17-50

A Guide to
**Diagnostic Clinical Chemistry**

# A Guide to
# Diagnostic Clinical Chemistry

R. N. WALMSLEY
MB BS, FRCPath, MCB, MAACB
*Director of Clinical Biochemistry, Flinders Medical Centre, Bedford Park,*
*South Australia,*
*Visiting Specialist (Chemical Pathology), Repatriation Hospital, Daw Park,*
*South Australia and Honorary Senior Lecturer in Clinical Biochemistry,*
*Flinders University of South Australia*

G. H. WHITE
BSc, PhD, MAACB
*Principal Clinical Biochemist, Flinders Medical Centre, Bedford Park,*
*South Australia and*
*Honorary Senior Lecturer in Clinical Biochemistry, Flinders University of South*
*Australia*

Blackwell Scientific Publications
MELBOURNE OXFORD LONDON EDINBURGH BOSTON

© 1983 by
Blackwell Scientific Publications
Editorial Offices:
99 Barry Street, Carlton
  Victoria 3053 Australia
Osney Mead, Oxford OX2 OEL
8 John Street, London WC1N 2ES
9 Forrest Road, Edinburgh EH1 2QH
52 Beacon Street, Boston
  Massachusetts 02108, USA

All rights reserved. No part of this
publication may be reproduced, stored
in a retrieval system, or transmitted,
in any form or by any means,
electronic, mechanical, photocopying,
recording or otherwise
without the prior permission of
the copyright owner.

First published 1983

Typeset by Abb-Typesetting Pty Ltd
Printed in Singapore by Richard Clay (S.E.Asia) Pte Ltd

DISTRIBUTORS

USA
Blackwell Mosby Book Distributors
11830 Westline Industrial Drive
St Louis, Missouri 63141

Canada
Blackwell Mosby Book Distributors
120 Melford Drive, Scarborough
Ontario M1B 2X4

Australia
Blackwell Scientific book Distributors
31 Advantage Road,
Highett, Victoria 3190

Cataloguing in Publication Data

Walmsley, R. N.
A guide to diagnostic clinical chemistry.

Includes index.
ISBN 0 86793 040 3.

1. Chemistry, Clinical. I. White, G. H.
II. Title.

616.07'56

# Contents

| | | |
|---|---|---|
| Preface | | vi |
| Common abbreviations | | viii |
| 1 | Introduction | 1 |
| 2 | Water and sodium | 2 |
| 3 | Potassium | 59 |
| 4 | Chloride | 80 |
| 5 | Anion gap | 89 |
| 6 | Acid-base and bicarbonate | 95 |
| 7 | Mineralocorticoids | 134 |
| 8 | Urea and creatinine | 150 |
| 9 | Calcium, phosphate and magnesium | 165 |
| 10 | Glucose | 205 |
| 11 | Proteins | 236 |
| 12 | Plasma enzymes | 262 |
| 13 | Liver function tests | 293 |
| 14 | Plasma lipids | 317 |
| 15 | Steatorrhoea and malabsorption | 334 |
| 16 | Iron | 351 |
| 17 | Porphyrins | 360 |
| 18 | Urate | 381 |
| 19 | Hypothalamic and anterior pituitary hormones | 388 |
| 20 | Thyroid hormones | 425 |
| 21 | Cortisol | 462 |
| 22 | Catecholamines and hypertension | 496 |
| 23 | Tumours: biochemical syndromes | 508 |
| Further reading | | 520 |
| Conversion factors for SI units | | 522 |
| Index | | 525 |

# Preface

The results of biochemical tests are of little value in the diagnosis and management of disease unless they can be interpreted and understood in terms of the underlying pathophysiology. However, when discussing the interpretative aspects many texts simply provide a list of the causes of abnormal tests, without demonstrating their relationship to the pathophysiology. This book is an attempt to rectify this deficiency, and is based on our experiences in teaching medical students, clinical chemists, technical staff and those who are preparing for professional examinations in medicine, chemical pathology and clinical chemistry.

Each chapter is designed to cover three closely related aspects of its subject. The first part discusses the relevant physiology related to the analyte in question. This section has purposely been kept brief because students using this book will have already acquired the basic knowledge needed, and should therefore need only a reminder of the principles involved. The second section deals with tests and procedures generally available for the investigation and elucidation of specific diagnostic and treatment problems. In this context the relative merits of the tests are critically discussed and reflect our own experiences rather than those found in the literature. The final section of each chapter deals with the pathophysiology of specific disease processes using case material from our own laboratory to illustrate how these processes express themselves in terms of clinical chemistry tests.

We consider that the use of actual patient cases has advantages, in that the quantitative changes in test results caused by disease and its treatment can be experienced at first hand, rather than having to rely on the 'increased, moderately increased' format found in most textbooks. The cases are selected to demonstrate the patterns of results that are most commonly encountered, it not being possible in a short book to illustrate the full range or subtlety of values that may be seen. The variety of cases appearing in individual chapters reflects the availability, and therefore the prevalence, of the particular diseases in our local population. However, we believe that the material covered is broadly representative of that seen by most general hospital laboratories. Obviously some disease processes will not be illustrated because of the scarcity of material. In these instances we have provided a short descriptive summary of the essential pathophysiology. Certain topics have been excluded, for example vitamins, various paediatric conditions and most inborn errors of metabolism, because they are not normally dealt with in our laboratory.

Each chapter is orientated towards either a single biochemistry test or a group of organ-related tests. We recognize that this approach may elicit criticism on the grounds that many analytes are closely interrelated, and that disturbed homeostasis of one metabolite is invariably associated with abnormalities in others. However, this division was considered appropriate, because in practice the difficult problems in interpretation usually arise when only one of a group of test results is found to be abnormal.

The reference ranges quoted throughout the text are those peculiar to our laboratory and should be used by the reader in this context. For some analytes the reference range quoted with patient cases occasionally changes; this reflects the usual laboratory practice of continually updating the reference values as a consequence of analytical method changes and subtle alterations in the population from which such ranges are derived. For visual clarity in the case examples we have identified tests by a short code (e.g. Amy. for amylase); these codes are used in our laboratory's computerized reporting system and should not be considered as standard abbreviations (see p. ix for full list).

A number of our colleagues have provided us with helpful criticisms and advice during the preparation of the text. For their particular help we wish to thank Drs G. D. Calvert, W. J. Riley, and especially Dr M. D. Guerin and Mrs B. Dilena who read the whole manuscript and provided invaluable comments. Special thanks are due to Mrs Joanna Fenton who typed the drafts and the final manuscript, and to Mr A. Bentley and Mr D. Jones of the Department of Medical Illustration and Media for Figs 9.2, 9.3, and the front cover illustration respectively.

*Adelaide, 1983*                                                                         R. N. *Walmsley*
                                                                                                      G. H. *White*

# Abbreviations

| | |
|---|---|
| [ ] | Concentration |
| AcAc | Acetoacetic acid |
| ACE | Angiotensin-converting enzyme |
| ACP | Acid phosphatase |
| ACTH | Adrenocorticotropic hormone |
| ADH | Antidiuretic hormone |
| ADP | Adenosine diphosphate |
| αFP | α-Fetoprotein |
| AHCO$_3$ | Actual bicarbonate |
| AIP | Acute intermittent porphyria |
| ALA | Aminolaevulinate |
| Alb. | Albumin |
| Aldo. | Aldosterone |
| ALT | Alanine aminotransferase |
| Amy. | Amylase |
| AP | Alkaline phosphatase |
| AST | Aspartate aminotransferase |
| α$_1$-AT | α$_1$-Antitrypsin |
| ATN | Acute tubular necrosis |
| ATP | Adenosine triphosphate |
| Bili. | Bilirubin |
| BJP | Bence Jones protein |
| BSP | Bromsulpthalein |
| Ca$^{2+}$ | Calcium ion |
| Ca | Total calcium |
| CAH | Chronic active hepatitis |
| CAT | Carnitine acyl transferase |
| CBG | Cortisol-binding globulin (transcortin) |
| CCF | Congestive cardiac failure |
| CEA | Carcinoembryonic antigen |
| CEP | Congenital erythropoietic porphyria |
| Chol. | Cholesterol |
| CK | Creatine kinase |
| Cl | Chloride |
| CNS | Central nervous system |
| COAD | Chronic obstructive airway disease |
| Cort. | Cortisol |

| | | |
|---|---|---|
| CPH | Chronic persistent hepatitis | |
| Creat. | Creatinine | |
| CRF | Chronic renal failure | |
| CSF | Cerebrospinal fluid | |

| | |
|---|---|
| $1,25\text{-}(OH)_2D_3$ | 1,25-dihydroxycholecalciferol |
| DIT | Diiodotyrosine |
| DOPA | Dihydroxyphenylalanine |

| | |
|---|---|
| $E_2$ | Oestradiol |
| EC | Erythropoietic coproporphyria |
| ECF | Extracellular fluid |
| ECV | Extracellular volume |
| EDTA | Ethylenediamine tetra-acetate (sequestrene) |
| EP | Erythropoietic protoporphyria |
| EPP | Electrophoretic pattern |
| EPG | Electrophoretogram |
| ESR | Erythrocyte sedimentation rate |

| | |
|---|---|
| Fe | Iron |
| $FE_{Na}$ | Fractional excretion of sodium |
| FFA | Free fatty acid |
| $\alpha FP$ | alpha-fetoprotein |
| FSH | Follicle-stimulating hormone |
| FTI | Free thyroxine index |

| | |
|---|---|
| GFR | Glomerular filtration rate |
| GGT | $\gamma$-Glutamyltransferase |
| GH | Growth hormone |
| GHRF | Growth hormone releasing factor |
| Glu. | Glucose |
| GnRH | Gonadotropin-releasing hormone |
| GTT | Glucose tolerance test |

| | |
|---|---|
| $H^+$ | Hydrogen ion |
| HBD | Hydroxybutyrate dehydrogenase |
| HC | Hereditary coproporphyria |
| $25\text{-}OHD_3$ | 25-Hydroxycholecalciferol |
| HDL | High density lipoprotein |
| HGPRT | Hypoxanthine guanine phosphoribosyl transferase |
| 5 HT | 5-Hydroxytryptamine (serotonin) |
| 5 HIAA | 5-Hydroxyindole acetic acid |
| HMMA | 4-Hydroxy-3-methoxymandelic acid (VMA) |
| HPA | Hypothalamic-pituitary-adrenal axis |
| HPL | Human placental lactogen |
| HVA | Homovanillic acid |

*Abbreviations*

| | |
|---|---|
| ICF | Intracellular fluid |
| IDDM | Insulin-dependent diabetes mellitus |
| IHD | Ischaemic heart disease |
| IM | Intramuscular |
| IV | Intravenous |
| IVV | Intravascular volume |
| | |
| K | Potassium |
| 17-KG | 17-Ketogenic steroids |
| | |
| LCAT | Lecithin:cholesterol acyl transferase |
| LD | Lactate dehydrogenase |
| LDL | Low density lipoprotein |
| LFT | Liver function tests |
| LH | Luteinizing hormone |
| LHRH | LH-releasing hormone |
| LRH | Low renin hypertension |
| | |
| MEN | Multiple endocrine neoplasia |
| Mg | Magnesium |
| MI | Myocardial infarction |
| MIT | Monoiodotyrosine |
| Mol. wt. | Molecular weight |
| | |
| Na | Sodium |
| NIDDM | Non insulin-dependent diabetes mellitus |
| 5'NT | 5'-Nucleotidase |
| $NAD^+$ | Nicotinamide-adenine dinucleotide |
| NADH | Nicotinamide-adenine dinucleotide (reduced) |
| | |
| OAF | Osteoclast activating factor |
| OGTT | Oral glucose tolerance test |
| 17-OHCS | 17-Hydroxycorticosteroids |
| β-OHB | β-Hydroxybutyric acid |
| OP | Osmotic pressure |
| Osmo. | Osmolality |
| 17-Oxo. | 17-Oxosteroids |
| | |
| PBG | Porphobilinogen |
| PCT | Porphyria cutanea tarda |
| $P_{CO_2}$ | Partial pressure of carbon dioxide |
| PG | Prostaglandin |
| PIF | Prolactin inhibitory factor |
| $P_{O_2}$ | Partial pressure of oxygen |
| $PO_4$ | Inorganic phosphate |
| PRA | Plasma renin activity |
| PRF | Prolactin releasing factor |
| PRL | Prolactin |

| | | |
|---|---|---|
| PRPP | Phosphoribosyl pyrophosphate | |
| PRU | Prerenal uraemia | |
| PTH | Parathyroid hormone | |
| | | |
| RBF | Renal blood flow | |
| | | |
| SI | Système International | |
| SIAD | Syndrome of inappropriate antidiuresis | |
| SIADH | Syndrome of inappropriate secretion of ADH | |
| SD | Standard deviation | |
| | | |
| $T_3$ | Triiodothyronine | |
| $T_4$ | Thyroxine | |
| TBG | Thyroxine-binding globulin | |
| TBP | Thyroxine-binding protein | |
| TBPA | Thyroxine-binding prealbumin | |
| TCA | Tricarboxylic acid cycle | |
| Te | Testosterone | |
| TIBC | Total iron binding capacity | |
| TP | Total protein | |
| TRH | Thyrotropin-releasing hormone | |
| Trig. | Triglyceride | |
| TSH | Thyroid-stimulating hormone | |
| | | |
| UDP | Uridine diphosphate | |
| UV | Ultraviolet light | |
| | | |
| VIP | Vasoactive intestinal peptide | |
| VLDL | Very low density lipoprotein | |
| VMA | Vanillyl mandelic acid (HMMA) | |
| VP | Variegate porphyria | |
| | | |
| ↑ | Increase | |
| ↑↑ | Large increase | |
| | | |
| ↓ | Decrease | |
| ↓↓ | Large decrease | |

# 1 Introduction to the interpretation of tests

Diagnostic clinical chemistry uses biochemical knowledge and techniques to assist in the diagnosis of human disease, to follow its progress and to monitor the effect of treatment. The practice of this discipline involves the equal participation of the medical practitioner and the clinical chemist. Until perhaps 20 years ago the limited biochemical knowledge and the small range of available tests usually kept the working relationship between the doctor and the clinical chemist a simple one; the doctor ordered the test and took the patient sample, the biochemist did the test and reported the result (Fig. 1.1).

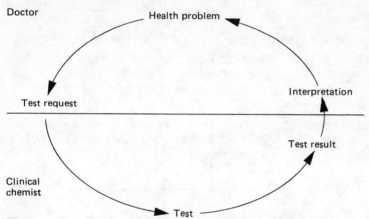

**Fig. 1.1** Until 20 years ago limited biochemical knowledge kept the working relationship between doctor and clinical chemist a simple one.

However, the explosion of new biochemical knowledge, techniques and instrumentation capabilities over the last two decades has brought a matching degree of complication to the early professional relationship. At the practical level it is now often the case that 10 or 15 different nursing, technical and other staff have interposed themselves into the once simple request–test–result chain of events. A typical organization for an average size clinical chemistry laboratory might be as shown in Fig. 1.2.

Before a result is interpreted for clinical purposes the doctor must be confident that the requested test has been performed on the correct patient, and that the result is unaffected by extraneous factors. The clinical chemist also wishes to be similarly confident.

# Chapter 1

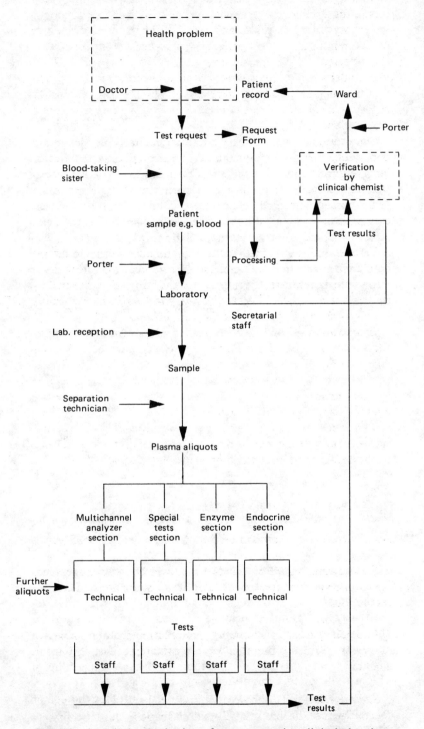

**Fig. 1.2** A typical organization of an average size clinical chemistry laboratory.

*Interpretation of tests*

With the growing complexity of laboratory organization there is an increasing likelihood of error at each step. With a typical modern laboratory system the clinician has to assume that the sample was (1) taken from the correct patient, (2) obtained in the proper way, (3) collected in the appropriate container, (4) unadulterated, (5) kept under suitable conditions until it reached the laboratory and that the appropriate results were recorded in the case-records. In reporting the test result the clinical chemist also has to make these assumptions.

These sections of a sample's route are not under the direct control of either the doctor or clinical chemist, and so they form an 'area of uncertainty'. The areas of 'potential certainty' are those of professional trust, i.e. the clinician's assumption that the requested test has been properly performed and the clinical chemist's assumption that his laboratory has been informed of all the factors that may deflect the test result from the true one.

To effectively monitor the 'area of uncertainty' it is necessary for both doctor and clinical chemist to be able to identify both gross and subtle data disasters, but to do this the clinician has to understand some aspects of clinical chemistry and the clinical chemist has to have knowledge of some aspects of medicine.

In addition to monitoring the validity of a patient's test results, the doctor and the clinical chemist must interact at two other levels, using the relevant knowledge of each other's profession to (1) interpret the results in the light of current biochemical and medical knowledge and (2) use new knowledge to improve the diagnosis and treatment of patients.

The following sections outline the minimum of analytical and statistical knowledge that both assists in identifying errors and aids in the meaningful interpretation of patient results.


## THE INVALID RESULT

An incorrect test result can be due to many factors, including:

*Sample handling errors* (1) Sample taken from wrong patient, (2) sample switched for another during manipulations, (3) incorrect report.

*Incorrect patient preparation* (1) Posture/physical activity, (2) stress, (3) drugs/therapy, (4) biological rhythms, (5) food/alcohol.

*Sample errors* (1) Incorrect storage, (2) effect of sampling, e.g. tourniquet, haemolysis, (3) contamination, e.g. IV therapy, (4) preservation, e.g. incorrect container, (5) interference, e.g. lipaemia.

Chapter 1    SAMPLE HANDLING ERROR

*Example*   A 72 year old man making a good recovery after surgery for fractured neck of femur.

| Date |      | 15/08 | 16/08 | 17/08 | 17/08* |        |              |
|------|------|-------|-------|-------|--------|--------|--------------|
| Plasma | Na   | 140   | 142   | 137   | 141    | mmol/l | (132–144)    |
|      | K    | 4.2   | 3.6   | 4.0   | 3.7    | mmol/l | (3.1–4.8)    |
|      | Cl   | 103   | 101   | 95    | 101    | mmol/l | (93–108)     |
|      | HCO$_3$ | 27  | 32    | 29    | 28     | mmol/l | (21–32)      |
|      | Urea | 12.0  | 10.4  | 14.7  | 9.6    | mmol/l | (3.0–8.0)    |
|      | Creat. | 0.16 | 0.12 | 0.59  | 0.11   | mmol/l | (0.06–0.12)  |

The creatinine result of 17/08 appeared incompatible with the previous results and the clinical picture. A repeat specimen was therefore requested and analysed.*

Many clinical chemistry laboratories have a system whereby all results are scrutinized before they are reported to the clinical staff. Obvious handling errors involving the switching of patient samples are easily detected before reporting, particularly if a cumulative patient record system is used which allows the new results to be easily compared with previous ones obtained for the patient. The above type of error is difficult to detect when the incorrect sample is also the first sample received from the patient. In this situation, the chances of detection are much improved if the clinician provides the laboratory with clinical information that is relevant to the tests requested.

*Example*   Admission samples on a 61 year old man (patient A) and a 46 year old man (patient B).

|        |         | Patient A | Patient B |        |              |
|--------|---------|-----------|-----------|--------|--------------|
| Plasma | Na      | 139       | 139       | mmol/l | (132–144)    |
|        | K       | 3.7       | 3.7       | mmol/l | (3.1–4.8)    |
|        | Cl      | 98        | 103       | mmol/l | (93–108)     |
|        | HCO$_3$ | 25        | 28        | mmol/l | (21–32)      |
|        | Urea    | 17.3      | 4.8       | mmol/l | (3.0–8.0)    |
|        | Creat.  | 0.58      | 0.09      | mmol/l | (0.06–0.12)  |
|        | CK      | 347       |           | U/l    | (30–140)     |
|        | HBD     | 360       |           | U/l    | (125–250)    |

Since there is no comparative set of data for either patient the above results would be reported. However, this is less likely if the laboratory is able to relate the results to the 'relevant' clinical information that prompted the test requests, i.e. patient A was admitted with chest pain and patient B was a new transfer patient with chronic renal failure, for dialysis.

With the appropriate clinical information the clinical chemist should realize that the renal function results for patient B are improbable and withhold the result. Following this, a simple investigation would reveal that plasma aliquots for patients A and B were switched prior to electrolyte and urea analysis; the enzyme results were valid for patient A.

The above types of error always carry potential danger for the patient if clinical action is taken on invalid results. Such mistakes can be minimized if test requests are accompanied by relevant clinical information. For this remedy to be effective the clinical chemist must have a clear understanding of the effect of disease and its treatment on biochemical parameters.

INCORRECT PATIENT PREPARATION

A 23 year old woman, 32 weeks pregnant, had an oral glucose tolerance test performed to investigate the possibility of gestational diabetes.

| Time | | 09.10 | 09.45 | 10.15 | 10.45 | 11.15 | |
|---|---|---|---|---|---|---|---|
| Plasma | Glu. | 3.7 | 6.3 | 7.6 | 6.4 | 4.3 | mmol/l |
| Urine | Glu. | <0.1 | <0.1 | <0.1 | <0.1 | <0.1 | % |
| Ketones | | ++ | | +++ | | − | |

The patient was fasted for 16 hours prior to the test, causing increased fat metabolism, resulting in ketonuria. There is, therefore, no diagnostic value in requesting the urinary ketone test under these conditions. Other tests, such as lipid studies, do require an overnight fast if the results are to be diagnostically helpful. Therefore it is worthwhile to check on the requirements of patient preparation before embarking on unfamiliar tests.

DRUG EFFECT ON PHYSIOLOGY

A 27 year old woman being investigated for anxiety and intermittent tachycardia.

| Plasma | $T_4$ | 176 | nmol/l | (60–160) |
|---|---|---|---|---|
| | FTI | 141 | units | (50–150) |
| | $T_3$ | 3.4 | nmol/l | (1.2–2.8) |

Many drugs have physiological effects that are reflected in altered biochemical tests. The above euthyroid patient was taking an oral contraceptive, the oestrogen content of which increased the plasma thyroid hormone levels by stimulating increased liver synthesis of the thyroid hormone transport protein (Chapter 20). The requesting doctor was unaware of this effect and the laboratory was unaware of the patient's drug history, resulting in the patient being referred to an endocrine clinic. Recent literature lists approximately 100 different effects that oral contraceptives can have on laboratory tests. Many other drugs can also have physiological or analytical effects on diagnostic tests. It is, therefore, important to include the patient's current drug history with the test request, as can also be seen in the following examples.

DRUG EFFECT ON A TEST

A 61 year old woman was admitted to hospital with a provisional diagnosis of Addison's disease. A blood sample for plasma electrolytes and cortisol was taken and the patient was admitted for a full work-up (12/10). An adrenal cortex stimulation test was performed the following day (only basal level shown).

| Date | | 12/10 | 13/10 | | |
|---|---|---|---|---|---|
| Plasma | Na | 115 | 115 | mmol/l | (132–144) |
| | K | 7.2 | 7.1 | mmol/l | (3.1–4.8) |
| | Cl | 85 | 90 | mmol/l | (93–108) |
| | $HCO_3$ | 20 | 14 | mmol/l | (21–32) |
| | Urea | 12.0 | 11.3 | mmol/l | (3.0–8.0) |
| | Creat. | 0.11 | 0.09 | mmol/l | (0.06–0.12) |
| | Cort. | 97 | 869 | nmol/l | (140–690) |

The plasma electrolyte and cortisol levels in the admission sample would support the provisional diagnosis, however the basal level cortisol prior to synacthen stimulation (13/10) would not. Investigation of the marked discrepancy revealed that the patient had been given prednisolone to provide steroid cover following admission. Prednisolone behaves virtually as cortisol in the assay used by the clinical chemistry laboratory. The synacthen test therefore had to be repeated.

DRUG EFFECT ON PHYSIOLOGY

An 11 year old boy. The clinical note accompanying the request for thyroid function studies stated: 'obese, mentally slow.'

| | | | |
|---|---|---|---|
| Plasma | $T_4$ | <10 | nmol/l |
| | TSH | <2.5 | mU/l |

The low $T_4$ and TSH led the clinical chemist to suggest on the patient report that the results suggested hypothyroidism secondary to anterior pituitary or hypothalamic disease. Laboratory concern at the lack of repeat or confirmatory tests led to the discovery that the patient was being prescribed $T_3$ for his obesity; the effect of $T_3$, in sufficient amounts, is to suppress both $T_4$ and TSH secretion.

*Interpretation of tests*

## INCORRECT STORAGE

A 62 year old lady attended the renal clinic as an outpatient. A blood sample was taken, stored at $+5°C$ in a refrigerator overnight and sent to the laboratory in the morning for separation and analysis.

| Plasma | Na | 145 | mmol/l | (132–144) |
|---|---|---|---|---|
| | K | 7.6 | mmol/l | (3.1–4.8) |
| | Cl | 98 | mmol/l | (93–108) |
| | $HCO_3$ | 25 | mmol/l | (21–32) |
| | Urea | 8.3 | mmol/l | (3.0–8.0) |
| | Creat. | 0.23 | mmol/l | (0.06–0.12) |
| | Ca | 2.41 | mmol/l | (2.15–2.55) |
| | $PO_4$ | 5.00 | mmol/l | (0.60–1.25) |
| | TP | 72 | g/l | (60–85) |
| | Alb. | 41 | g/l | (37–52) |
| | AP | 57 | U/l | (25–120) |

Prolonged contact between plasma and the red cells allows increased leakage of potassium and phosphate from the erythrocytes. Thus plasma $[K^+]$ can be markedly high without any visible evidence of haemolysis. Obviously storage times that are shorter than the above case can lead to more subtle alterations.

## SAMPLE CONTAMINATION

A 76 year old woman was admitted to hospital. A blood sample was taken after an IV saline drip had been set up.

| Time | | 20.00 | 21.45 | | |
|---|---|---|---|---|---|
| Plasma | Na | 148 | 140 | mmol/l | (132–144) |
| | K | 1.0 | 2.8 | mmol/l | (3.1–4.8) |
| | Cl | 125 | 88 | mmol/l | (93–108) |
| | $HCO_3$ | 16 | 35 | mmol/l | (21–32) |
| | Urea | 4.5 | 10.5 | mmol/l | (3.0–8.0) |
| | Creat. | 0.05 | 0.13 | mmol/l | (0.06–0.12) |

The venepuncture was made proximally to the IV drip line needle, thus the drip fluid diluted the blood sample taken; note the high $[Cl^-]$ and low $[K^+]$. As this was the first set of results the urea and creatinine levels would not be considered inappropriate for this patient. The second specimen taken later, from a site opposite to the drip arm, indicates the correct values.

### SPECIMEN CONTAMINATION

A 79 year old lady, receiving IV glucose and potassium. A blood sample was inadvertently taken from a site above the drip needle.

| Plasma | | | | |
|---|---|---|---|---|
| Na | 97 | mmol/l | (132–144) |
| K | >10.0 | mmol/l | (3.1–4.8) |
| Cl | 86 | mmol/l | (93–108) |
| $HCO_3$ | 14 | mmol/l | (21–32) |
| Urea | 2.5 | mmol/l | (3.0–8.0) |
| Creat. | 0.07 | mmol/l | (0.06–0.12) |
| Glu. | 62.0 | mmol/l | (3.0–5.5) |
| Ca | 1.32 | mmol/l | (2.05–2.55) |
| $PO_4$ | 0.51 | mmol/l | (0.06–1.25) |
| TP | 48 | g/l | (60–85) |
| Alb. | 25 | g/l | (30–50) |
| AP | 72 | U/l | (25–100) |
| ALT | 6 | U/l | (<35) |
| Bili. | 9 | µmol/l | (<20) |

The dilutional effect is obvious, with a number of analytes, particularly sodium, the proteins and calcium being markedly affected. The drip content is clearly shown by the glucose level.

### INAPPROPRIATE SAMPLE

A routine postrenal dialysis blood sample was taken from a 56 year old male patient.

| Time | | 12.20 | 13.45 | | |
|---|---|---|---|---|---|
| Plasma | Na | 137 | 138 | mmol/l | (132–144) |
| | K | 3.1 | 3.4 | mmol/l | (3.1–4.8) |
| | Cl | 97 | 102 | mmol/l | (93–108) |
| | $HCO_3$ | 5 | 20 | mmol/l | (21–32) |
| | Urea | 1.6 | 11.3 | mmol/l | (3.0–8.0) |
| | Creat. | 0.13 | 0.46 | mmol/l | (0.06–0.12) |

The first sample was inadvertently taken from the venous return side of the dialyser and reflects the plasma composition immediately after it has been dialysed and prior to its having mixed with the patient's remaining blood.

WRONG PRESERVATIVE

A blood sample was taken by a nurse from a 42 year old woman. The sample was aliquoted into a lithium-heparin containing tube for electrolyte, calcium, protein and other assays, and into a tube with potassium-EDTA preservative for an ESR estimation.

| Plasma | Na | 141 | mmol/l | (132–144) |
|---|---|---|---|---|
| | K | 10.3 | mmol/l | (3.1–4.8) |
| | Cl | 94 | mmol/l | (93–108) |
| | $HCO_3$ | 28 | mmol/l | (21–32) |
| | Urea | 5.5 | mmol/l | (3.0–8.0) |
| | Creat. | 0.09 | mmol/l | (0.06–0.12) |
| | TP | 77 | g/l | (64–84) |
| | Alb. | 46 | g/l | (37–52) |
| | Ca | 0.56 | mmol/l | (2.15–2.55) |
| | $PO_4$ | 0.95 | mmol/l | (0.6–1.25) |
| | AP | <20 | U/l | (25–120) |
| | Bili. | 10 | μmol/l | (<20) |

These odd results arose because the nurse thought that there was inadequate blood in the lithium-heparin tube and decanted some from the K-EDTA tube. This is reflected by the high [$K^+$], low [Ca] (due to complexing with EDTA), and the low AP, due to EDTA complexing $Mg^{2+}$ from one of the analytical reagents ($Mg^{2+}$ is required as a cofactor in the AP assay).

**Summary**

Many data disasters can be recognized before the results leave the laboratory. The detection rate of errors and invalid results is markedly improved if (1) doctors ensure that test requests are accompanied by the relevant clinical and drug history and (2) clinical chemists can relate biochemical parameters to disease processes and the effects of treatment.

INTERPRETATION OF A RESULT

Once confident that a test result is valid both the doctor and the clinical chemist wish to make an interpretation; the former as an aid to patient management, the latter to detect errors by ensuring that the result has reasonable harmony with the clinical expectation.

Clinical chemistry is a quantitative discipline, most tests yielding numerical results that can be related to an appropriate reference range. This numerical aspect gives the superficial impression that deciding if a result is normal or abnormal is a simple matter of comparison. However, an understanding of the following brief sections is crucial to obtaining meaningful diagnostic information from a test result.

The interpretation of a test result has three stages: (1) analytical, (2) statistical, (3) clinical. Much literature and data exist for each of these areas; the following sections provide the barest essentials for the practical task of diagnostic interpretation.

## Analytical interpretation

The following two points are to be considered.

### ANALYTICAL ACCURACY (OR INACCURACY)

Some analytes can be estimated by a variety of techniques. Although the various methods will estimate the analyte with differing degrees of accuracy, this is not too important, providing the patient result is always compared with a reference range that has been calculated from results determined by the same method.

Doctors should verify that the reference ranges quoted to them have been determined by the laboratory, using their current methods.

### ANALYTICAL IMPRECISION (OR PRECISION)

The analytical imprecision of a method describes the spread of results obtained when the same sample is repeatedly tested over a period of time. For example the sodium concentration of a plasma sample measured 20 times on consecutive days may yield results within the range 137–143 mmol/l (mean $\pm 2$ SD). This imprecision reflects the sum of small changes in many factors, particularly the ageing and renewal of reagents, differences in pipetting, instrument behaviour, calibration, graphing techniques, etc. Analytical imprecision differs for every method, and for the same method as used in different laboratories, and so laboratories should determine for themselves the imprecision of each of their methods (Table 1.1).

**Table 1.1** An indication of analytical imprecision for some commonly used analytes at levels within the reference range

| Rule of Thumb | | | Authors' Laboratory (2 SD) | |
|---|---|---|---|---|
| ~2.5% | Plasma | Na | 3 | mmol/l |
| | | K | 0.1 | mmol/l |
| | | Cl | 2 | mmol/l |
| | | TP | 2 | g/l |
| | | Ca | 0.06 | mmol/l |
| | | Urate | 0.01 | mmol/l |
| ~5.0% | Plasma | Alb. | 2 | g/l |
| | | $PO_4$ | 0.04 | mmol/l |
| | | $T_4$ | 6 | nmol/l |
| ~10.0% | Plasma | $HCO_3$ | 2 | mmol/l |
| | | Urea | 0.5 | mmol/l |
| | | Amy. | 20 | U/l |
| | | AP | 8 | U/l |
| | | ALT | 3 | U/l |
| | | CK | 5 | U/l |
| | | Glu. | 0.4 | mmol/l |
| | | Cort. | 10 | nmol/l |
| ~15.0% | Plasma | Bili. | 3 | µmol/l |
| | | Creat. | 0.02 | mmol/l |

Because a test is usually performed on a patient sample once and not many times, the single value obtained masks the fact that a spread of values would be obtained if the measurement was repeated a number of times. If the analytical imprecision for plasma $[Na^+]$ is 3 mmol/l (2 SD), then a result for $[Na^+]$ of 134 mmol/l actually indicates that there is a 95% probability of the true value lying in the range $134 \pm 3$ (131–137) mmol/l.

Analytical imprecision assumes importance for the interpretation of results either when they lie close to the reference range or clinical decision limits or when there is a need to decide if there is a significant difference between two results.

Doctors should ensure that their laboratory provides them with analytical imprecisions, determined in-house, for all tests offered.

*Example* A 56 year old woman with hypercalcaemia due to malignancy is being treated with IV calcitonin in order to lower her plasma calcium level. Commencing with 17/01, decide for each day whether the treatment is having the desired effect. Also consider whether significant changes occur in the other biochemical parameters (use Table 1.1).

| Date | | 17/01 | 18/01 | 19/01 | 20/01 | 21/01 | |
|---|---|---|---|---|---|---|---|
| Plasma | Ca | 3.23 | 3.04 | 2.98 | 3.00 | 2.92 | mmol/l |
| | $PO_4$ | 0.78 | 0.76 | 0.95 | 0.70 | 0.55 | mmol/l |
| | TP | 71 | 66 | 69 | 67 | 69 | g/l |
| | Alb. | 34 | 33 | 34 | 35 | 36 | g/l |
| | AP | 273 | 239 | 230 | 230 | 235 | U/l |

It should always be borne in mind that the above style is a convenient means of representing the following:

| Date | | 17/01 | 18/01 | 19/01 | 20/01 | 21/01 | |
|---|---|---|---|---|---|---|---|
| Plasma | Ca | 3.17–3.29 | 2.98–3.10 | 2.96–3.04 | 2.94–3.06 | 2.86–2.98 | mmol/l |
| | $PO_4$ | 0.74–0.82 | 0.72–0.80 | 0.91–0.99 | 0.66–0.74 | 0.51–0.59 | mmol/l |
| | TP | 69–73 | 64–68 | 67–71 | 65–69 | 67–71 | g/l |
| | Alb. | 32–36 | 31–35 | 32–36 | 33–37 | 34–38 | g/l |
| | AP | 265–281 | 221–247 | 222–238 | 222–238 | 227–243 | U/l |

If test results are thought of in terms of their analytical imprecision it is a simpler matter to decide whether a result is significantly different from a reference range limit or another result.

**The statistical interpretation**

A normal or an abnormal test result is equally informative. However, the problem lies in deciding into which category the individual result fits. To aid this decision the patient result is generally compared with a relevant reference range provided by the clinical chemistry laboratory.

THE REFERENCE RANGE

If an analyte is repeatedly measured over a period of time in an individual, the results are generally found to vary. In addition to the variation caused by analytical imprecision, there is biological variation which is due to many factors, particularly biochemical individuality, diet, physical state, biological rhythms, etc. The biological variation is large for some biochemical parameters, insignificant for others, and may vary markedly from individual to individual. For these reasons it would be ideal to determine the reference range for all biochemical parameters for each person, so that they may be used when the individual falls ill. As this is clearly not feasible, population reference ranges are used instead.

A population reference range is usually determined by measuring the analyte of interest in a large number (should be > 100) of people who are thought to be healthy. This reference population must be representative in relevant respects of those to whom the test may be applied for diagnostic purposes, i.e. age, sex, stage of pregnancy, etc. From the values obtained the reference range is often simply calculated as the mean ± 2 SD; this means that the reference range encompasses 95% (the actual value is 95.4%, but 95% is used for simplicity) of the values obtained for that reference population. Therefore, by definition 5% of any healthy group subsequently tested will fall outside the range (2.5% below and 2.5% above assuming a Gaussian distribution) and be classified as abnormal (false positives).

If the same analyte is now estimated in a large number of patients known to have a disease that causes abnormally altered levels of the analyte, a second 'disease' reference range could be determined. However, for most analytes there is considerable overlap between the 'well' and 'unwell' range (Fig. 1.3). This can arise for several reasons, particularly: (1) the definition of a reference range, (2) the biochemical parameter is only moderately altered by the disease, (3) a number of individuals will have the disease in mild form at the time of testing, (4) the patient's 'personal' reference range (own biological variation) lies well within the population reference range so that an abnormal result for the patient is normal in relation to the population, (5) the distribution of values in a normal population is non-Gaussian for many analytes.

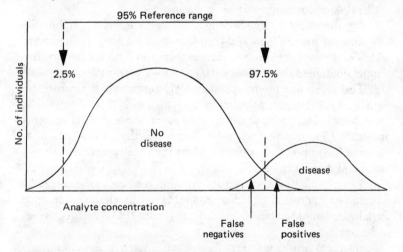

**Fig. 1.3** A typical distribution of test results obtained by screening subjects for a particular disease, e.g. plasma [urate] as a screening test for gout. Note that if the upper reference limit is moved left or right, the false results on one side improve only at the expense of the other.

The upper reference limit for the healthy population therefore encloses some diseased patients (*false negatives*), and excludes some healthy individuals (*false positives*). If an individual result (value ± analytical imprecision) is remote from a reference range limit then the classification of normal or abnormal is simple. As the result approaches a reference range limit, the interpretation generally becomes increasingly dependent on the clinical findings and possibly the results of other tests.

However, in practice the biological variation is not separately identified but forms an inherent part of the overall variation (analytical imprecision + biological variation) that is reflected by both reference ranges and individual results.

*Example* A 44 year old male laboratory technician donated a blood sample for various reference range determinations. A raised plasma [Ca] was incidentally noted. Plasma [Ca] was estimated on three further occasions, using blood samples obtained without the use of a tourniquet.

| Day | | 1 | 2 | 3 | | |
|---|---|---|---|---|---|---|
| Plasma | Ca | 2.63 | 2.61 | 2.58 | mmol/l | (2.15–2.55) |
| | $PO_4$ | 1.04 | 1.11 | 1.09 | mmol/l | (0.06–1.25) |
| | TP | 77 | 77 | 76 | g/l | (64–84) |
| | Alb. | 39 | 38 | 39 | g/l | (37–52) |
| | AP | 86 | 91 | 87 | U/l | (25–120) |

The results do not clearly establish hypercalcaemia. In this case several considerations may aid a decision:

(1) Are the given reference ranges appropriate, i.e. for ambulant subjects for plasma [Ca] and [Alb.]?

(2) Will the average result ± analytical imprecision fall within the upper reference limit (2.60 ± 0.04 = 2.56–2.64 mmol/l)?

(3) The reference range approximately represents the mean ± 2 SD of the healthy population studied, i.e. 2.35 ± 0.2 mmol/l, which includes ~95% of the group studied. Mean ± 3 SD encompasses 99.7% of the group, i.e. some healthy individuals had plasma [Ca] up to 2.65 mmol/l. Thus the results from the above case could be normal.

(4) The plasma [Ca] could be abnormal in relation to the subject's 'unknown', personal biological range. Other biochemical data, if available, should be considered. In this case [$PO_4$] and Alb. (see Chapter 7).

Equivocal results may remain as such, since clinical chemistry tests are an aid only and any diagnostic or treatment decision must remain a clinical one.

*Example* A 56 year old woman complained of increasing tiredness. Amongst other tests, a thyroid function test was performed.

| | | | |
|---|---|---|---|
| Plasma | $T_4$ | 51 nmol/l | (60–160) |
| | TSH | <2.5 mU/l | (<5.5) |

This lady was considered to be euthyroid (normal TSH, no strong clinical evidence) and one of the 2.5% of people whose $T_4$ can be expected to fall below the reference limit. This was confirmed by the TSH value.

*Interpretation of tests*

*Example* A 61 year old lady consulted her general practitioner complaining of recent weight gain, occasional constipation and tiredness. Thyroid function tests were performed.

Plasma $T_4$    87 nmol/l    (60–160)
TSH   >40 mU/l    (<5.5)

Although the plasma $T_4$ result was well within the reference range, a plasma TSH level was estimated because of the several thyroid associated clinical symptoms present in the patient. The response by the patient's own anterior pituitary (↑TSH) clearly indicates that the circulating $T_4$ level is below the patient's 'own' reference range (primary hypothyroidism), although it is within that of the population.

*Example* A 54 year old Australian aboriginal woman had thyroid function tests performed as part of a clinical investigation of her cardiac disease.

Plasma $T_4$    47 nmol/l    (60–160)
TSH   <2.5 mU/l    (<5.5)

Australian aboriginals often have markedly lower levels of circulating thyroxine-binding globulin than Australians of European origin. The low TBG is reflected in low plasma $T_4$ levels. In this patient, there was no clinical evidence of thyroid or pituitary disease. Therefore the above reference range is inappropriate for this particular patient.

**Summary**

(1) Clinical chemistry results must be compared with a reference range that is relevant for the individual whose result is being interpreted. Factors to consider include: race, sex, age, physique, menstrual and other biological rhythms, postmenopause, diet, etc.
(2) A reference range is a guide only and describes the values found for about 95% of a small group of individuals who are thought to be healthy.

# SELECTION AND DIAGNOSTIC VALUE OF A TEST

## TEST SELECTION

Two distinct styles of laboratory organization have developed during the last decade.

*Discretionary* testing allows the clinician to select individual or small groups of tests that are considered to be relevant to the management of the patient.

*Profile* testing is a second approach, by which large batteries of tests are carried out, usually as a screening procedure. However, a laboratory organized for the latter situation usually still has to perform the large analytically related block of tests, even if only one or two specific tests were requested. Additional to yielding unwanted data that either has to be reported or suppressed, the profile or screening approach produces the difficult problem of the unexpected abnormal result. Apart from the presence of disease, such results will arise because of the definition of a reference range, that is $\sim 5\%$ of normal subjects tested will be classified as abnormal. It can be shown that:

| Number of different tests performed on a single sample from an individual | Percentage chance of an abnormal result (%) |
|---|---|
| 1 | 5 |
| 2 | 10 |
| 5 | 23 |
| 10 | 40 |
| 20 | 64 |

Therefore, as with all clinically unexpected test results, the unexpected abnormal result produced by a test screen should be confirmed with a repeat test before further diagnostic or other action is taken.

## DIAGNOSTIC VALUE OF A TEST

Ideally a test should be able to confirm the presence or absence of disease, but unfortunately this is rarely possible because of the marked overlap of test results that occurs between healthy and ill people (Fig. 1.3). Thus, the usual definition of a reference range (95% of normals) ensures that 5% of normal individuals will be misclassified by a particular test ($\sim 2.5\%$ below the lower reference limit, $\sim 2.5\%$ above the upper reference limit). Similarly, a number of individuals with a disease will be misclassified and give normal values for the relevant test. These and other factors ensure that no diagnostic tests can perfectly classify healthy and ill individuals.

*Interpretation of tests*

The sensitivity of a test numerically describes its ability to detect a disease when it is present, whilst specificity indicates the capability of the test to reflect the absence of the disease in subjects free of the disease. The sensitivity and specificity of a test is dependent on the selected reference or action limit (Fig. 1.3), but is also affected by the population tested, as well as biological and analytical variation.

What is the chance of a normal (or abnormal) result reflecting the sought for normal (or abnormal) condition?

*Example* Consider 100 subjects known to have primary hypothyroidism by clinical and other test (e.g. thyroid uptake, etc.) criteria. A plasma $T_4$ is measured for each subject and 95 results are found to fall below the lower reference limit for $T_4$. Plasma $T_4$ can be said to have 95% sensitivity (5% false negatives).

If plasma $T_4$ is now measured in 100 individuals known to be free of thyroid disease, and 95 results fall within the reference range, then the test has 95% specificity (5% false positives).

Although the test seems very effective, it is being used in an artificial situation, because the 'prevalence' of the disease in the above case is 50% (100 normal individuals, 100 thyroid disease patients), whilst primary hypothyroidism has a prevalence in the general population of $\sim 1\%$. If a thyroid laboratory instituted a massive screening programme for the disease and measured plasma $T_4$ in 100 000 of the general population, without prior clinical intervention, the following findings could be predicted.

| | |
|---|---|
| Numbers of the population with primary hypothyroidism | 1000 |
| $T_4$ has 95% sensitivity. Therefore | |
| Number of positive tests | 950 |
| Number of (false) negative tests (5%) | 50 |
| Number of subjects without thyroid disease | 99 000 |
| $T_4$ has 95% specificity. Therefore | |
| Number of negative tests | 94 050 |
| Number of (false) positive tests (5%) | 4950 |

Therefore the total number of positive tests (950+4950) is 5900. Of these, 950 have primary hypothyroidism.

Therefore the predictive value of a positive test is:

$$\frac{\text{No. of 'positive' tests given by subjects with the disease}}{\text{Total no. of positive tests}} \times 100$$

$$= \frac{950}{(950+4950)} \times 100 = 16.1\%$$

The total number of negative tests (50+94 050) = 94 100.
Therefore the predictive value of a negative test is:

$$\frac{\text{No. of negative tests given by non-diseased subjects}}{\text{Total no. of negative tests}} \times 100 =$$

$$\frac{94\,050}{94\,050 + 50} \times 100 = 99.9\%$$

In the above case, with a disease prevalence of 1%, less than two out of 10 positive tests will be due to subjects with the disease sought.

If all those subjects with positive tests (i.e. low $T_4$) are retested (5900):

| | |
|---|---|
| Number with primary hypothyroidism (remember 50 gave negative tests) | 950 |
| Number with positive tests (95%) | 902 |
| Number of negative tests (5%) | 48 |
| Number without the disease | 4950 |
| Number with negative tests (95%) | 4702 |
| Number with positive tests (5%) | 248 |

Predictive value of $T_4$ for a positive test $\quad \dfrac{902}{902 + 248} \times 100 = 78.4\%$

Predictive value of $T_4$ for a negative test $\quad \dfrac{4702}{4702 + 48} \times 100 = 99\%$

Therefore by selecting the population tested, the predictive value has markedly risen (from 16.1 to 78.4%), so that now almost eight out of 10 abnormal results (i.e. low $T_4$) will be due to subjects with primary hypothyroidism. This is simply because the disease prevalence has increased in the group of samples being retested.

| Disease prevalence % | Predictive value of a positive test % |
|---|---|
| 1 | 16.1 |
| 16.1 | 78.4 |

In the above situation a different test might be applied (e.g. plasma TSH) instead of repeating the same test, in which case the predictive value of the second test is also improved.

Therefore as the prevalence of a condition increases, so does the predictive value of a test. Many diseases have a prevalence of much less than 1%, with a consequently very poor predictive value.

The disease prevalence is crucial to the diagnostic effectiveness of the test. By careful clinical assessment and selection the doctor can markedly improve the prevalence of a disease in the samples submitted for analysis, and in so doing he improves the predictive value of the results he receives.

**Summary**

The clinician is always concerned with the individual patient, and although the concepts discussed above are statistically based, their incorporation into the background thinking when considering an individual result greatly assists in extracting the maximum diagnostic value.

**Clinical interpretation**

Once the valid clinical chemistry result has been interpreted both analytically and statistically, it may then be clinically interpreted. Both clinician and clinical chemist need to do this, the former for the management of the patient, the latter as a last check that the result, by reasonably fitting clinical expectation, can be a valid result for that patient. Clinical interpretation and pathophysiology for both clinician and clinical chemist forms the subject matter of the following chapters.

# 2 Water and sodium

Water is distributed between the body fluid compartments according to their initial relative osmotic pressures, and these in turn are determined by the number, not the weight, of solute particles present in each compartment. For example, there is a greater weight of albumin (40 g/l) in plasma than sodium (3.2 g/l), but there are more than 200 times as many sodium ions per litre as there are albumin molecules, as is directly reflected by calculating the molar concentration, i.e.

Albumin: mol.wt. ~ 65 000            Sodium: mol.wt. 22.99

$$\text{mmol/l} = \frac{\text{wt./l}}{\text{mol.wt.}} = \frac{40}{65\,000} = \sim 0.6 \text{ mmol/l} \qquad \frac{3.2}{22.99} = 139 \text{ mmol/l}$$

Sodium therefore contributes far more to the plasma osmotic pressure than does albumin, and in fact is responsible for about 48% of the total osmotic pressure (see p. 28). Potassium fulfils a similar role within the cell. The intimate association between water distribution and the major ions (sodium, potassium, chloride) requires that their clinical and diagnostic significance should be considered together, but for convenience the following sections will discuss the relevant homeostatic and pathophysiological aspects of water and sodium, whilst subsequent chapters will cover those of potassium and chloride.

## WATER BALANCE

The volume of water taken in and lost by the body depends on diet, activity and the environment. Average values for a healthy adult are shown in Table 2.1. The minimum intake of water that is necessary to maintain normal body function is the amount required for (1) renal elimination of the solutes produced by metabolism and (2) replacement of the losses from the skin and lungs.

In the normal subject approximately 600 mmol of solute have to be excreted by the kidneys each day. Since the kidney can concentrate tubular fluid to a maximum of 1200 mmol/kg the minimum amount of water required for solute excretion is 600/1200 = 0.5 l

**Table 2.1** Average daily water balance for a healthy 70 kg adult

|  | Input |  | Output |
|---|---|---|---|
| Oral fluids | 1400 | Urine | 1500 |
| Food | 700 | Lung | 500 |
| Metabolic oxidation | 400 | Skin | 400 |
|  |  | Faeces | 100 |
|  | 2500 |  | 2500 |

(500 ml). A similar water loss occurs via the skin and lungs, but these insensible losses can be markedly increased in an air-conditioned environment ($\sim$ 1000 ml/24 h). Metabolic water is produced during energy production and the metabolism of carbohydrates, fats and proteins. The minimum quantity of exogenous water required, therefore, lies in the range 500–1000 ml/24 h.

## WATER HOMEOSTASIS

Intracellular and extracellular water content depends on (1) fluid shifts across cell membranes, (2) loss of fluid from the body and (3) fluid intake.

### Cell membrane shifts

Water moves freely across the cell membranes in response to changes in the osmotic pressures of the two adjoining compartments. Sodium and potassium ions on the other hand tend to remain in their respective compartments under the influence of the 'sodium pump'. Sodium is the major cation of the extracellular fluid (ECF) and therefore the osmotic pressure of the ECF is largely governed by its sodium concentration. Changes in sodium concentration will result in osmolality changes (see p. 28) and thus in fluid movement between the ECF and intracellular fluid compartment (ICF).

$\uparrow$ECF sodium (or $\downarrow$ECF water)→$\uparrow$ECF osmolality→water flow from ICF to ECF space (cellular dehydration)

$\downarrow$ECF sodium (or $\uparrow$ECF water)→$\downarrow$ECF osmolality→water shift from ECF to ICF space (cellular overhydration)

### Loss of body fluids

Although water is lost from the skin, lungs and gut, the major organ controlling water loss from the body is the kidney. The cells of the collecting ducts passing through the renal medulla (where osmolality may reach as high as 1200 mmol/kg) are normally impermeable to water. In the presence of antidiuretic hormone (ADH), they become permeable and water is reabsorbed from the duct lumen into the renal medulla along the osmotic gradient. ADH is synthesized in the hypothalamus, passed down neural tracts and is stored in, and released into the blood from, the posterior pituitary. The formation and secretion of this peptide is regulated by osmoreceptors situated in the hypothalamus and baroreceptors situated in the right atrium and carotid sinus.

The osmoreceptors respond to changes in osmolality:

(1) ↑ECF osmolality→↑ADH release→renal water reabsorption to dilute ECF. Urine is concentrated
(2) ↓ECF osmolality→↓ADH release→↓water reabsorption and a dilute urine

The baroreceptors respond to intravascular volume (IVV) changes:

(1) ↓IVV→↑ADH release→↑renal water reabsorption to restore IVV
(2) ↑IVV→↓ADH→↓renal water reabsorption to reduce IVV

The osmoreceptors will respond to a change in ECF osmolality of around 2%. The baroreceptors on the other hand require a 10% change in IVV volume, before they respond. However, the baroreceptors can in certain circumstances over-ride the osmoreceptors, i.e. ↓IVV in the presence of ↓osmolality (stimulus for ADH cut off) will result in the secretion of ADH and water conservation. Thus the control mechanism will attempt to maintain IVV and, therefore, the blood circulation and its associated functions, at the expense of an abnormal ECF osmolality.

**Fluid intake**

The major mechanism controlling water intake is thirst. The thirst centre, situated in the hypothalamus, appears to be controlled by the same factors that control ADH. A decrease in blood volume or an increase in plasma osmolality will stimulate thirst and increase the oral intake of fluid.

**Control of extracellular volume**

In the normal person water homeostasis and extracellular volume (ECV) is, in the final analysis, dictated by the extracellular sodium content. An increase in total extracellular sodium will result in an increase in extracellular water.

↑ECF sodium content→↑ECF [Na$^+$]→↑ECF osmolality→
(1) thirst→↑oral fluid input
(2) ↑secretion of ADH→↑renal water reabsorption
(3) water shift out of cells into the ECF
1+2+3→↑ECV

A decrease in extracellular sodium content promotes the reverse situation.

↓ECF sodium content→↓ECF [Na$^+$]→↓ECF osmolality→
(1) ↓ADH secretion→↑renal water loss
(2) water shift from the ECF into the cells
1+2→↓ECV

Thus ECV is finally controlled by total body sodium balance.

When there are changes in the ECF sodium concentration the homeostatic mechanisms, mainly through ADH, attempt to keep the ECF osmolality as near normal as possible. This is done at the expense of the extracellular volume (ECV) and is an attempt to minimize compartmental movements of water and changes in the intracellular environment. In states of excessive extracellular fluid overload or depletion however, the control of vascular volume can, in defence of cardiovascular status, override the maintenance of a normal plasma osmolality.

## SODIUM HOMEOSTASIS

Sodium balance, which mainly depends on the renal regulation of sodium excretion, can generally be considered to be determined by the intravascular volume (IVV).

↓IVV→↑renal retention of sodium
↑IVV→↑renal loss of sodium

### Renal handling of sodium

Every 24 hours approximately 25 000 mmol of sodium are filtered by the kidneys. However, due to tubular reabsorption less than 1% of this sodium appears in the urine (100–200 mmol/24 h).

PROXIMAL TUBULE

Approximately 70% of the filtered sodium is reabsorbed in this segment by an energy-dependent process. The tubular fluid remains iso-osmotic to plasma due to the concurrent and osmotically equivalent reabsorption of water.

ASCENDING LOOP OF HENLE

Twenty to 30% of the filtered sodium is reabsorbed in this area. This reabsorption occurs mainly from the thick diluting segment and is thought to be consequent to the active reabsorption of chloride ions. The segment is impermeable to water and thus the remaining tubular fluid becomes hypotonic. This is a necessary process in the renal diluting mechanism. The reabsorbed sodium chloride is trapped in the renal interstitium and along with urea, which is reabsorbed in the collecting duct, contributes to the high medullary osmolality that is necessary for the process of urinary concentration (part of the countercurrent multiplying system).

DISTAL TUBULE

Five to 10% of the filtered sodium is reabsorbed by the distal convoluted tubule, which also secretes potassium into the tubular fluid. Although this reabsorption of $Na^+$ is under the influence of aldosterone and is also associated with the secretion of $K^+$ there is not a 1:1 molar exchange of these two ions.

COLLECTING DUCT

A final regulation of sodium balance occurs in this part of the nephron under the influence of aldosterone. The postulated natriuretic hormone ('third factor') probably also plays a part here.

The three main factors concerned in the renal handling of sodium are: (1) glomerular filtration rate (GFR), (2) aldosterone and (3) 'third factor'.

**Glomerular filtration rate**

A fall in GFR will result in a reduced tubular flow and an increased sodium conservation, and *vice versa*. The exact role of the GFR in sodium balance is unclear as many of the factors which control the filtration rate (e.g. ↑ or ↓ renal blood flow) also influence other renal sodium transport mechanisms.

**Aldosterone**

Decreased renal blood flow stimulates the production of aldosterone, through the renin-angiotensin system.

(1) ↓renal blood flow→↑renin secretion (renal juxtaglomerular apparatus)
(2) renin→conversion of angiotensinogen to angiotensin I
(3) angiotensin I converted to angiotensin II by converting enzyme (present mainly in lung)
(4) angiotensin II →(a) vasoconstriction, (b) ↑secretion of aldosterone
(5) aldosterone→↑distal renal tubular reabsorption of $Na^+$ and excretion of $K^+$. Aldosterone therefore promotes sodium retention and potassium depletion

**Third factor**

If GFR and aldosterone secretion rate remain constant an increase in the IVV will result in decreased proximal tubular and collecting duct sodium reabsorption. The reverse occurs if the IVV is decreased. The mechanism by which this occurs has been called the third factor. Cross perfusion experiments in dogs suggest that with volume expansion there is secretion of natriuretic hormone. Not all authorities accept this and suggest that the mechanism may involve changes in intrarenal haemodynamics.

Although it has been stated that body water is controlled by the extracellular sodium content and that sodium balance is controlled by the IVV, it must be recognized that these two mechanisms are closely related and act concurrently.

## DISORDERS OF PLASMA SODIUM CONCENTRATION

As indicated earlier plasma sodium concentration is finally dependent on the regulation of the body water. If the absolute amount of body sodium remains constant then a positive water balance (intake > output) will result in a low plasma $[Na^+]$ (total body sodium diluted). On the other hand a negative water balance (output > intake) will result in hypernatraemia (total body sodium concentrated). In this context water refers to free water, i.e. water free of solute.

In the normal subject the major factor controlling water balance is thirst, and since it is possible for a person to drink 10–20 litres of water a day it follows that hypernatraemia should not occur unless an adequate intake is prevented. Therefore hypernatraemia is usually associated with an inadequate water intake in the face of normal or increased losses from the body, e.g. too old, too young, too sick to drink.

*Hyponatraemia* infers that water is present in the ECF in excess of sodium. Therefore in order to maintain a normal plasma sodium concentration, the subject must be able to excrete sufficient free water from the body to maintain a normal ratio between the sodium and water content of the ECF. This is done by diluting the urine and excreting water in excess of sodium (i.e. $Na^+$ excreted $\lesssim 140$ mmol per litre of water excreted).

This excess urinary water (free water) may be calculated as follows:

$$\text{Free water} = V - \frac{U\ Osmo. \times V}{P\ Osmo.}$$

V = volume per unit time, U Osmo. = urinary osmolality, P Osmo. = plasma osmolality.

The above equation calculates the amount of water, free of any solute, excreted by the kidney. The $\frac{U\ Osmo. \times V}{P\ Osmo.}$ part of the equation calculates the quantity of total solute cleared by the kidney or, to put it another way, the amount of fluid isotonic to plasma that is excreted by the kidney. From the electrolyte point of view the above equation underestimates the 'electrolyte-free water' clearance because urea is a major contributor to the urinary osmolality and 'urea containing water' is electrolyte-free. Also urea freely crosses cell membranes and, therefore, does not cause intracellular water shifts. Hence to calculate the electrolyte-free water clearance the following equation should be used:

$$\text{Electrolyte-free water} = V - \frac{U([Na^+]+[K^+]) \times V}{P[Na^+]}$$

where $U([Na^+]+[K^+])$ = urinary concentration of $Na^+ + K^+$, $P[Na^+]$ = plasma concentration of $Na^+$.

A fall in the ECF osmolality of around 2–3% will result in the normal subject, in the excretion of a dilute urine (↑free water) with an osmolality of less than 100 mmol/kg. As it is possible for normal subjects to take in 10–20 l of free water a day and excrete it via the kidneys (generation of free water) without becoming hyponatraemic it follows that most patients who have hyponatraemia must have a disability of urinary free water clearance.

*Urinary dilution* or *free water clearance* depends on three major factors:
(1) Adequate delivery of fluid to the diluting segment of the ascending loop of Henle.
(2) Dilution of the urine in the ascending loop of Henle by the reabsorption of NaCl (this segment is impermeable to water).

(3) Absence of the action of ADH in the collecting duct.

These factors may be disturbed in the following situations and can result in free water retention and hyponatraemia if free water input is excessive.
(1) Inadequate delivery of fluid to diluting segment—↓GFR and ↑proximal reabsorption—shock, hypovolaemia.
(2) Inadequate dilution at diluting segment—loop diuretics (frusemide), thiazide diuretics, osmotic diuresis.
(3) ↑ADH action in collecting ducts—↑secretion of ADH in the face of ↓ECF osmolality, e.g. (a) volume depletion, (b) stress—physical, psychogenic, e.g. postoperative state, (c) drugs, e.g. barbiturates, opiates, chlorpropamide, tolbutamide, clofibrate, vincristine, vinblastine, tegretol, (d) augmented action of ADH—chlorpropamide, panadol, (e) circulation of ADH-like substance—syndrome of inappropriate antidiuresis (SIAD), (f) oxytocin (large doses).

## LABORATORY INVESTIGATION

In the investigation of dehydration, overhydration, hypernatraemia and hyponatraemia the following estimations are useful:

*Plasma*
Electrolytes, urea and creatinine, osmolality.

*Urine*
Osmolality, electrolytes.

**Plasma electrolytes**

PLASMA SODIUM

Plasma sodium levels do not give any information about the total body sodium content. They are only useful insofar as they classify the patient as hyponatraemic, normonatraemic or hypernatraemic.

PLASMA POTASSIUM

Plasma potassium levels may give some indication of associated conditions, e.g. high levels are associated with acidosis, hypoaldosteronism and renal failure, whereas low levels may be associated with hyperaldosteronism or diuretic therapy.

## PLASMA BICARBONATE

Bicarbonate levels usually provide a useful guide to acid-base status, but in this context care must be exercised (see Chapter 6).

## PLASMA UREA AND CREATININE

Increased urea levels indicate renal insufficiency consequent to a reduced GFR. This may occur in the presence of an otherwise normally functioning kidney (prerenal uraemia) or in the presence of intrinsic renal disease (renal uraemia, see p. 156). It is difficult to differentiate between the two on the basis of plasma urea levels alone, but a plasma urea concentration greater than say 20 mmol/l is more likely to result from intrinsic renal disease than a decrease in GFR, whilst levels between 8 and 20 mmol/l may reflect either condition. A low plasma urea may be taken to reflect overhydration providing the patient is not a neonate, does not have severe liver disease and has a normal protein intake. However, overhydration can be present in a patient with a normal or increased plasma urea level if the patient's original level was high or 'high normal'.

Similar care is required for interpreting plasma creatinine in the presence of altered hydration states. Plasma concentrations of creatinine are related to muscle mass and, therefore, plasma creatinine levels will generally be lower in patients of small build (children, elderly females) than those of an average size adult.

### Plasma and urinary osmolality

Osmotic pressure (OP) is the force that must be applied to prevent water moving from a compartment containing pure water to an adjoining one containing a solute solution through a membrane that is impermeable to the solute molecules. Osmotic pressure is dependent on the number of solute particles (ions or molecules) that are dissolved in the solvent. If no opposing force is applied in the above situation, the water will continue to enter the solute compartment and dilute the solute until the osmotic pressures of both compartments are equivalent. In the body, the semipermeable membranes are formed by cell walls and tissues, across which ions and water, but not proteins, can pass, whilst in some situations some ions cannot permanently pass either (ion pumps).

The traditional unit of measurement of osmotic pressure is the osmol, defined as the osmotic pressure associated with 1 mole of solute dissolved in 1 litre of water. The SI unit of the osmol is the mol. Two terms may be used to express OP.

(1) Osmolarity: expresses the solute concentration per volume of solution. Unit: mmol/l.
(2) Osmolality: expresses the solute concentration per weight of solvent. Unit: mmol/kg water.

Because the 'mole' describes the amount of a substance in terms of the number of molecules present, (i.e. 1 mole $= 6.024 \times 10^{23}$ molecules (Avogadro's number) and moles of a substance $=$ (wt. of substance in g)/(mol. wt.)) the use of such units in measuring plasma analytes allow the osmotic pressure of a plasma sample to be easily calculated if the direct estimation is not available. In making such a calculation it should be remembered that each positive ion present in plasma has to be matched by a negative ion to maintain electroneutrality. Not all these negative ions are measured in the diagnostic laboratory (usually only $Cl^-$ and $HCO_3^-$) and, so as to allow for this, the values obtained for the measured cations ($Na^+$, $K^+$) need to be multiplied by a factor of two. This obviously does not apply to non-electrolytes such as glucose and urea.

|  |  |  | mmol /kg |
|---|---|---|---|
| Plasma Na | 140 | mmol/l | 280 |
| K | 4.0 | mmol/l | 8 |
| Urea | 5.0 | mmol/l | 5 |
| Glu. | 4.0 | mmol/l | 4 |
| Total osmotic pressure |  |  | 297 |

(Calculated OP (mmol/kg) $= 2([Na^+] + [K^+]) + [Urea] + [Glu.])$

This calculation does not take account of the contribution made by osmotically active particles such as proteins, calcium, magnesium, etc. However such estimates give a rough guide to plasma osmolality and are particularly useful in situations where the water content of plasma has been markedly altered (hyperlipidaemia, hyperproteinaemia).

It is important to distinguish between plasma total osmotic pressure, which reflects the presence of all the osmotically active particles present, and tonicity (effective osmotic pressure), which is the part of the total osmotic pressure in plasma that is due to those osmotically active molecules that are unable to freely cross the cell membrane ($Na^+$, $K^-$, $Cl^-$, $HCO_3^-$, glucose, etc.). High concentrations of freely mobile molecules such as urea or ethanol increase the total (measured) osmolality of plasma, but do not affect its tonicity.

The tonicity of the ECF is of clinical importance because of its effect on cell hydration; a hypertonic ECF causes cellular dehydration, whilst a hypotonic ECF overhydrates cells.

The total (measured) plasma osmolality and tonicity usually parallel each other, but there can be important discrepancies. For example:

Plasma  Na    136  mmol/l
        K     6.0  mmol/l
        Urea  78.1 mmol/l
        Glu.  5.0  mmol/l
        Osmo. 368  mmol/kg (275–295)
Calculated tonicity = $2([Na^+]+[K^+])+[Glu.]$
                     289 mmol/kg

The tonicity is normal and, therefore, there will be no water movement between cells and the ECF, the additional osmotically active particles (urea) being equally distributed between both compartments.

OSMOLAL GAP

The osmolal gap is the difference between the measured osmolality and the calculated osmolality of plasma.

Measured osmolality (mmol/kg) $-\{2\times([Na^+]$ mmol/l $+[K^+]$ mmol/l$)+[Urea]$ mmol/l $+[Glu.]$ mmol/l$\}$

The osmolal gap gives a measure of low molecular weight substances that are present in the plasma, other than $Na^+$, $K^+$, urea and glucose, which contribute to the plasma osmolality, e.g. ethanol, methanol, ethylene glycol, drugs, mannitol. The normal osmolal gap is approximately 10 mmol/l.

*Example* Patient admitted after a heavy bout of drinking.

Plasma  Na         144   mmol/l   (132–144)
        K          3.6   mmol/l   (3.2–4.8)
        Cl         100   mmol/l   (93–108)
        $HCO_3$    29    mmol/l   (23–33)
        Urea       2.8   mmol/l   (3.0–8.0)
        Glu.       5.0   mmol/l   (3.0–5.5)
        Osmo.      371   mmol/kg  (275–295)
        Osmo. gap  67    mmol/kg  (<10)

The plasma alcohol was 2.9 g/l (0.29% or 63 mmol/l) and thus accounted for the osmolal gap. The calculated tonicity is 300 mmol/kg (ethanol penetrates cell membranes) and is another example of plasma tonicity markedly differing from plasma osmolality.

The normal plasma osmolality is around $285\pm10$ mmol/kg. There is no 'normal value' for urinary osmolality; it depends on the state of hydration and can range from 50 to 1200 mmol/kg. In water and electrolyte disturbances, osmolality measurements have a valuable part to play in diagnosis and treatment but only if samples for plasma and 'spot' urines are taken *at the same time*. Random urinary osmolalities are useless unless they form part of a dynamic function test (e.g. water deprivation test). In the presence of normal renal, adrenal and hypothalamic-pituitary function the following relationships apply:

### Euhydration

(1) ↑fluid intake

$$\left.\begin{array}{l}\text{Plasma Osmo.} = N \\ \text{Urine Osmo.} = \downarrow\end{array}\right\} \quad \frac{\text{U Osmo.}}{\text{P Osmo.}} = <1$$

(2) ↓fluid intake

$$\left.\begin{array}{l}\text{Plasma Osmo.} = N \\ \text{Urine Osmo.} = \uparrow\end{array}\right\} \quad \frac{\text{U Osmo.}}{\text{P Osmo.}} = >1$$

*Diarrhoea, Vomiting*

### Overhydration

For example, infusions of large amounts of 5% glucose. Glucose is metabolized immediately and the process is therefore tantamount to infusing pure water.

$$\left.\begin{array}{l}\text{Plasma Osmo.} = \downarrow \\ \text{Urine Osmo.} = \downarrow\downarrow\end{array}\right\} \quad \frac{\text{U Osmo.}}{\text{P Osmo.}} = <1$$

Urine will approach maximum dilution (50–100 mmol/kg)

### Dehydration

Unconscious patient who has not taken fluid for some hours but who will still be losing fluid from the skin, lungs and kidney.

$$\left.\begin{array}{l}\text{Plasma Osmo.} = \uparrow \\ \text{Urine Osmo.} = \uparrow\uparrow\end{array}\right\} \quad \frac{\text{U Osmo.}}{\text{P Osmo.}} = >1$$

Urine will approach maximum concentration (1000–1200 mmol/kg)

In chronic renal failure the kidney has difficulty in concentrating and diluting urine due to nephron damage and osmotic (urea) diuresis, and so the urine osmolality tends to be much the same as the plasma osmolality.

$$\frac{\text{U Osmo.}}{\text{P Osmo.}} \simeq 1$$

The plasma osmolality in renal failure will depend on the state of hydration and the plasma urea level.

**Urine electrolytes**

SODIUM

The 24 hour urinary sodium output is dependent on dietary intake, and so this measurement is valueless unless sodium intake is known. However, sodium concentrations in 'spot urines', if looked at in conjunction with the plasma sodium levels, are helpful procedures in evaluating patients with (1) volume depletion, (2) oliguria, (3) hyponatraemia.

*Note* Patients on diuretic therapy will have high spot urinary sodium concentrations ($>20$ mmol/l), depending on the duration of therapy. With long term therapy these patients may come back into sodium balance (escape phenomenon).

Volume depletion

In a subject with a depleted extracellular volume a spot urinary sodium can indicate whether sodium is being lost from the kidney (hypoaldosteronism, salt-losing nephritis, diuretics, etc.) or from an extrarenal source (diarrhoea, etc.). If the loss is extrarenal the volume depletion will stimulate the kidney to conserve sodium and the spot [$Na^+$] will be low ($<10$ mmol/l). If there is renal salt loss the concentration will be high ($>20$ mmol/l).

Oliguria

Oliguria is generally defined as a urinary volume of less than 400 ml/24 h and may be due to prerenal causes (dehydration or blood loss), or to acute tubular necrosis (acute renal failure). It is important to differentiate between these two conditions as their treatment is markedly different.

In prerenal uraemia (in the absence of intrinsic renal disease) the urinary spot sodium concentration will be low ($<10$ mmol/l) as aldosterone and other factors act to increase sodium retention and so repair any ECV deficit. If there is tubular damage (acute tubular necrosis) some sodium will escape reabsorption by these tubules and the spot value will be high ($>20$ mmol/l).

Hyponatraemia

In the presence of normal renal function hyponatraemia due to extrarenal loss of sodium will be associated with a low spot urinary sodium concentration (<10 mmol/l). When hyponatraemia is associated with a high urinary spot [Na$^+$] (>20 mmol/l) one of the following causes should be suspected: (1) renal salt-losing states, (2) syndrome of inappropriate secretion of ADH, (3) diuretic therapy, (4) urinary loss of complex anions, e.g. HCO$_3^-$ during active vomiting (see below).

Fractional excretion of sodium (FE$_{Na^+}$)

This calculation estimates the amount of sodium that is filtered by the glomeruli but then escapes the tubular reabsorption process

$$FE_{Na^+} = \frac{U[Na^+] \times P[Creat.]}{P[Na^+] \times U[Creat.]} \times 100\%$$

Where U[Na$^+$] = urinary sodium concentration, P[Na$^+$] = plasma sodium concentration, P[Creat.] = plasma creatinine concentration, U[Creat.] = urinary creatinine concentration.

This estimation is a more sensitive technique for the differentiation of the causes of oliguria (prerenal or renal) than a spot urinary sodium estimation.

(1) FE$_{Na^+}$ < 1%—prerenal uraemia (dehydration)
(2) FE$_{Na^+}$ > 1%—acute tubular necrosis

## CHLORIDE

Changes in spot urine chloride concentrations generally parallel those of sodium and provide the same diagnostic information. However a spot urinary chloride estimation is useful in those situations where hypovolaemia (stimulus for increased renal retention of NaCl) is associated with a renal loss of a poorly reabsorbable anion (e.g. HCO$_3^-$, SO$_4^{2-}$, HPO$_4^{2-}$). This may arise in two conditions:

(1) *Hypovolaemia associated with metabolic alkalosis* During vomiting the continual loss of gastric HCl causes the generation of excess HCO$_3^-$. If the glomerulus filters more HCO$_3^-$ than can be reabsorbed by the proximal tubule the excess appears in the urine as NaHCO$_3$, the HCO$_3^-$ being poorly absorbed by the distal nephron (thus urinary [Na$^+$]↑). Chloride is avidly taken up by the proximal nephron, so that if the vomiting also caused hypovolaemia the urinary [Cl$^-$] will be low (<10 mmol/l). Therefore in cases of severe vomiting, a spot urinary chloride value will give a better indication of the patient's state of hydration than the urinary [Na$^+$]. Typically, spot urinary results during the vomiting phase with hypovolaemia are:

Urine $[Na^+]$ = >20 mmol/l
$FE_{Na^+}$ = >1%
$[Cl^-]$ = <10 mmol/l

The above results should be compared with those found in acute tubular necrosis when generally urinary $[Cl^-]$ = >20 mmol/l.

(2) *Proximal renal tubular acidosis (RTA)* In this disease the proximal renal tubule is unable to reabsorb bicarbonate. The consequent loss of $HCO_3^-$ and its obligatory $Na^+$ in the urine leads to sodium depletion and dehydration. In this situation the filtered chloride is reabsorbed by the distal nephron, so that urinary $[Cl^-]$ = <10 mmol/l. The urinary $[Na^+]$ will be >20 mmol/l because of the obligatory loss with $HCO_3^-$.

## Laboratory investigation of salt and water disturbances

The investigation of salt and water disturbances can be often assisted by assessing the osmolality and $[Na^+]$ values obtained from both plasma and urine, as shown in Figs 2.1, 2.2, and 2.3.

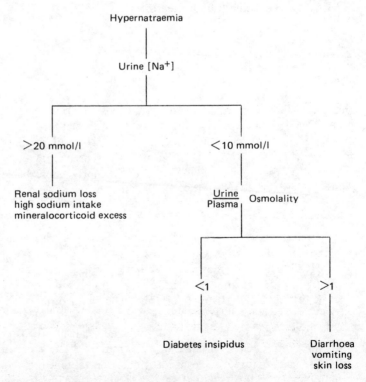

**Fig. 2.1** Laboratory investigation of hypernatraemia.

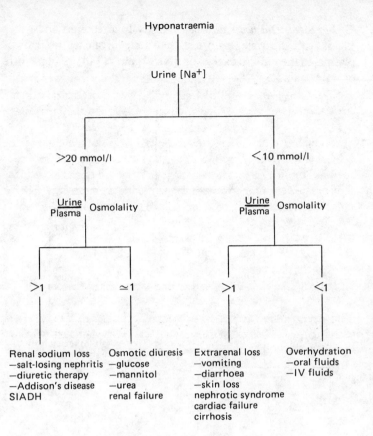

**Fig. 2.2** Laboratory investigation of hyponatraemia.

## HYPERNATRAEMIA

A high plasma sodium concentration does not necessarily mean that the total body sodium content is increased, but infers that the extracellular sodium is excessive relative to water (i.e. $\gtrsim 140$ mmol/l. This can occur if the total body sodium is normal, increased or decreased. As noted earlier hypernatraemia is usually associated with an insufficient water intake.

### Classification

(1) *Decreased total body sodium*  Extracellular sodium and water deficiency, with the water loss being greater than its equivalent, in terms of plasma, to the sodium loss, i.e. $[Na^+]$ loss $\lesssim 140$ mmol per litre of water loss.

(2) *Normal total body sodium*  Extracellular water deficiency associated with minimal or no sodium deficit.

(3) *Increased total body sodium* Extracellular sodium and water excess with the water gain less than its equivalent, in terms of plasma, to the sodium excess, i.e. [Na$^+$] gain $\gtrsim$140 mmol per litre of water gain.

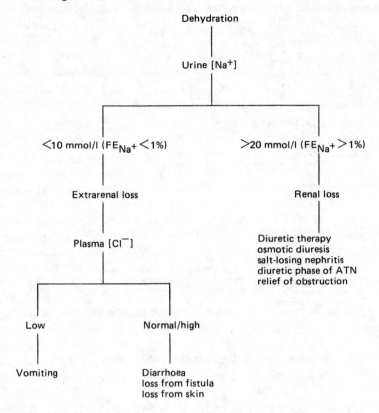

Fig. 2.3 Laboratory investigation of dehydration.

## Decreased sodium and water

LOSS OF HYPOTONIC FLUID

Extrarenal

(1) Vomiting/diarrhoea, (2) excessive sweating, (3) dialysis.

Renal

Osmotic diuresis (glucose, urea, mannitol).

## Normal sodium and decreased water

PURE WATER DEPLETION

Extrarenal

*Decreased intake* (1) Unconscious, (2) thirst centre dysfunction, (3) mechanical obstruction, (4) inadequate IV therapy.

*Increased loss* (1) Fever/thyrotoxicosis, (2) hot dry environment, (3) non-humidified ventilators.

Renal

Diabetes insipidus.

## Increased sodium and increased water

*High sodium intake* (1) Seawater ingestion, (2) high sodium diet (accidental in infants feed), (3) iatrogenic (hypertonic saline, sodium bicarbonate).

*Low sodium output* Mineralocorticoid excess.

*Note* None of the body fluids mentioned (urine, gut secretions, etc.) have a sodium concentration greater than that of plasma and, therefore, in these conditions hypernatraemia will occur only if water intake is inadequate to replace the losses (see Table 2.2).

**Table 2.2** Approximate sodium concentration of some body fluids (mmol/l)

| Plasma | Sweat | Gastric juice | Pancreatic/ bile fluid | Small bowel | Diarrhoea fluid |
|---|---|---|---|---|---|
| 140 | 40 | 60 | 140 | 100 | 60 |

## Case examples/pathophysiology

RENAL SODIUM AND WATER LOSS

Osmotic diuresis due to diabetes mellitus in a 29 year old male.

Chapter 2

| Plasma | Na | 158 mmol/l | (132–144) | Urine | Na | 25 mmol/l |
| --- | --- | --- | --- | --- | --- | --- |
| | Urea | 16 mmol/l | (3.0–8.0) | | Osmo. | 450 mmol/kg |
| | Glu. | 43 mmol/l | (3.0–5.5) | $\dfrac{\text{Urine}}{\text{plasma}}$ Osmo. | | 1.14 |
| | Osmo. | 393 mmol/kg | (275–295) | | | |

Pathophysiology

See Fig. 2.4.

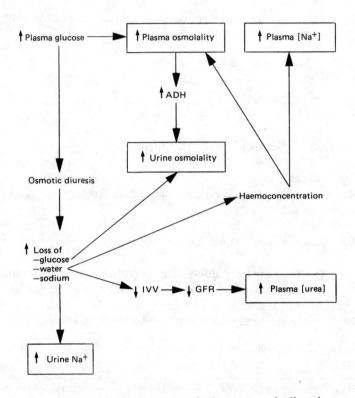

Fig. 2.4 Mechanism of hypernatraemia due to osmotic diuresis.

EXTRARENAL SODIUM AND WATER LOSS

A 53 year old admitted to hospital after persistent vomiting for the previous 18 hours.

| Plasma | Na | 150 mmol/l | (132–144) | Urine | Osmo. | 710 mmol/kg |
| --- | --- | --- | --- | --- | --- | --- |
| | K | 2.8 mmol/l | (3.1–4.8) | | Na | 5 mmol/l |
| | Cl | 90 mmol/l | (98–108) | | Creat. | 10 mmol/l |
| | $HCO_3$ | 37 mmol/l | (23–33) | $\dfrac{\text{Urine}}{\text{plasma}}$ Osmo. | | 1.75 |
| | Urea | 25 mmol/l | (3.0–8.0) | | | |
| | Creat. | 0.22 mmol/l | (0.06–0.12) | | | |
| | Osmo. | 405 mmol/kg | (275–295) | $FE_{Na}$ | | 0.07 % |

## Pathophysiology

See Fig. 2.5. Note that the urinary $[Na^+]$ is <20 mmol/l, suggesting that all the filtered bicarbonate is being reabsorbed by the renal tubules (see p. 33).

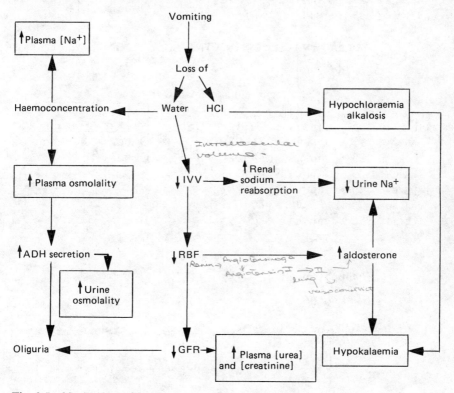

**Fig. 2.5** Mechanism of hypernatraemia due to vomiting.

### RENAL WATER LOSS—DIABETES INSIPIDUS

In this condition the absence of ADH results in a constantly dilute urine (50–200 mmol/kg). The urine flow may be of the order of 5–10 l/24 h, but if the patient is able to replace this orally there will be no dehydration; under these circumstances the plasma osmolality and sodium concentration will be within normal limits. If the patient is unable to replace fluid losses (ill or unconscious) he will become hypertonically dehydrated.

|  |  | Balanced fluid intake | Inadequate fluid intake |  |  |
|---|---|---|---|---|---|
| Plasma | Na | 144 | 153 | mmol/l | (132–144) |
|  | Urea | 5.0 | 12.5 | mmol/l | (3.0–8.0) |
|  | Osmo. | 285 | 345 | mmol/kg | (275–295) |
| Urine | Osmo. | 120 | 250 | mmol/kg |  |
|  | Vol. | 6 | 4 | l/24 h |  |
| Urine/plasma | Osmo. | 0.42 | 0.72 |  |  |

If a patient with diabetes insipidus becomes severely dehydrated there is a slight increase in urinary osmolality and a decrease in urinary volume due to (1) some residual ADH secretion (occasional) or (2) decreased GFR causing an increased water (and sodium) reabsorption by the proximal tubule.

EXTRARENAL LOSS AND INADEQUATE INTAKE OF WATER

A 46 year old unconscious woman with fever.

| Plasma | Na | 165 | mmol/l | (132–144) | Urine | Na | 5 | mmol/l |
| --- | --- | --- | --- | --- | --- | --- | --- | --- |
| | K | 3.2 | mmol/l | (3.1–4.8) | | Osmo. | 900 | mmol/kg |
| | Cl | 123 | mmol/l | (98–108) | $\dfrac{\text{Urine}}{\text{Plasma}}$ | Osmo. | 2.43 | |
| | $HCO_3$ | 25 | mmol/l | (23–33) | | | | |
| | Urea | 20.6 | mmol/l | (3.0–8.0) | | | | |
| | Creat. | 0.21 | mmol/l | (0.06–0.12) | | | | |
| | Osmo. | 370 | mmol/kg | (275–295) | | | | |

Pathophysiology

See Fig. 2.6.

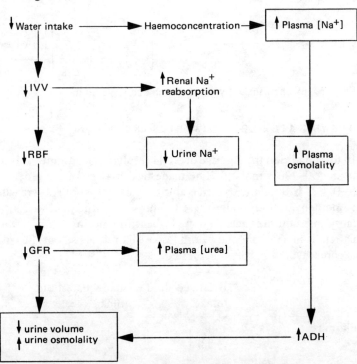

**Fig. 2.6** Mechanism of hypernatraemia due to inadequate intake of water.

*Note* In pure water deficiency the water loss is 'shared' between the intracellular and extracellular compartments (osmotic equilibrium). Therefore, if a person whose normal total body water is 40 l loses 1 l of pure water, his extracellular (and plasma) volume will decrease by 1/40 (2.5%). If the pure water loss is large it should be remembered that replacement therapy should be given slowly, otherwise the rapid return of water to depleted intracellular areas such as brain cells will cause damage. On the other hand, if our normal subject loses 1 l of isotonic fluid from his ECF there will be no cellular shift of water and the loss will be exclusive to the extracellular compartment. The extracellular fluid volume will then decrease by 1/15 (15 l of water in ECF) or 6.7%. In this case, replacement therapy should be rapid so that the circulating blood volume can be restored. Therefore, isotonic fluid losses have usually more serious consequences (i.e. decreased blood volume and reduced blood pressure) than pure fluid losses.

SALT OVERLOAD

A psychiatrically disturbed patient who consumed half a bottle of salt tablets ($\sim 25$ g of NaCl).

| Plasma | Na | 157 | mmol/l | (132-144) | Urine Na | 75 | mmol/l |
|---|---|---|---|---|---|---|---|
| | K | 4.8 | mmol/l | (3.2-4.8) | Osmo. | 678 | mmol/kg |
| | Cl | 126 | mmol/l | (98-108) | | | |
| | $HCO_3$ | 26 | mmol/l | (23-33) | | | |
| | Urea | 4.1 | mmol/l | (3.0-8.0) | | | |
| | Creat. | 0.10 | mmol/l | (0.06-0.12) | | | |
| | Osmo. | 332 | mmol/kg | (275-295) | | | |

With pure salt (sodium) gain, as in other hypertonic states, water moves out of the cells into the hypertonic extracellular fluid causing cellular dehydration. This water shift may be extensive and result in severe extracellular fluid overload and pulmonary oedema. This contrasts with the other types of hypertonic situations (pure water loss, water and salt loss) in which the extracellular fluid is contracted and the patient is dehydrated.

PRIMARY ALDOSTERONISM

In primary aldosteronism there is an increased sodium retention by the distal tubule, raising the serum osmolality, which in turn releases ADH. This causes concurrent water reabsorption equivalent to the retained sodium and the end result may be a normal plasma sodium concentration. Occasionally, the water retention may be less than equivalent to the sodium and a slightly high plasma sodium concentration will result. Generally, the plasma sodium levels are in the upper half of, or just above, the reference range (see p. 138 for a full discussion of this condition).

|        |         | Patient A | Patient B |        |           |
|--------|---------|-----------|-----------|--------|-----------|
| Plasma | Na      | 141       | 146       | mmol/l | (132–144) |
|        | K       | 2.9       | 2.8       | mmol/l | (3.2–4.8) |
|        | Cl      | 92        | 99        | mmol/l | (93–108)  |
|        | $HCO_3$ | 35        | 36        | mmol/l | (23–33)   |
|        | Urea    | 4.8       | 6.0       | mmol/l | (3.0–8.0) |

As might be expected from the action of aldosterone the most common plasma electrolyte abnormality in this condition is hypokalaemic alkalosis (see Chapter 3) but surprisingly this is not seen in every case. Overall however, the total body sodium (and water) is increased and body potassium is depleted.

It would be reasonable to expect the retention of sodium and water to result in oedema. This is not the case because as the extracellular volume increases two events occur:
(1) Sodium begins to be excreted by the kidney due to the increased IVV. This is termed the 'escape phenomenon' (? third factor, p. 25).
(2) ADH production is cut off (volume expansion) and renal reabsorption of water is decreased.

As these two factors come into play the urinary osmolality and sodium concentration will reflect the oral input of sodium and water.

**Consequences of the hypertonic state**

Hypertonicity of the extracellular fluid, depending on its duration, may cause the following.

Fluid shifts

Water moves from the intracellular to the extracellular fluid causing cellular dehydration. In the brain the resulting tissue shrinkage may result in tearing of the blood vessels that are attached to the rigid calvarium, causing haemorrhage.

Hyperglycaemia

Hypertonicity may be associated with hyperglycaemia due to a decreased rate of insulin secretion. The exact cause of this is unknown.

Hyperkalaemia

Extracellular hypertonicity results in movement of potassium out of cells. This may represent an attempt by the cell to maintain the intracellular/extracellular potassium gradient.

### Formation of 'idiogenic molecules'

It has been shown experimentally in animals that prolonged hypertonic states cause brain cells to increase their intracellular concentration of osmotically active particles. This has the effect of returning the intracellular fluid content and volume back to normal, thus protecting the organ. Up to 50% of this intracellular increase in osmolality is due to cellular uptake of $Na^+$, $K^+$ and glucose from the extracellular fluid. The other 50% is attributed to particles that are generated within the cell (idiogenic molecules). There is also good evidence to suggest that this process also occurs in humans subjected to prolonged hypertonicity. This mechanism has important implications with regard to therapy, i.e. sudden rehydration of a hypertonically dehydrated subject with hypotonic solutions may result in severe cerebral oedema. It is not known how long it takes for these idiogenic molecules to disappear and, therefore, rehydration with hypotonic fluids should take place slowly over 1–3 days.

## Principles of therapy

### Pure water loss

If the water loss is from the ECF, then water moves from the ICF to return the ECF osmolality towards normal, and so the water loss is equally distributed between the two compartments. Therefore, severe hypovolaemia and shock tends not to occur unless the dehydration is severe (e.g. plasma $[Na^+] > 170$ mmol/l). Therapy is aimed at replacing the fluid deficit with hypotonic fluids over a period of 1–3 days. Replacement must not be rapid as this will suddenly lower the extracellular tonicity, resulting in large fluid shifts into the cells and possible cerebral oedema (see idiogenic molecules above). Fluids which may be used are:
(1) Tap water. If this can be taken orally this is the best method of rehydration.
(2) 5% glucose. This fluid is isotonic (osmolality $\sim 300$ mmol/kg), but after infusion the glucose is quickly metabolized and thus it is equivalent to infusing pure water.
(3) 4% glucose in 1/5 normal saline. This solution is also isotonic (osmolality $\sim 300$ mmol/kg). Note that in the case of intravenous saline solutions 'normal' means that the solution contains 0.9% NaCl, i.e. $\sim 150$ mmol of sodium/l and an osmolality of $\sim 300$ mmol/kg.
(4) Half-normal saline. This fluid is hypotonic with regard to plasma. Hypotonic fluids have the disadvantage of producing haemolysis when infused intravenously.

## Water and salt loss

If the fluid lost from the ECF has the same tonicity as that of the ICF (isotonic), then there will be no fluid shift between the two compartments and the volume loss will be sustained by the ECF alone. Since the consequences of a reduced ECF are potentially serious (reduced circulation and renal function) the loss of isotonic fluid requires rapid repair. Normal saline is the fluid of choice and may be administered quickly since there will be no water movement into the brain.

Usually most fluid losses (gastrointestinal, sweat, osmotic diuresis) are hypotonic. In these cases, the loss could be considered as being partly isotonic and partly of pure water, e.g. a 1 litre loss of fluid of 70 mmol $Na^+/l$ could be considered as comprising (1) 500 ml isotonic (140 mmol $Na^+/l$) and (2) 500 ml pure water.

Therapy can then be considered in two parts.

(1) *Expansion of extracellular volume* This must be done rapidly to prevent the serious consequences of hypovolaemia (e.g. acute tubular necrosis) and to improve renal function. Normal saline is the fluid of choice but if the patient is in a state of shock plasma or whole blood infusions may be necessary.

(2) *Replacement of pure water loss* After blood pressure and renal function have improved (increasing urine output) the pure water deficit may be corrected over the next one to three days as outlined above.

### Pure salt gain

This is uncommon but may occur due to injudicious infusion of hypertonic saline solutions (e.g. twice normal saline). The danger in these cases is a rapid extracellular volume expansion (water drawn out of cells) and a subsequent pulmonary oedema. Therapy is aimed at removing the excess salt and correcting the hypertonicity with 5% glucose infusions. The removal of salt can be accomplished by the use of a potent diuretic (e.g. frusemide). This treatment will aggravate the hypertonicity as the diuresis fluid will contain sodium in a concentration less than the plasma level (i.e. more water is being lost than sodium). Therefore at the same time as the diuresis, an infusion of 5% glucose will be necessary. If this therapy is ineffective dialysis may be required.

## HYPONATRAEMIA

A low plasma sodium concentration infers that the plasma sodium content is relatively less than its normal equivalent water content. However, this situation can occur in the presence of low, normal or increased body sodium. In almost all cases of true hyponatraemia there is a defect in the urinary excretion of free water, due either to excessive ADH (or ADH-like substance) or to intrarenal mechanisms.

## Classification

Hyponatraemia may occur under the following circumstances.

(1) *Hypotonic dehydration* Decreased total body sodium associated with a less than equivalent loss of extracellular water, i.e. $Na^+$ loss $\gtrsim 140$ mmol per litre of water lost.

(2) *Normal total body sodium* associated with (a) euhydration, normal extracellular water, (b) overhydration, increased extracellular water.

(3) *Overhydration* Increased total body sodium associated with a more than equivalent gain of extracellular water, i.e. $Na^+$ gain $\lesssim 140$ mmol per litre of water gained.

## Case examples/pathophysiology

### Hyponatraemia with decreased body sodium and water

Always due to fluid loss being replaced by salt-poor fluids.

Extrarenal sodium loss
*Skin*  (1) Sweating, (2) burns.
*Gut*  (1) Vomiting, (2) diarrhoea, (3) fistula.

Renal sodium loss
(1) Osmotic diuresis, (2) diuretic therapy, (3) mineralocorticoid deficiency, (4) salt-losing nephritis, (5) diuretic phase of acute tubular necrosis, (6) renal tubular acidosis.

EXTRARENAL SODIUM LOSS

Prolonged vomiting in a 48 year old man with a suspected intestinal obstruction.

Chapter 2

| Plasma | Na | 128 | mmol/l | (132–144) | Urine | Na | 6 | mmol/l |
|---|---|---|---|---|---|---|---|---|
| | K | 2.1 | mmol/l | (3.2–4.8) | | Osmo. | 650 | mmol/kg |
| | Cl | 79 | mmol/l | (98–108) | Urine/Plasma | Osmo. | 2.39 | |
| | HCO$_3$ | 40 | mmol/l | (23–33) | | | | |
| | Urea | 3.0 | mmol/l | (3.0–8.0) | | | | |
| | Osmo. | 272 | mmol/kg | (275–295) | | | | |

Fluid loss from the gut will, in the first instance, result in hypernatraemia (water loss greater than salt loss). The thirst mechanism is stimulated and fluids are taken orally. However, the underlying illness often makes patients disinclined to eat, and so very little salt is replaced. Thus the water deficiency is corrected more rapidly than that of sodium, resulting in hyponatraemia.

(1) water and sodium loss with pure water (or salt-poor fluid) replacement →↓ECF sodium →↓ECF osmolality→
    (a) fluid shift into cells (cellular overhydration)     →↓ECV and ↓IVV
    (b) ↓ADH secretion →↑ renal water loss
(2) ↓IVV→ (a) ↑renal sodium retention →↓ urinary sodium (<10 mmol/l), (b) ↑ADH secretion →↑renal water reabsorption → ↑urine osmolality
(Recall from p. 22 that volume contraction can override hypo-osmolality with regard to ADH control.)

Gastric fluid is rich in chloride ions and thus vomiting will result in hypochloraemia. The loss of hydrogen ion from the stomach results in a metabolic alkalosis (↑plasma [HCO$_3^-$]). The hypokalaemia reflects (1) loss in the vomitus, (2) renal loss due to the aldosteronism that is secondary to the contracted ECV, (3) return of potassium into the cells due to the alkalosis.

RENAL SODIUM LOSS–ADDISON'S DISEASE

Biochemical results on a blood sample taken on admission of a 61 year old woman with subsequently diagnosed Addison's disease.

| Plasma | Na | 122 | mmol/l | (132–144) | Urine | Na | 30 | mmol/l |
|---|---|---|---|---|---|---|---|---|
| | K | 6.2 | mmol/l | (3.1–4.8) | | Osmo. | 350 | mmol/kg |
| | HCO$_3$ | 17 | mmol/l | (23–33) | Urine/Plasma | Osmo. | 1.28 | |
| | Urea | 12 | mmol/l | (3.0–8.0) | | | | |
| | Osmo. | 273 | mmol/kg | (275–295) | | | | |

In primary adrenal failure the lack of aldosterone results in salt wasting from the distal renal tubules (↓Na$^+$-K$^+$ exchange). This will lead to hyponatraemia and a low plasma osmolality, causing a cessation of ADH secretion. Water will then be lost by the kidney, bringing the plasma [Na$^+$] and osmolality back to normal. However, the salt loss is continuous and, therefore, so will be the water loss. If these events are followed to their logical conclusion the patient would soon become totally depleted of both sodium and water. This does not happen because as the IVV decreases ADH will eventually be secreted, regardless of the plasma osmolality, causing renal water conservation (baroreceptors over-riding osmoreceptors, p. 22). The decreased IVV also stimulates proximal tubular sodium reabsorption, resulting in some sodium conservation.

The high plasma urea reflects the lowered GFR due to the dehydration (prerenal uraemia). The high [K$^+$] is due to decreased distal tubular secretion and the concurrent acidosis. The low bicarbonate (acidosis) is due to depressed H$^+$ secretion (aldosterone deficiency) and renal insufficiency (prerenal uraemia).

*Water and sodium*

SALT-LOSING NEPHRITIS

A 25 year old man with polycystic kidneys.

| Plasma | Na | 124 | mmol/l | (132–144) | Urine | Na | 65 | mmol/l |
|---|---|---|---|---|---|---|---|---|
| | K | 6.0 | mmol/l | (3.1–4.8) | | Osmo. | 310 | mmol/kg |
| | HCO$_3$ | 12 | mmol/l | (23–33) | $\dfrac{\text{Urine}}{\text{Plasma}}$ Osmo. | | 1.01 | |
| | Urea | 33 | mmol/l | (3.0–8.0) | | | | |
| | Osmo. | 305 | mmol/kg | (275–295) | | | | |

In early chronic renal failure the salt intake and excretion is usually normal and the patient is in sodium balance. As the disease progresses disturbances of sodium and water homeostasis may occur.

In some cases there may be gradual sodium retention as GFR falls, resulting in ECV expansion and hypertension. On the other hand, as in the above case, there may be a slow progressive sodium loss producing sodium depletion, decreased plasma sodium concentration and a contracted extracellular volume. In the above case, the urinary sodium greatly exceeds 20 mmol/l in the face of a low plasma sodium. The exact cause of this sodium-losing nephritis is unclear, but it is probably related to malfunction of the remaining nephrons and osmotic (urea) diuresis. This condition is most commonly associated with analgesic nephropathy, polycystic kidneys, medullary cystic disease and pyelonephritis.

OSMOTIC DIURESIS

Diabetes mellitus in an adult.

| | | | | | | | |
|---|---|---|---|---|---|---|---|
| Plasma Na | 123 | mmol/l | (132-144) | Urine Na | | 91 | mmol/l |
| K | 7.8 | mmol/l | (3.2-4.8) | Osmo. | | 655 | mmol/kg |
| $HCO_3$ | 3 | mmol/l | (23-33) | $\dfrac{\text{Urine}}{\text{Plasma}}$ Osmo. | | 2.04 | |
| Urea | 16 | mmol/l | (3.0-8.0) | | | | |
| Glu. | 30 | mmol/l | (2.5-5.5) | | | | |
| Osmo. | 320 | mmol/kg | (275-295) | | | | |

Osmotic diuresis results in the renal loss of both sodium and water, but relatively more plasma water is lost than sodium (i.e. $\leq 140$ mmol $Na^+$ lost per litre of water lost). This would result in hypernatraemia, but in the early stages of the disease the resulting thirst usually causes the patient to drink pure water or salt-poor fluids, which produces hyponatraemia. If the diabetic crisis progresses to coma the patient will cease fluid replacement and hypernatraemic dehydration may occur. The raised plasma urea concentration in this patient reflects the decreased GFR due to dehydration. The hyperkalaemia is due to the lack of insulin and the consequent metabolic acidosis ($K^+$ moves out from the cells).

**Hyponatraemia with normal body sodium and normal or increased water**

Euhydrated

*Pseudohyponatraemia*  Hyperlipidaemia, hyperproteinaemia.

*Excess extracellular solute*  Hyperglycaemia.

Overhydrated

Salt-poor water intake in association with: (1) stress—physical (pain, postoperative), mental, (2) hypovolaemia—shock, haemorrhage, diuretic therapy, (3) antidiuretic drugs—chlorpropamide, tolbutamide, barbiturates, opiates, vinblastine, vincristine, carbamazepine, clofibrate, cyclophosphamide, (4) glucocorticoid deficiency, (5) myxoedema, (6) SIADH, (7) transurethral resection of the prostate, (8) compulsive water drinking

EUHYDRATION: FACTITIOUS HYPONATRAEMIA

Preanaesthetic electrolyte check on a 58 year old man. Plasma noted to be very lipaemic by laboratory.

| | | | |
|---|---|---|---|
| Plasma Na | 120 | mmol/l | (132-144) |
| K | 3.8 | mmol/l | (3.2-4.8) |

| | | | |
|---|---|---|---|
| Cl | 87 | mmol/l | (93–108) |
| HCO₃ | 20 | mmol/l | (23–33) |
| Urea | 4.5 | mmol/l | (3.0–8.0) |
| Chol. | 21 | mmol/l | (2.5–7.3) |
| Trig. | 81 | mmol/l | (0.3–1.7) |

The concentration of plasma sodium is always expressed in terms of plasma volume, although the sodium ions are present only in the plasma water. Plasma normally comprises about 93% water. The water content of plasma can be markedly reduced if large quantities of lipid or protein are present. This leads to artefactually lowered estimations of solutes such as sodium because the ion measured is related to the volume of plasma taken for analysis. In these conditions hyponatraemia does not occur until the plasma concentration of protein or triglyceride is greater than about 120 g/l and about 50 mmol/l respectively.

EUHYDRATION: EXCESSIVE EXTRACELLULAR SOLUTE

A 26 year old diabetic man.

| | | | | |
|---|---|---|---|---|
| Plasma | Na | 127 | mmol/l | (132–144) |
| | K | 7.0 | mmol/l | (3.2–4.8) |
| | Cl | 93 | mmol/l | (93–108) |
| | HCO₃ | 12.5 | mmol/l | (23–33) |
| | Urea | 9.0 | mmol/l | (3.0–8.0) |
| | Creat. | 0.14 | mmol/l | (0.06–0.12) |
| | Glu. | 41.5 | mmol/l | (2.5–5.5) |
| | Osmo. | 330 | mmol/kg | (275–295) |

Non-electrolyte solutes that are confined to the extracellular space create a concentration gradient which favours the movement of water out of cells and the retention in the ECF of administered fluid. Various empirical formulae have been used to roughly calculate a 'corrected' $[Na^+]$ (i.e. if no water movement had occurred), e.g.

$[Na^+]$ corrected $= [Glu.]/4 + [Na^+]$ measured

It should be noted that hyponatraemia associated with diabetes mellitus may be due to several factors, i.e. extracellular solute, hyperlipidaemia, osmotic diuresis, oral fluid replacement.

OVERHYDRATION: IATROGENIC WATER OVERLOAD

Excessive intravenous 5% dextrose infusion in a patient aged 79 years with a fractured neck of femur.

Chapter 2

| Plasma | Na | 118 | mmol/l | (132–144) | Urine | Na | 5 | mmol/l |
| --- | --- | --- | --- | --- | --- | --- | --- | --- |
| | K | 3.1 | mmol/l | (3.2–4.8) | | Osmo. | 150 | mmol/kg |
| | Cl | 78 | mmol/l | (98–108) | $\dfrac{\text{Urine}}{\text{Plasma}}$ Osmo. | | 0.58 | |
| | HCO$_3$ | 27 | mmol/l | (23–33) | | | | |
| | Urea | 2.5 | mmol/l | (3.0–8.0) | | | | |
| | Osmo. | 258 | mmol/kg | (275–295) | | | | |

This condition is usually associated with inappropriate intravenous therapy, where salt-poor fluids are given rapidly or over a long period of time. In the case of dextrose infusions the glucose is immediately metabolized and therefore the process is equivalent to infusing pure water (see Fig. 2.7).

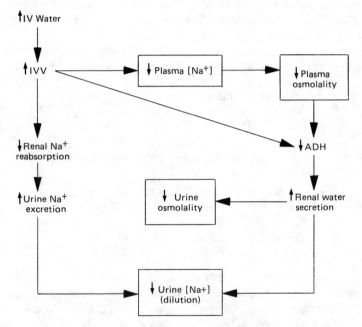

**Fig. 2.7** Mechanism of hyponatraemia due to excessive IV 5% dextrose infusion.

OVERHYDRATION: COMPULSIVE WATER DRINKING

Hyponatraemia in this situation is uncommon but when it occurs the biochemistry and pathophysiology is similar to the above case.

It must be remembered that renal excretion of a large water load is limited by two factors:
(1) The diluting capacity of the renal tubules (maximal dilution = 50 mmol/kg).
(2) Total solute available for excretion. On a high protein diet approximately 1700 mmol of solute is excreted daily. Therefore, if

a subject is able to dilute his urine to 50 mmol/kg he will be able to excrete 1700/50 or 34 l/day. On a medium protein diet about 1200 mmol of solute is excreted per day. The maximum urinary volume would then be 1200/50 or 24 l/day, i.e. water overload would not occur unless more than 24 l of fluid a day was drunk. However, if the subject stops eating and takes only glucose he will have only about 200 mmol of solute available for excretion per day, and therefore he would only be able to excrete 200/50 or 4 l/day. It is in this situation that the compulsive water drinker gets overloaded with fluid and hyponatraemic.

## OVERHYDRATION: SYNDROME OF INAPPROPRIATE ADH SECRETION (SIADH)

In this syndrome there is continued and inappropriate secretion of ADH in the face of hypo-osmolality and a normal or increased volume of the extracellular fluid.

*Example* A 64 year old man with carcinoma of the lung.

| Plasma | Na | 122 | mmol/l | (132–144) | Urine | Na | 60 | mmol/l |
| | K | 4.4 | mmol/l | (3.2–4.8) | | Osmo. | 531 | mmol/kg |
| | Cl | 94 | mmol/l | (98–108) | Urine/Plasma | Osmo. | 2.01 | |
| | $HCO_3$ | 22 | mmol/l | (23–33) | | | | |
| | Urea | 3.5 | mmol/l | (3.0–8.0) | | | | |
| | Creat. | 0.06 | mmol/l | (0.06–0.12) | | | | |
| | Osmo. | 264 | mmol/kg | (275–295) | | | | |

Pathophysiology

See Fig. 2.8.

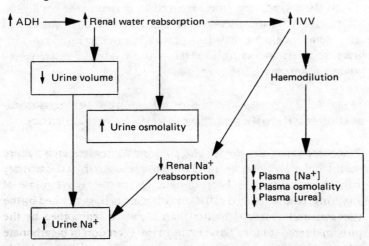

**Fig. 2.8** Mechanism of hyponatraemia in the syndrome of inappropriate secretion of ADH (SIADH).

For a definitive diagnosis of SIADH the following criteria should be satisfied, but usually point (7) is assumed if all other criteria are met.
(1) Decreased plasma osmolality (low plasma [$Na^+$]).
(2) Urine osmolality high relative to plasma.
(3) High urinary sodium concentration (>30 mmol/l).
(4) No evidence of hypovolaemia.
(5) Normal pituitary, adrenal, renal, cardiac and hepatic function.
(6) Absence of therapeutic agents.
(7) Increased plasma ADH concentration (inappropriate to plasma osmolality).
(8) Response to water restriction by increasing the plasma osmolality and sodium concentration.

Causes

*Malignancy*   Bronchogenic carcinoma, brain neoplasm, renal carcinoma, lymphoma, intestinal carcinomas, etc.

*Cerebral disorders*   Infections, trauma, tumours.

*Pulmonary disorders*   Pneumonia, tuberculosis, pneumothorax.

*Miscellaneous disorders*   Acute intermittent porphyria, Guillain-Barré syndrome.

A syndrome with similar biochemical features to SIADH may occur in the following conditions: (1) therapy with antidiuretic agents—chlorpropamide, tolbutamide, etc., (2) thiazide diuretic therapy, (3) glucocorticoid deficiency, (4) hypothyroidism
This latter syndrome differs from SIADH in that the condition will resolve when the offending drug is discontinued or the endocrine deficiency is treated. It is probably better to refer to all of these disorders including SIADH as the syndrome of inappropriate antidiuresis (SIAD).

## OVERHYDRATION: POSTOPERATIVE STATE

Postcholecystectomy, on 5% glucose intravenous therapy.

| Plasma | | | | Urine | | |
|---|---|---|---|---|---|---|
| Na | 121 | mmol/l | (132–144) | Na | 6 | mmol/l |
| K | 4.6 | mmol/l | (3.2–4.8) | Osmo. | 410 | mmol/kg |
| Cl | 90 | mmol/l | (98–108) | Urine/Plasma Osmo. | | 1.56 |
| $HCO_3$ | 22 | mmol/l | (23–33) | | | |
| Urea | 3.5 | mmol/l | (3.0–8.0) | | | |
| Osmo. | 263 | mmol/kg | (275–295) | | | |

(1) stress reaction (surgical operation)→ (a) ↑ADH secretion→

water retention and concentrated urine, (b) ↑aldosterone secretion→renal $Na^+$ retention →↓urine $[Na^+]$
(2) IV salt poor fluid + ↑ renal water reabsorption→haemodilution →↓plasma $[Na^+]$

## OVERHYDRATION: TRANSURETHRAL RESECTION OF PROSTATE

Transurethral resection of the prostate for benign prostatic hypertrophy in a 66 year old man. During surgery the bladder was irrigated with a 1.5% (200 mmol/l) glycine solution.

|  |  | Preop. | Postop. |  |  |
|---|---|---|---|---|---|
| Plasma | Na | 139 | 123 | mmol/l | (132–144) |
|  | K | 3.9 | 4.8 | mmol/l | (3.2–4.8) |
|  | Cl | 100 | 91 | mmol/l | (98–108) |
|  | $HCO_3$ | 27 | 22 | mmol/l | (23–33) |
|  | Creat. | 0.08 | 0.07 | mmol/l | (0.06–0.12) |
|  | Osmo. | 281 | 272 | mmol/kg | (275–295) |
|  | TP | 64 | 54 | mmol/l | (60–85) |
|  | Alb. | 35 | 29 | mmol/l | (30–50) |

Occasionally during this procedure substantial amounts of the irrigating solution may enter the bloodstream via open veins or sinuses. Plasma sodium levels as low as 110–115 mmol/l have been described in such cases. The increase in $[K^+]$ probably reflects slight haemolysis.

## Hyponatraemia with increased body sodium and water

### CAUSES

*Oedematous states* Congestive cardiac failure, nephrotic syndrome, cirrhosis of liver.

*Renal failure* Acute, chronic.

### Oedematous states

Hyponatraemia is occasionally associated with congestive cardiac failure, nephrosis and cirrhosis of the liver.
Nephrotic syndrome in a 20 year old male.

| Plasma | Na | 122 | mmol/l | (132–144) | Urine | Na | 8 | mmol/l |
|---|---|---|---|---|---|---|---|---|
|  | Urea | 4.1 | mmol/l | (3.0–8.0) |  | Osmo. | 400 | mmol/kg |
|  | Alb. | 22 | g/l | (32–50) | Urine/Plasma | Osmo. | 1.50 |  |
|  | Osmo. | 266 | mmol/kg | (275–295) |  |  |  |  |

In oedematous states the pathophysiology of hyponatraemia is unclear. It may be due to inadequate delivery of fluid to the distal diluting segment either as a result of a low GFR, or it may result from increased proximal tubular reabsorption. However, the primary cause appears to be retention of sodium and water by the kidney, with water retention predominating.

In the case of hypoalbuminaemia, the following sequence of events has been suggested:

(1) ↓plasma albumin →↓plasma oncotic pressure→loss of fluid from vascular space →↓IVV and oedema
(2) ↓IVV→↑renal reabsorption of salt and water (water predominating) →↓plasma [$Na^+$] and ↑urine osmolality

In congestive cardiac failure (CCF) the primary cause of hyponatraemia is renal retention of sodium and water, possibly due to altered haemodynamics through the kidney.

CCF→↓cardiac output →↓effective circulating volume (↓IVV) →↑renal reabsorption of water and sodium, etc

It is important to remember that the most common cause of hyponatraemia in oedematous states is diuretic therapy.

Renal failure

In advanced renal failure, the patient is unable to dilute or concentrate his urine due to the intrinsic renal disease. As there is a reduction in glomerular filtrate he is unable to adequately secrete water (and sodium) loads. With a creatinine clearance of 0.1 ml/s the patient may only be able to secrete 1.5 l of water a day and any input in excess of this amount will result in overhydration and hyponatraemia. Retention of sodium occurs for similar reasons. Hyponatraemia may also occur in renal disease due to a salt-losing lesion.

CONSEQUENCES OF THE HYPOTONIC STATE

Hypotonicity of the ECF causes water to move from the ECF into the intracellular fluid (ICF), with consequent cellular overhydration. The organ at most risk is the brain as it is contained within a rigid structure. With acute lowering of the ECF tonicity (e.g. rapid intravenous infusions of hypotonic solutions) there may be a rapid expansion of the brain, resulting in cerebral oedema, increased intracranial pressure and irreversible brain damage. If on the other hand, ECF hypotonicity develops slowly the consequences are less dramatic and patients may only complain of lethargy and listlessness.

The clinical manifestations will depend on the cause of the hypotonic state:
(1) *Normal total body sodium with increased water* This situation occurs in acute and chronic water overload (e.g. infusions) and in SIADH. The extracellular volume is normal or expanded and the symptoms are due to the hypotonicity alone.

(2) *Decreased body sodium and water* This condition is associated with sodium loss (renal or extrarenal) and ECF volume contraction. Thus the clinical manifestations will reflect hypotonicity and hypovolaemia (hypotension, tachycardia, decreased GFR).

(3) *Increased total body sodium and water* (a) Oedematous states: the main clinical findings are those of the original condition (e.g. cardiac failure, nephrosis, cirrhosis) and the symptoms due to hypotonicity are usually not evident. (b) Fluid overload in renal failure: water overload in patients with acute renal failure may occur because of inappropriate intravenous therapy. Symptoms reflecting hypotonicity will depend on the rate and severity of the fall in the plasma sodium concentration.

**Principles of therapy**

(1) NORMAL TOTAL BODY SODIUM AND INCREASED WATER

Acute water overload

This may present as a medical emergency due to a rapid expansion of the brain volume. Therapy is aimed at preventing and reversing the flow of water into cells and may be achieved by the infusion of hypertonic saline. Care should be taken to prevent an excessive expansion of the ECF as this may result in pulmonary oedema and cardiac failure.

Chronic water overload

Clinical manifestations are less severe in this situation and dramatic therapy is usually not required. If the hyponatraemia is due to intravenous therapy, then slowing the rate or stopping the infusion is usually sufficient. Management of SIADH depends on the severity of the hyponatraemia. In mild to moderate cases, decreasing fluid intake ($<500$ ml/d) usually resolves the situation. If the hyponatraemia is severe and is producing symptoms of hypotonicity, it may be necessary to increase the ECF tonicity with hypertonic saline infusions. At the same time a potent diuretic should be given (e.g. frusemide) to decrease the extracellular volume.

## (2) DECREASED BODY SODIUM AND WATER

The main problem is sodium depletion and extracellular fluid contraction (hypovolaemia). Treatment involves infusion of normal (isotonic) saline until the deficit is repaired. Hypertonic saline solutions, although physiologically desirable, are usually not used because once renal function is improved the kidneys can resolve the water and sodium deficits.

## (3) INCREASED BODY SODIUM AND WATER

When hyponatraemia occurs in oedematous states (not due to diuretic therapy) it usually infers serious malfunction of the primary organs and these patients are often refractory to diuretic therapy; they may require severe fluid restriction or, in the case of cirrhosis and nephrosis, infusion of plasma expanders (e.g. albumin) in an attempt to increase the effective blood volume and thus the GFR.

# NORMONATRAEMIA

Sodium and water disturbances associated with a normal plasma sodium concentration may occur under the following circumstances:
(1) Overhydration with retention of sodium and water in proportional amounts (i.e. $\sim 140$ mmol of $Na^+$ retained per litre of water retained).
(2) Dehydration with proportional loss of sodium and water.

### Sodium and water excess—overhydration

*Oedematous states*   (1) Congestive cardiac failure, (2) nephrotic syndrome, (3) cirrhosis.

*Inappropriate intravenous therapy*

*Haemodialysis*

*Mineralocorticoid excess*

*Renal failure*

### Sodium and water depletion—dehydration

In clinical practice the commonest type of dehydration is the isotonic (normonatraemic) variety.

Pathogenesis

(1) loss of hypotonic fluid (renal, gastrointestinal, sweat) →$H_2O$ loss > $Na^+$ loss → hypertonic dehydration → thirst
(2) thirst → pure $H_2O$ intake (tap water) → replacement of $H_2O$ deficit → normonatraemia

If the patient with hypotonic fluid losses cannot take water, the hypertonic dehydration will persist, whilst if excessive water is ingested hypotonic dehydration will occur. Thus isotonic dehydration infers (a) the patient is water and salt depleted and (b) the patient has ingested water.

Extrarenal loss

*Skin*   (1) Sweat, (2) burns.

*Respiratory tract*   (1) Overbreathing, (2) pyrexia.

*Gut*   (1) Vomiting, (2) diarrhoea, (3) fistula loss.

Renal loss

(1) Osmotic diuresis, (2) diuretic therapy, (3) salt-losing nephritis, (4) diuretic phase of acute tubular necrosis, (5) renal tubular acidosis.

**Case example/pathophysiology**

Small gut obstruction with vomiting in an adult.

| Plasma | Na | 137 | mmol/l | (132–144) | Urine | Na | 5 | mmol/l |
|---|---|---|---|---|---|---|---|---|
| | K | 4.6 | mmol/l | (3.2–4.8) | | Creat. | 16 | mmol/l |
| | Cl | 107 | mmol/l | (98–108) | | Osmo. | 680 | mmol/kg |
| | $HCO_3$ | 20 | mmol/l | (23–33) | Urine/Plasma | Osmo. | 2.26 | |
| | Urea | 12 | mmol/l | (3.0–8.0) | | | | |
| | Creat. | 0.17 | mmol/l | (0.06–0.12) | | $Fe_{Na}$ | 0.04% | |
| | Osmo. | 301 | mmol/kg | (275–295) | | | | |

Pathophysiology

(1) water and sodium loss → dehydration and sodium depletion → ↓IVV
(2) ↓IVV → (a) ↓GFR → ↑plasma [urea] and [creatinine], (b) ↑ADH → ↑ urine osmolality (U/P > 1), (c) ↑renal $Na^+$ reabsorption → ↓urine $Na^+$ ($FE_{Na}$ = <1%)

*Note* Vomiting from above the small intestine (e.g. pyloric stenosis) results in loss of acid (HCl) causing a metabolic alkalosis. Fluid from the small gut contains a large amount of alkali ($HCO_3^-$) and loss from this region (e.g. severe diarrhoea) will result in a metabolic acidosis. Loss from both gut and stomach simultaneously (gut obstruction) will result either in a normal plasma bicarbonate (loss of $H^+ + HCO_3^-$) or a slight acidosis due to the renal impairment associated with dehydration.

**Consequences of isotonic dehydration**

Loss of hypotonic fluid followed by pure water replacement is tantamount to the loss of an isotonic fluid from the extracellular space. Thus there will be no intracellular-extracellular fluid shifts. The clinical manifestations will therefore be due to the extracellular volume depletion (hypovolaemia), i.e. hypotension, tachycardia, decreased GFR, shock. A serious consequence of a severe longstanding hypovolaemia is the development of acute tubular necrosis.

**Principles of therapy**

In most instances, all that is required is the intravenous infusion of normal saline, the aim being to restore the blood volume and improve renal function. In severe cases of dehydration it is important to check the urine electrolytes (U[$Na^+$], $FE_{Na}$) before commencing intravenous fluids. If acute tubular necrosis has occurred therapy has to be modified.

# 3 Potassium

## BODY DISTRIBUTION

Potassium is the major cation within the cell, its high concentration there being maintained by the $Na^+$-$K^+$ pump. Potassium is required for normal neuromuscular action because of its role in determining membrane polarization.

| | | |
|---|---|---|
| Total body potassium | ~3500 mmol | |
| Intracellular | ~3400 mmol | (130–159 mmol/l) |
| Extracellular | ~75 mmol | (3–5 mmol/l) |

Approximate potassium concentration of body fluids is given in Table 3.1.

**Table 3.1** Approximate potassium concentration of body fluids (mmol/l)

| Plasma | Sweat | Gastric juice | Pancreatic/bile fluid | Small bowel fluid | Diarrhoea fluid |
|---|---|---|---|---|---|
| 4 | 10 | 10 | 5 | 5 | 40 |

## HOMEOSTASIS

See Fig. 3.1. Maintenance of a normal plasma potassium level requires (1) adequate intake, (2) normal distribution between intracellular and extracellular fluid, (3) normal excretion and losses.

### Intake

The daily potassium requirements of 1–2 mmol/kg body weight a day are easily met by the average diet (~100 mmol/day). Plasma abnormalities due to inappropriate intake appear only when there are gross excesses or deficiencies in the diet and then they are only transitory, unless there is a concomitant defect in renal secretion.

Chapter 3

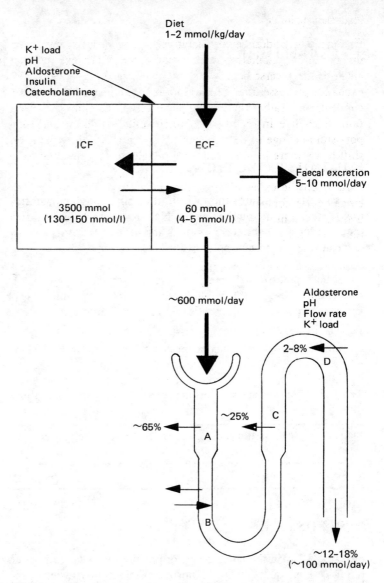

**Fig. 3.1** Homeostasis and renal handling of potassium. A, proximal tubule; B, descending loop of Henle; C, thick ascending loop of Henle; D, distal tubule and collecting duct. The amount of $K^+$ absorbed or excreted is given as a percentage of the amount filtered at the glomerulus.

### Intracellular-extracellular distribution

The gradient between intracellular and extracellular potassium is maintained by a $Na^+$-$K^+$ ATPase pump. The exact mechanism is unclear but sodium ions, which diffuse into cells along a concentration gradient, are pumped out in exchange for potassium ions.

Factors influencing internal distribution are given below.

### Acid-base balance

Extracellular acidosis is associated with movement of potassium out of cells in exchange for hydrogen ions. With extracellular alkalosis the reverse occurs. Recent evidence suggests that these exchange processes may be in response to the plasma bicarbonate concentration rather than to the plasma hydrogen ion concentration. A change in blood pH of 0.10 units results in a plasma potassium change of the order of 0.3–1.3 mmol/l, the potassium shift being more pronounced for a metabolic acidosis than for a respiratory one.

As might be expected from the above, any alterations in the extracellular concentration of potassium will influence cellular hydrogen ion movement. Hypokalaemia causes hydrogen ions to move into cells in exchange for potassium ions, resulting in an intracellular acidosis and an extracellular alkalosis. Hyperkalaemia promotes the reverse situation.

### Aldosterone

Mineralocorticoids promote potassium uptake by the renal distal tubule cells. Uptake is also increased in epithelial cells, particularly those of the salivary glands, sweat glands and colonic mucosa. The resulting increase in intracellular potassium concentration in the renal tubule cells causes potassium secretion into the tubule lumen to be enhanced. Aldosterone therefore increases the urinary excretion of potassium. The effect of aldosterone on muscle and liver cells is controversial.

### Insulin

Cellular uptake of potassium is promoted by insulin, whilst insulin secretion is stimulated by hyperkalaemia.

### Catecholamines

Infusions of adrenalin result in an initial rise in plasma $[K^+]$ (? release from liver), followed by a sustained fall due to $K^+$ entry into cells, particularly muscle. This effect is mediated through the $\beta 2$ adrenergic receptors (salbutamol, a specific $\beta 2$ agonist, produces the same effect, but $\beta 1$ and $\alpha$ agonists do not).

## Excretion

GUT

Under normal circumstances the gastrointestinal loss of less than

10 mmol/day is not important. However in two situations losses via the gut do assume significance.
(1) Prolonged diarrhoea or vomiting may enhance losses to an extent that makes them clinically significant.
(2) When urinary potassium secretion is decreased, as in renal failure, the gut provides another pathway of excretion. In these circumstances, losses may approach 35% of oral intake.

RENAL

Eighty-five to 90% of filtered potassium is reabsorbed in the proximal tubule and the ascending limb of the loop of Henle. The major portion of potassium appearing in the urine is derived from distal tubular secretion. There is a loose association between sodium reabsorption and potassium and hydrogen secretion. This is not a one-for-one molar process because sodium uptake is often greater than the combined hydrogen and potassium secretion. Also sodium uptake occurs mainly in the early part of the tubule, whilst potassium secretion is predominant in the distal portion (Fig. 3.1).

Distal tubular potassium secretion is influenced by the following.

Potassium intake

High intake (or hyperkalaemia) stimulates renal tubular cell uptake and secretion of potassium. This may be due to stimulation of the $Na^+$-$K^+$ ATPase pump, but it should be recalled that hyperkalaemia is also a potent stimulus to aldosterone production.

Acid-base status

Extracellular acidosis, as described above, causes potassium to move out of cells, including those that line the renal tubules. These potassium depleted tubule cells then secrete less potassium into the urine. Alkalosis has the opposite effect and increases potassium excretion. These effects are more pronounced in metabolic disturbances. In metabolic acidosis, however, the degree of renal excretion of potassium is dependent on the duration of the disturbance. In the initial phase there is a decreased potassium excretion, but in the chronic phase there is increased excretion which may result in mild to moderate potassium depletion. The exact cause of this increased excretion is unclear, but it may be due to an elevated distal tubular fluid flow, e.g. excretion of acid anions accompanied by sodium ions.

### Fluid delivery to distal tubule

Increased fluid delivery to the tubule lumen (diuretics, presence of poorly absorbable anions, osmotic diuresis) increases potassium secretion. Equilibration of the potassium between the distal tubular cell and the tubular fluid is rapid and, therefore, if the rate of fluid flow past the secretory site is fast more potassium will be excreted per unit time.

### Mineralocorticoids

Aldosterone stimulates distal tubular $Na^+$ reabsorption and $K^+$ secretion and hence increases renal excretion of potassium. This mechanism is probably directly related to the intracellular $K^+$ concentration, i.e.

aldosterone $\rightarrow \uparrow Na^+$-$K^+$ exchange at the peritubular side of the cell $\rightarrow \uparrow$ intracellular $[K^+] \rightarrow \uparrow$ luminal secretion of $K^+$ (along a concentration gradient)

Of the above factors probably only aldosterone and distal tubular flow rate influence the physiological excretion of potassium.

### Potassium loading

During the first 4–6 hours following a potassium load (oral or IV) ~50% of the potassium appears in the urine. Of the 50% retained by the body about 80% is translocated intracellularly, i.e.

$\uparrow K^+ \rightarrow$ (1) $\uparrow$ aldosterone $\rightarrow \uparrow$ renal $K^+$ secretion, (2) $\uparrow$ insulin $\rightarrow \uparrow$ cellular $K^+$ uptake

These cellular mechanisms protect the body against hyperkalaemia.

## LABORATORY INVESTIGATION

#### PLASMA POTASSIUM

Plasma potassium concentrations do not reflect the total body levels, although hypokalaemia is usually associated with a potassium deficit. Plasma levels, however, are of major importance in regard to therapy because severe deviations from the norm are dangerous, regardless of the state of the total body level.

## PLASMA BICARBONATE

In most situations associated with abnormal plasma potassium levels the bicarbonate concentration tends to vary inversely with the plasma potassium concentration. There are two general reasons for this:

(1) In metabolic alkalosis (high plasma bicarbonate), there is movement of potassium into cells (in exchange for $H^+$) and also an increased secretion of potassium into the urine. This results in hypokalaemia. The reverse situation occurs in acidosis.

(2) Severe hypokalaemia *per se* can be the cause of a metabolic alkalosis, i.e. movement of hydrogen ions into cells in exchange for potassium ions, and increased renal secretion of hydrogen ion. Conversely hyperkalaemia could theoretically result in metabolic acidosis, but this is probably rare. In most situations where hyperkalaemia and metabolic acidosis coexist, the primary disease usually causes both acidosis and hyperkalaemia, e.g. diabetic ketoacidosis, acute renal failure.

If the above reciprocal relationships do not hold for a given potassium ($K^+$) and bicarbonate ($HCO_3^-$) level the following should be considered:

$\downarrow K^+$ and $\downarrow HCO_3^-$: diseases associated with potassium loss and metabolic acidosis, e.g. diarrhoea, renal tubular acidosis

$\uparrow K^+$ and $\uparrow HCO_3^-$: respiratory acidosis partially compensated by a metabolic alkalosis

## PLASMA UREA AND CREATININE

A high level of plasma urea (or creatinine) associated with hyperkalaemia suggests renal failure or hypovolaemia (prerenal uraemia). Hypovolaemia is usually not associated with hyperkalaemia, unless there is a concurrent acidosis or impairment of renal potassium secretion due to either reduced GFR or tubular damage.

Acute renal failure is invariably accompanied by hyperkalaemia because of the major reduction in overall renal function that occurs. In chronic renal failure hyperkalaemia rarely arises if the GFR remains above approximately 0.3 ml/s. If hyperkalaemia is present in this situation it is nearly always due to either an accompanying acidosis or distal tubule dysfunction.

## URINARY POTASSIUM

Twenty-four hour urinary potassium secretion rates are of little value unless the dietary intake is known. However, potassium levels on 'spot' urines are helpful where the question to be answered is whether the route of loss is renal or extrarenal.

Renal loss  $[K^+] > 20$ mmol/l
Extrarenal loss  $[K^+] < 20$ mmol/l

In most circumstances the route of loss is evident from the clinical picture and there is no need for laboratory confirmation. However, hypokalaemia may result from longstanding subtle gut losses (e.g. chronic laxative abuse) and in these circumstances spot urinary concentrations are helpful during the initial clinical work-up.

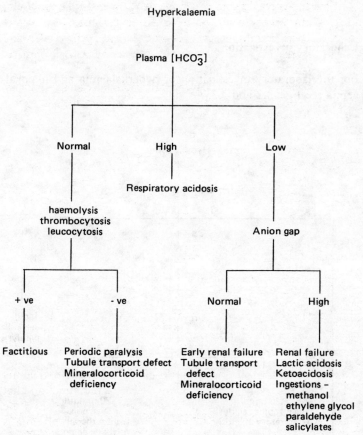

Fig. 3.2 Scheme for the laboratory investigation of hyperkalaemia.

URINARY CHLORIDE

Urinary 'spot' chloride concentrations can sometimes be helpful in identifying the causes of hypokalaemic metabolic alkalosis. A value less than 10 mmol/l suggests previous diuretic therapy or loss from the upper gastrointestinal tract (e.g. vomiting), whilst a value greater than 20 mmol/l is suggestive of current diuretic therapy or mineralocorticoid excess.

OTHER INVESTIGATIONS

Depending on the clinical picture and information gained from the plasma and urinary electrolytes some of the following investigations may also be indicated: (1) plasma glucose, ketone bodies, lactate, (2) hormone investigations — renin, aldosterone, cortisol.

**Laboratory investigation**

For the laboratory investigation of hyperkalaemia and hypokalaemia see Figs 3.2 and 3.3.

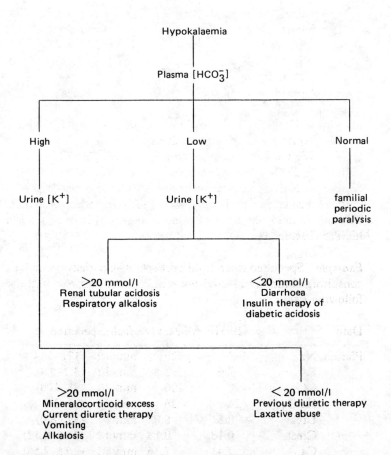

Fig. 3.3 Scheme for the laboratory investigation of hypokalaemia.

# HYPERKALAEMIA

Hyperkalaemia may be caused by the following situations.

*Factitious* (1) Improper collection, (2) haematological disorders.

*Increased input* Oral, intravenous therapy.

*Altered distribution* (1) Acidaemia, (2) insulin deficiency, (3) hyperkalaemic periodic paralysis.

*Reduced excretion* (1) Renal failure, (2) mineralocorticoid deficiency, (3) renal tubule transport defect.

**Case examples/pathophysiology**

IMPROPER COLLECTION

Hyperkalaemia due to improper collection may be due to:
(1) Haemolysis, particularly if the blood sample is forcibly expelled from the syringe.
(2) Incorrect specimen tube, e.g. potassium — EDTA as used for lipid or ESR analysis.
(3) Prolonged use of a tourniquet — local anoxia causes potassium to seep out of cells.
(4) Heel and finger prick collections — combination of local tissue damage (squeezing) and red cell damage.
(5) Allowing red cells to remain in contact with plasma for a long period ($> 2h$) — potassium seeps out of cells as the energy-dependent $Na^+$-$K^+$ pump system of the red cells runs down.

In these cases, excepting the use of a tourniquet when the hyperkalaemia is mild, the level of plasma potassium is usually fairly high (6–10 mmol/l).

*Example* Specimen taken from a preoperative patient attending a consulting clinic (29/11) and not sent to the laboratory until the following day.

| Date |        | 29/11 | 30/11 | (fresh specimen) |             |
|------|--------|-------|-------|--------|-------------|
| Plasma | Na   | 133   | 135   | mmol/l | (122–144)   |
|        | K    | 7.6   | 3.8   | mmol/l | (3.2–4.8)   |
|        | Cl   | 98    | 100   | mmol/l | (98–108)    |
|        | HCO$_3$ | 24 | 26    | mmol/l | (23–33)     |
|        | Urea | 8.3   | 8.0   | mmol/l | (3.0–8.0)   |
|        | Creat. | 0.13 | 0.13 | mmol/l | (0.06–0.12) |
|        | Ca   | 2.41  | 2.36  | mmol/l | (2.15–2.55) |
|        | PO$_4$ | 5.00 | 1.10 | mmol/l | (0.60–1.25) |

If whole blood is allowed to stand for long periods prior to plasma separation both potassium and phosphate will diffuse out of the red cells resulting in hyperkalaemia and hyperphosphataemia.

HAEMATOLOGICAL DISORDERS

Large numbers of abnormal white cells (leukaemia) and platelets (thrombocytosis) may, on standing or during clotting, release significant amounts of potassium into the plasma (or serum). This type of hyperkalaemia is termed pseudohyperkalaemia. It can be overcome by collecting the blood into heparin and separating the plasma without delay.

Intravascular haemolysis (haemolytic disease) may also be associated with a factitious hyperkalaemia.

It is not often appreciated that the plasma potassium concentration in whole blood packs, when first removed from refrigeration, may be of the order of 10–20 mmol/l. Potassium returns to the cells as the blood is warmed.

*Example*  Inpatient with pulmonary tuberculosis who had a persistent hyperkalaemia (5.2–6.4 mmol/l). The patient had a total white cell count of 35 000/cmm and a platelet count of 1 300 000/cmm. The laboratory performed assays on serum. A repeat assay on plasma suggested pseudohyperkalaemia as the cause of the high serum potassium level in this patient.

|       | Serum | Plasma |        |             |
|-------|-------|--------|--------|-------------|
| Na    | 138   | 139    | mmol/l | (135–145)   |
| K     | 6.4   | 4.6    | mmol/l | (3.5–5.0)   |
| Cl    | 95    | 98     | mmol/l | (98–108)    |
| $HCO_3$ | 27  | 25     | mmol/l | (24–32)     |
| Urea  | 8.0   | 8.5    | mmol/l | (3.0–8.0)   |
| Creat.| 0.09  | 0.09   | mmol/l | (0.06–0.14) |

INCREASED INPUT

With normal renal function fairly large amounts of potassium can be administered (orally or intravenously) without any danger, other than transient hyperkalaemia. However, if renal function is poor a persistent and potentially dangerous hyperkalaemia may result from inappropriate potassium administration.

## DISTURBED INTERNAL DISTRIBUTION – ACIDOSIS

A 68 year old man with chronic obstructive airways disease.

| Plasma | Na | 133 | mmol/l | (132-144) | pH | 7.19 | (7.35-7.45) |
|---|---|---|---|---|---|---|---|
| | K | 5.2 | mmol/l | (3.2-4.5) | $P_{CO_2}$ | 14.7 kPa | (4.5-6.0) |
| | Cl | 94 | mmol/l | (98-108) | | | |
| | $HCO_3$ | 42 | mmol/l | (23-33) | | | |
| | Urea | 8.5 | mmol/l | (3.0-8.0) | | | |

In acidaemia $H^+$ tends to move from the plasma into cells in exchange for $K^+$, resulting in hyperkalaemia. There is also decreased renal potassium secretion. Respiratory acidosis partially compensated by a metabolic alkalosis is a situation where the hyperkalaemia is associated with a high bicarbonate.

The magnitude of hyperkalaemia in respiratory acidosis is dependent on the duration of the disturbance. The rise in plasma potassium becomes evident within several hours and then may disappear as compensation (increased plasma $HCO_3^-$) occurs. Chronic respiratory acidosis is therefore not always associated with hyperkalaemia.

## DISTURBED INTERNAL DISTRIBUTION – INSULIN DEFICIENCY

Adult with diabetic ketoacidosis.

| Plasma | Na | 132 | mmol/l | (132-144) |
|---|---|---|---|---|
| | K | 6.5 | mmol/l | (3.2-4.8) |
| | Cl | 89 | mmol/l | (98-108) |
| | $HCO_3$ | 6 | mmol/l | (23-33) |
| | Urea | 14.5 | mmol/l | (3.0-8.0) |
| | Glu. | 64 | mmol/l | (3.0-5.5) |
| Blood pH | | 7.02 | | (7.35-7.45) |

The hyperkalaemia of diabetic acidosis is due to three mechanisms:
(1) Insulin deficiency (a) decreases cellular potassium uptake, (b) increases protein breakdown. Potassium is closely associated with the protein in cells and, therefore, protein breakdown (starvation, diabetes mellitus) is associated with potassium release.
(2) Acidaemia.
(3) Hypertonicity — increased extracellular tonicity (↑glucose) results in moderate hyperkalaemia, possibly due to increased intracellular $[K^+]$ (cell dehydration) with net efflux of potassium.

Although these patients are hyperkalaemic, they are invariably potassium deficient because the potassium being released from cells is subsequently lost in the urine.

## DISTURBED INTERNAL DISTRIBUTION – HYPERKALAEMIC PERIODIC PARALYSIS

A rare familial disorder that is associated with intermittent attacks of muscle paralysis and hyperkalaemia. It is thought that during or before the attacks potassium leaks out of muscle cells. The exact aetiology is unknown.

## DECREASED RENAL EXCRETION – RENAL FAILURE

Analgesic nephropathy in an adult female.

| Plasma | Na | 142 | mmol/l | (132–144) |
|---|---|---|---|---|
| | K | 5.5 | mmol/l | (3.2–4.8) |
| | Cl | 101 | mmol/l | (98–108) |
| | $HCO_3$ | 15 | mmol/l | (23–33) |
| | Urea | 63 | mmol/l | (3.0–8.0) |
| | Creat. | 0.49 | mmol/l | (0.06–0.12) |

In chronic renal failure hyperkalaemia does not usually occur unless there is an acidosis or there is severe renal impairment preventing adequate secretion (creatinine clearance less than ~0.3 ml/s). Early in the disease process the plasma potassium is kept within normal limits by (1) renal secretion, there being an adaptive increase in secretion by the remaining nephrons and (2) secretion into the gut.

Acute renal failure (acute tubular necrosis) on the other hand, due to the rapid onset and severity of the disease process, is invariably associated with hyperkalaemia in its early stages.

## DECREASED RENAL EXCRETION – MINERALOCORTICOID DEFICIENCY

Adult female with Addison's disease.

| Plasma | Na | 118 | mmol/l | (132–144) |
|---|---|---|---|---|
| | K | 5.6 | mmol/l | (3.2–4.8) |
| | Cl | 91 | mmol/l | (98–108) |
| | $HCO_3$ | 21 | mmol/l | (23–33) |
| | Urea | 16 | mmol/l | (3.0–8.0) |
| | Creat. | 0.14 | mmol/l | (0.06–0.12) |

(1) ↓distal tubular $Na^+$ reabsorption and $K^+$ secretion→ (a) $K^+$ retention→hyperkalaemia, (b) $Na^+$ loss→$Na^+$ depletion→ (i) hyponatraemia, (ii) dehydration →↑plasma [Urea]
(2) Acidosis (↓[$HCO_3^-$]) is due to (a) ↓distal tubular $H^+$ secretion, (b) ? dehydration (renal function impairment), (c) ? hyperkalaemia

For other causes of mineralocorticoid deficiency see Chapter 7.

DECREASED RENAL EXCRETION – DEFECTS IN TUBULAR TRANSPORT

Adult on long term spironolactone therapy for hypertension.

| Plasma | Na | 115 | mmol/l | (132–144) |
|---|---|---|---|---|
| | K | 7.2 | mmol/l | (3.2–4.8) |
| | Cl | 85 | mmol/l | (98–108) |
| | $HCO_3$ | 20 | mmol/l | (23–33) |
| | Urea | 19 | mmol/l | (3.0–8.0) |
| | Creat. | 0.13 | mmol/l | (0.06–0.12) |

The potassium-sparing diuretic spironolactone acts by inhibiting the action of aldosterone on the distal renal tubule. Thus the sequence of events will be the same as that for mineralocorticoid deficiency (see above). Other potassium-sparing diuretics which may also be associated with hyperkalaemia are amiloride and triamterene.

Hyperkalaemia due to a defect at the level of renal tubular transport may be seen in the following conditions:
(1) Pseudohypoaldosteronism — inborn defect of tubular transport where the tubule is insensitive to aldosterone.
(2) Tubular disease—renal transplant, systemic lupus erythematosis, interstitial nephritis, amyloidosis.

## Consequences of hyperkalaemia

*Neuromuscular*  Weakness, parathesia, paralysis.

*Gastrointestinal*  Nausea, vomiting, pain, ileus.

*Cardiovascular*  Conduction defects, arrhythmias, cardiac arrest.

## Principles of therapy

Therapy will depend on the cause and severity of the hyperkalaemia. If the hyperkalaemia is due to intracellular shifts (e.g.

acidaemia), treatment of the primary disorder is all that is required. On the other hand, severe hyperkalaemia causing symptoms (cardiac conduction defects, neuromuscular signs, etc.) requires rapid correction. Treatment is aimed at (a) inducing potassium to move into cells and/or (b) increasing net excretion.

### Cellular uptake

This may be increased by (1) $HCO_3^-$ infusions or (2) infusion of insulin and glucose.

These measures give a rapid but transient fall in the plasma level and they therefore should be coupled with methods to increase excretion from the body.

### Excretion

Increased excretion from the body may be carried out by (1) cation exchange resins, e.g. $Na^+$ polystyrene sulfonate (Resonium-A), orally or by enema (1 g of resin will bind approximately 1 mmol of $K^+$) or (2) dialysis against a potassium-free dialysate.

## HYPOKALAEMIA

Hypokalaemia may be caused by the following.

*Inadequate intake*

*Disturbed internal distribution* (1) Insulin therapy, (2) alkalaemia (3) hypokalaemic periodic paralysis, (4) salbutamol.

*Gastrointestinal loss*

*Renal loss* (1) Diuretics, (2) renal disease, (3) mineralocorticoid excess.

*Haemodialysis.*

**Case examples/pathophysiology**

INADEQUATE INTAKE

Adult on total parenteral nutrition.

| Plasma | Na | 137 | mmol/l | (132-144) | Urinary spot | K | 4 mmol/l |
| --- | --- | --- | --- | --- | --- | --- | --- |
| | K | 2.3 | mmol/l | (3.2-4.8) | | | |
| | Cl | 95 | mmol/l | (98-108) | | | |
| | HCO$_3$ | 35 | mmol/l | (23-33) | | | |
| | Urea | 5.0 | mmol/l | (3.0-8.0) | | | |
| | Creat. | 0.10 | mmol/l | (0.06-0.12) | | | |

Inappropriate intravenous therapy can result in hypo- as well as hyperkalaemia. The metabolic alkalosis, if due only to potassium deficiency, suggests that the potassium depletion is of the order of 500-1000 mmol since deficiencies of less than 500 mmol have been found not to cause alkalosis. The low (<20 mmol/l) urinary spot potassium level suggests that there is no excessive loss from the kidney. Potassium depletion is rarely due to an inadequate oral intake, except where there is a concurrent increased potassium excretion from the body (e.g. diuretic therapy).

### DISTURBED INTERNAL DISTRIBUTION: INSULIN THERAPY

Blood specimen taken after 4 hours of insulin therapy in a patient with diabetic ketoacidosis.

| Plasma | Na | 143 | mmol/l | (132-144) | Urine spot | K | 2 mmol/l |
| --- | --- | --- | --- | --- | --- | --- | --- |
| | K | 2.7 | mmol/l | (3.2-4.8) | | | |
| | Cl | 113 | mmol/l | (98-108) | | | |
| | HCO$_3$ | 18 | mmol/l | (23-33) | | | |
| | Urea | 16 | mmol/l | (3.0-8.0) | | | |
| | Creat. | 0.18 | mmol/l | (0.06-0.12) | | | |

On admission to hospital most patients with diabetic ketoacidosis have hyperkalaemia and an associated potassium depletion (see hyperkalaemia). Insulin therapy causes the potassium to move back into the cells, so that there is a danger of severe hypokalaemia if potassium supplements are not given (Hyperchloraemia, see Chapter 4).

### DISTURBED INTERNAL DISTRIBUTION: ALKALAEMIA

Male aged 72 years who, over a long period, ingested increasingly large amounts of alkali (magnesium carbonate and sodium bicarbonate) in preference to seeking medical help for a gastric ulcer.

## Chapter 3

| Plasma | Na | 151 | mmol/l | (132–144) | Urine spot | K | 54 mmol/l |
|---|---|---|---|---|---|---|---|
| | K | 1.9 | mmol/l | (3.2–4.8) | | | |
| | Cl | 81 | mmol/l | (98–108) | | | |
| | HCO$_3$ | 61 | mmol/l | (23–33) | | | |
| | Urea | 9.5 | mmol/l | (3.0–8.0) | | | |
| | Creat. | 0.15 | mmol/l | (0.06–0.12) | | | |

Metabolic alkalosis induces hypokalaemia by two mechanisms: (1) transfer of $K^+$ into cells in exchange for $H^+$, (2) increased renal excretion of $K^+$.

The latter mechanism is the major cause of the potassium depletion. This is manifested by the large amount of potassium in the urine, i.e. in the presence of normal renal function hypokalaemia, due to inadequate intake or increased cellular uptake, would be associated with an urinary potassium concentration of less than 10–20 mmol/l.

### DISTURBED INTERNAL DISTRIBUTION: HYPOKALAEMIC PERIODIC PARALYSIS

Adult male who suffers frequent bouts of muscle weakness.

| Plasma | Na | 145 | mmol/l | (132–144) | Urine spot | |
|---|---|---|---|---|---|---|
| | K | 2.2 | mmol/l | (3.2–4.8) | K | 6 mmol/l |
| | Cl | 109 | mmol/l | (98–108) | | |
| | HCO$_3$ | 27 | mmol/l | (23–33) | | |
| | Urea | 6.0 | mmol/l | (3.0–8.0) | | |
| | Creat. | 0.08 | mmol/l | (0.06–0.12) | | |

This is a rare inherited disease characterized by intermittent attacks of muscle weakness and paralysis associated with hypokalaemia. The aetiology is unknown. The low urinary potassium level suggests that the potassium must be moving from the extracellular to the intracellular space.

### DISTURBED INTERNAL DISTRIBUTION: SALBUTAMOL

Salbutamol infusion given to a pregnant woman in an attempt to prevent premature labour.

| | | Preinfusion | 30 min of infusion | 4 h of infusion | | |
|---|---|---|---|---|---|---|
| Plasma | Na | 137 | 136 | 134 | mmol/l | (132–144) |
| | K | 4.1 | 2.7 | 2.2 | mmol/l | (3.2–4.8) |
| | Cl | 108 | 106 | 104 | mmol/l | (98–108) |
| | HCO$_3$ | 18 | 18 | 18 | mmol/l | (23–33) |
| | Urea | 3.4 | 3.3 | 3.1 | mmol/l | (3.0–8.0) |

In this case the hypokalaemia results from $K^+$ uptake by the cells due to β2 receptor stimulation (salbutamol is a specific β2 agonist). The low $[HCO_3^-]$ probably reflects a moderate compensated respiratory alkalosis (hyperventilation), which is not uncommon during late pregnancy.

## GASTROINTESTINAL LOSS: VOMITING

Psychogenic vomiting in a young adult. Dehydrated.

| Plasma | | | | Urine spot | | |
|---|---|---|---|---|---|---|
| Na | 140 | mmol/l | (132–144) | Urine spot K | 15 | mmol/l |
| K | 2.4 | mmol/l | (3.2–4.8) | | | |
| Cl | 84 | mmol/l | (98–108) | | | |
| HCO₃ | 39 | mmol/l | (23–33) | | | |
| Urea | 15.5 | mmol/l | (3.0–8.0) | | | |
| Creat. | 0.14 | mmol/l | (0.06–0.12) | | | |

Vomiting results in:
(1) loss of HCl→alkalosis
(2) loss of $K^+$ (small amount in vomitus)
(3) loss of water →dehydration→↑plasma [urea]

The hypokalaemia is due to:
(1) small loss in vomitus
(2) cellular uptake due to alkalaemia (loss of acid from stomach)
(3) increased renal secretion: alkalosis and secondary aldosteronism due to the hypovolaemia of dehydration

With rehydration, and providing renal function is normal, the urinary potassium concentration will decrease (normovolaemia suppresses aldosterone secretion).

## GASTROINTESTINAL LOSS: FROM LARGE GUT

A 58 year old adult with prolonged laxative abuse causing potassium loss from the intestine.

| Plasma | | | | Urine spot | | |
|---|---|---|---|---|---|---|
| Na | 138 | mmol/l | (132–144) | Urine spot K | 13 | mmol/l |
| K | 2.8 | mmol/l | (3.2–4.8) | | | |
| Cl | 86 | mmol/l | (98–108) | | | |
| HCO₃ | 35 | mmol/l | (23–33) | | | |
| Urea | 5.5 | mmol/l | (3.0–8.0) | | | |
| Creat. | 0.08 | mmol/l | (0.06–0.12) | | | |

The normal daily loss of potassium in the faeces is of the order of 5–10 mmol. This amount is increased by diarrhoea and can result in significant potassium depletion. The low urinary spot potassium in this case indicates renal conservation and, therefore, an extrarenal cause for the potassium deficiency.

In severe diarrhoea with associated dehydration the resulting hypovolaemia will stimulate increased aldosterone secretion. In such a situation there may also be potassium loss from the kidney and, therefore, an increased urinary potassium concentration (>20 mmol/l).

Other conditions where there may be an excessive loss of potassium from the large bowel are: (1) Villous adenoma of colon—because of increased potassium levels in tumour secretions, and (2) Ureterosigmoidostomy—secretion of $K^+$ and $HCO_3^-$ by the colon in exchange for $Na^+$ and $Cl^-$ (see p. 85).

INCREASED RENAL SECRETION: DIURETIC THERAPY

Patient with congestive cardiac failure on thiazide therapy.

| Plasma | Na | 135 | mmol/l | (132–144) | Urine spot | K | 26 mmol/l |
|---|---|---|---|---|---|---|---|
| | K | 2.6 | mmol/l | (3.2–4.8) | | | |
| | Cl | 88 | mmol/l | (98–108) | | | |
| | $HCO_3$ | 37 | mmol/l | (23–33) | | | |
| | Urea | 15.0 | mmol/l | (3.0–8.0) | | | |
| | Creat. | 0.13 | mmol/l | (0.06–0.12) | | | |

Several mechanisms may be involved, e.g.

(1) ↑distal tubular flow rate → ↑$K^+$ secretion
(2) volume depletion → secondary aldosteronism → ↑$K^+$ secretion

All diuretics, except those of the potassium-sparing variety, have been shown to be capable of causing hypokalaemia. However, it was shown in a large study that less than 5% of patients on non potassium-sparing diuretic therapy and not on potassium supplements showed hypokalaemia. Hypokalaemia appeared to be most common when non-potassium-sparing diuretics are used in oedematous conditions which already have an associated secondary aldosteronism (↑$K^+$ excretion). Hypertensive patients on long-term diuretic therapy (>6 months) may occasionally be hypokalaemic. However, studies have shown that the majority of these patients have a normal total body potassium. It has therefore been suggested that the hypokalaemia in these cases is due to an altered internal distribution of potassium.

INCREASED RENAL SECRETION: RENAL DISEASE

Patient with renal tubular acidosis (RTA).

| Plasma | Na | 139 | mmol/l | (132–144) | Urine spot K 25 mmol/l |
|---|---|---|---|---|---|
| | K | 2.9 | mmol/l | (3.2–4.8) | |
| | Cl | 116 | mmol/l | (98–108) | |
| | $HCO_3$ | 11 | mmol/l | (23–33) | |
| | Urea | 5.3 | mmol/l | (3.0–8.0) | |
| | Creat. | 0.06 | mmol/l | (0.06–0.12) | |

Decreased $H^+$-$Na^+$ exchange by the renal tubule results in chronic $Na^+$ depletion and hypovolaemia. The increased distal tubular flow rate ($\downarrow Na^+$ reabsorption reduces water removal from tubule lumen) and aldosteronism (hypovolaemia) will result in increased $K^+$ secretion by the distal tubules.

Other mechanisms may also be involved. In Type 1 RTA the distal tubular permeability to potassium may be increased, thereby favouring enhanced potassium diffusion from the tubular cell into the lumen. RTA is one of the situations in which hypokalaemia is found in association with a low plasma bicarbonate concentration. Other causes are: (1) severe diarrhoea, (2) ureterosigmoidostomy, (3) respiratory alkalosis (rare), (4) insulin therapy of diabetic ketoacidosis (see p. 73).

The commonest situation in which renal disease is associated with hypokalaemia is in the postdialysis state, i.e. plasma potassium is removed by dialysis and is also lowered as $K^+$ moves back into the cells as the acidaemia is corrected. Other renal conditions associated with hypokalaemia are the diuretic phase of acute tubular necrosis and diuresis following relief of obstruction.

## INCREASED RENAL SECRETION: MINERALOCORTICOID EXCESS

Patient with hypertension due to primary aldosteronism.

| Plasma | Na | 138 | mmol/l | (132–144) | Urine spot K 36 mmol/l |
|---|---|---|---|---|---|
| | K | 2.5 | mmol/l | (3.2–4.8) | |
| | Cl | 89 | mmol/l | (98–108) | |
| | $HCO_3$ | 34 | mmol/l | (23–33) | |
| | Urea | 5.0 | mmol/l | (3.0–8.0) | |
| | Creat. | 0.12 | mmol/l | (0.06–0.12) | |

Aldosterone induces increased $K^+$ secretion by the distal renal tubule and thus increased renal potassium excretion. Diagnosis depends on the demonstration of an aldosterone excess in the presence of a low plasma renin activity (see p. 141).

Other causes of mineralocorticoid excess can be found in Chapter 7.

## Chapter 3 HAEMODIALYSIS

*Example* A 48 year old woman with chronic renal failure due to analgesic nephropathy who was being managed with haemodialysis. The dialysis fluid concentration was: Na=137 mmol/l, K=1.0 mmol/l, Cl=102 mmol/l, Ca=1.6 mmol/l, Mg=0.25 mmol/l, acetate=40 mmol/l.

|        |        | Predialysis | Postdialysis |        |              |
|--------|--------|-------------|--------------|--------|--------------|
| Plasma | Na     | 145         | 140          | mmol/l | (132–144)    |
|        | K      | 4.6         | 2.6          | mmol/l | (3.2–4.8)    |
|        | Cl     | 105         | 95           | mmol/l | (98–108)     |
|        | $HCO_3$ | 20         | 25           | mmol/l | (23–33)      |
|        | Urea   | 35          | 15           | mmol/l | (3.0–8.0)    |
|        | Creat. | 1.16        | 0.56         | mmol/l | (0.06–0.12)  |

There are two reasons for the postdialysis hypokalaemia: (1) loss of $K^+$ into the low $[K^+]$ dialysis fluid, (2) acetate ions from the dialysis fluid enter the plasma by crossing the dialysis membrane and are then converted to $HCO_3^-$. The resultant reduction in the acidaemia allows the extracellular $K^+$ to pass into the tissue cells.

### Consequences of hypokalaemia

Impaired neuromuscular function

*Skeletal muscle*   Weakness, paralysis.

*Cardiac muscle*   Conduction defects, arrhythmias.

*GIT muscle*   Paralytic ileus.

Kidney

Impaired concentrating ability, interstitial nephritis (long standing $K^+$ depletion).

Central nervous system

Psychiatric disturbance (occasionally).

### Principles of therapy

Potassium is an intracellular ion and therefore plasma levels give a poor indication of total body levels. Therapy will depend on whether there is a change in internal distribution or a depletion of stores. Thus in hypokalaemic alkalosis therapy should first be aimed at correction of the alkalaemia.

The method of potassium therapy will depend on the severity of the depletion and the status of the patient. Mild deficiencies may be corrected by oral potassium. It is important to remember that (1) 1 g of KCl contains only 13 mmol of potassium and (2) 50% of the administered potassium will be lost in the urine.

If the patient is unable to take potassium orally, or the depletion is severe, intravenous therapy may be given. As potassium is not immediately taken up by the cells there is a danger of severe hyperkalaemia if it is infused too quickly. Therefore, the potassium-containing solution should not have a concentration greater than 80 mmol/l or be infused at a rate greater than 40 mmol/h.

# 4 Chloride

Chloride is the major anion present in the extracellular fluid. The control of chloride in the body is closely related to that of sodium.

**Intake**

100–200 mmol/day, mainly as sodium chloride.

**Absorption**

Occurs in the small intestine. The mechanism of chloride uptake is unclear, but it appears to depend on an exchange process with the bicarbonate anion, whilst the accompanying sodium exchanges for a hydrogen ion.

**Plasma-erythrocyte shift (Chloride shift)**

Carbon dioxide ($CO_2$) that is derived as an end product from cellular metabolism, diffuses from tissues through the plasma and into red cells where the $CO_2$ concentration is relatively low. Within the erythrocyte the $CO_2$ is combined with water to form carbonic acid ($H_2CO_3$), promoted by the action of carbonic anhydrase. The acid then dissociates into a bicarbonate ion ($HCO_3^-$) and a hydrogen ion ($H^+$). The $H^+$ is buffered by haemoglobin and the $HCO_3^-$ diffuses from the red cell into the plasma in exchange for $Cl^-$. The reverse reactions occur when the erythrocyte reaches the lung and its $CO_2$ content exceeds that of the alveoli. Thus arterial and venous plasmas will differ slightly (2–3 mmol/l) in their plasma chloride concentrations (see p. 104).

**Excretion**
*Extrarenal*   (1) Sweat <5 mmol/day, (2) faeces <5 mmol/day.

*Renal*   100–200 mmol/l, varying with diet. 99% of the $Cl^-$ in the glomerular filtrate is reabsorbed by the renal tubule.

## PROXIMAL TUBULE

$Na^+$ is actively reabsorbed with $Cl^-$ passively following.

## ASCENDING LOOP OF HENLE

In the thick segment $Cl^-$ is actively reabsorbed with $Na^+$ following. The reabsorbed $Na^+$ and $Cl^-$ from this portion of the nephron plays an important role in maintaining the high renal medullary osmolality the kidney needs in order to concentrate urine. The loop diuretics (frusemide, ethacrynic acid) act on this part of the nephron by preventing the reabsorption of $Cl^-$.

## DISTAL TUBULE, COLLECTING DUCT

A small amount of $Cl^-$ reabsorption occurs.

## Control of excretion

The control of absorption and excretion of chloride appears to be similar to that of sodium.

↑ (↓) blood volume → ↓ (↑) reabsorption of chloride

However, there are some differences between the renal reabsorption of $Na^+$ and $Cl^-$ that may result in dissimilar urinary levels of the two ions. For example, $Na^+$ can be reabsorbed independently of $Cl^-$ in the distal tubule, in exchange for $K^+$ or $H^+$ under the influence of aldosterone.

## Plasma chloride

Plasma levels of chloride vary with, and to a great extent depend on, the plasma concentrations of sodium and bicarbonate.

Plasma ↓ $Na^+$ associated with ↓ $Cl^-$
↑ $Na^+$ usually associated with ↑ $Cl^-$
↑ $HCO_3^-$ associated with ↓ $Cl^-$
↓ $HCO_3^-$ may be associated with ↑ $Cl^-$

## Chapter 4 LABORATORY INVESTIGATION

### PLASMA SODIUM

Except in situations where there are disturbances of acid-base homeostasis, the plasma chloride will parallel changes in the plasma sodium concentration.

### PLASMA BICARBONATE

Metabolic alkalosis ($\uparrow$ [$HCO_3^-$]), whether primary or secondary, is always associated with hypochloraemia. In a metabolic acidosis the plasma chloride may be elevated (normal anion-gap acidosis) or normal (increased anion-gap acidosis).

### PLASMA POTASSIUM

Hyperkalaemia is usual in states of hyperchloraemic acidosis unless there is a concomitant loss of potassium, e.g. potassium loss in: diarrhoea, ureterosigmoidostomy, renal tubular acidosis.

Hypokalaemia is often associated with metabolis alkalosis (hypochloraemia).

### URINARY CHLORIDE

'Spot' urinary chloride concentrations are useful in classifying metabolic alkalosis into the saline responsive and saline unresponsive types (see p. 126).

*Urine [$Cl^-$] < 10 mmol/l*   Saline responsive—vomiting, diuretic therapy, chloride diarrhoea, ingestion of alkali.

*Urine [$Cl^-$] > 20 mmol/l*   Saline unresponsive—mineralocorticoid excess, Bartter's syndrome, severe potassium deficiency.

In some situations, the urinary chloride concentration is a more useful tool in the assessment of volume depletion than the urinary sodium concentration. For example, during the active phase of vomiting, the loss of acid causes bicarbonate to be generated by the stomach wall at a rate greater than it can be reabsorbed by the proximal renal tubule. This results in the excretion of bicarbonate (along with sodium ions) in the urine. The urinary chloride concentration at this stage is low because of (1) low plasma concentration (loss in vomitus), (2) increased tubular reabsorption of NaCl. Therefore during vomiting (from above the pylorus), it is generally found that the urinary concentration of sodium is high (>20 mmol/l) and that of chloride is low (<20 mmol/l). This emphasizes the necessity for measuring both urinary sodium and chloride in attempting to assess the state of volume depletion during vomiting. A similar situation would arise during gastric drainage.

## HYPERCHLORAEMIA

Hyperchloraemia may be associated with the following.

*Hypernatraemia*

*Metabolic acidosis* (normal anion gap)  (1) renal tubular acidosis, (2) carbonic anhydrase inhibitors, (3) diarrhoea, (4) ureterosigmoidostomy, (5) aldosterone deficiency, (6) HCl, $NH_4Cl$, arginine HCl, lysine HCl therapy, (7) early uraemic acidosis, (8) treated diabetic ketoacidosis.

*Respiratory alkalosis*

**Case examples/pathophysiology**

HYPERTONIC DEHYDRATION

Inadequate fluid intake in a comatose adult.

| Plasma | Na | 153 | mmol/l | (132–144) |
|---|---|---|---|---|
| | K | 3.7 | mmol/l | (3.2–4.8) |
| | Cl | 117 | mmol/l | (98–108) |
| | $HCO_3$ | 27 | mmol/l | (23–33) |
| | Urea | 15 | mmol/l | (3.0–8.0) |

↓ fluid intake → haemoconcentration → ↑ $[Cl^-]$ and ↑ $[Na^+]$

RENAL TUBULAR ACIDOSIS

A 57 year old male who presented with severe muscle weakness which promptly responded to potassium therapy. He was diagnosed as having renal tubular acidosis.

| Plasma | Na | 137 | mmol/l | (132-144) | Urine pH | 7.0 | |
|---|---|---|---|---|---|---|---|
| | K | 1.7 | mmol/l | (3.2-4.8) | Na | 43 | mmol/l |
| | Cl | 120 | mmol/l | (98-108) | K | 92 | mmol/l |
| | $HCO_3$ | 8 | mmol/l | (23-33) | | | |
| | Urea | 8.4 | mmol/l | (3.0-8.0) | | | |
| | Creat. | 0.18 | mmol/l | (0.06-0.12) | | | |
| Anion gap | | 11 | mEq/l | (7-17) | | | |

The primary biochemical lesion in distal, or Type I, renal tubular acidosis is an inability of the distal renal tubule to secret $H^+$ in exchange for luminal $Na^+$.

$\downarrow Na^+-H^+$ *exchange* →
(1) retention of $H^+$ → $\downarrow [HCO_3^-]$ (metabolic acidosis) → inability to acidify urine (pH > 5.3)
(2) renal loss of $Na^+$ with its accompanying acid anions ($Na^+.A^-$) → sodium depletion

*Sodium depletion* →   volume depletion → ↑ tubular reabsorption of NaCl (↑ aldosterone, ? third factor)
Therefore sodium lost as $Na^+.A^-$ is replaced by NaCl (acid anions replaced by $Cl^-$) resulting in ↑ plasma $[Cl^-]$.

The hypokalaemia is caused by increased renal secretion of $K^+$ due to the hyperaldosteronism.

The other type of renal tubular acidosis (Type II, proximal) is due to an inability of the proximal renal tubule to reabsorb filtered bicarbonate. In this disease the biochemical features are similar to those seen in the above case with one important exception— during the period of systemic acidaemia ($\downarrow$ plasma $[HCO_3^-]$) the urinary pH is < 5.3. In this disorder the distal tubular $H^+$ secreting mechanism is intact, so that a $H^+$ gradient can be maintained across the distal tubular cells. However, during bicarbonate therapy the plasma $[HCO_3^-]$ increases and it begins to appear in the urine, resulting in an increasing urinary pH.

GASTROINTESTINAL DISEASE

Severe acute diarrhoea in a young adult.

| Plasma | Na | 136 | mmol/l | (132-144) |
|---|---|---|---|---|
| | K | 2.5 | mmol/l | (3.2-4.8) |
| | Cl | 112 | mmol/l | (98-108) |
| | $HCO_3$ | 14 | mmol/l | (23-33) |
| | Urea | 2.2 | mmol/l | (3.0-8.0) |
| Anion gap | | 12 | mEq/l | (7-17) |

(1) diarrhoea → (a) loss of $Na^+$ → $Na^+$ depletion, (b) loss of $HCO_3^-$ → ↓ $[HCO_3^-]$, (c) loss of $H_2O$ → hypovolaemia
(2) hypovolaemia → ↑ renal reabsorption of $Na^+$ and $Cl^-$
(3) $Na^+$ and $Cl^-$ is reabsorbed by the kidney in a 1:1 ratio and returned to the extracellular fluid where the ratio is ∼ 1.4:1. Thus the proportion of chloride will rise and lead to a hyperchloraemia. Also any NaCl taken orally is retained, giving a further increase in the proportion of $Cl^-$.

Diarrhoea is often associated with hypokalaemia (loss of potassium in diarrhoea fluid and in the urine due to aldosteronism consequent to the hypovolaemia). Other causes of hyperchloraemic acidosis associated with hypokalaemia are: (1) ureterosigmoidostomy, (2) renal tubular acidosis, (3) carbonic anhydrase inhibitor (acetazolamide).

## URETEROSIGMOIDOSTOMY

Transplantation of the ureters into the colon (e.g. total cystectomy for carcinoma of the bladder) results in urine being exposed to the absorptive processes of the large bowel. The following sequence of events occurs:

(1) urinary $Cl^-$ is reabsorbed in exchange for $HCO_3^-$ → ↓ plasma $[HCO_3^-]$ and ↑ plasma $[Cl^-]$
(2) urinary $Na^+$ is reabsorbed in exchange for $K^+$ → hypokalaemia

## EARLY URAEMIC ACIDOSIS

Renal failure due to analgesic abuse in a 57 year old woman.

| | | | | |
|---|---|---|---|---|
| Plasma Na | 143 | mmol/l | (132–144) |
| K | 5.0 | mmol/l | (3.2–4.8) |
| Cl | 119 | mmol/l | (98–108) |
| $HCO_3$ | 15 | mmol/l | (23–33) |
| Urea | 26 | mmol/l | (3.0–8.0) |
| Creat. | 0.30 | mmol/l | (0.06–0.12) |
| Anion gap | 14 | mEq/l | (7–17) |

In early renal failure distal tubular failure may be proportionally greater than glomerular dysfunction. In this case acid anions may be excreted normally (glomerulus) but hydrogen ions retained (distal tubule), producing a situation similar to renal tubular acidosis. Hyperkalaemia associated with hyperchloraemic acidosis may occur in the following conditions: early uraemic acidosis, $NH_4Cl$ ingestion, therapy with arginine and lysine HCl, treated diabetic ketoacidosis, mineralocorticoid deficiency.

### TREATED DIABETIC KETOACIDOSIS

Male aged 59 years admitted with diabetic ketoacidosis. Treated with normal saline and insulin infusion.

| Time | | 09.15 | 20.00 | | |
|---|---|---|---|---|---|
| Plasma | Na | 141 | 139 | mmol/l | (132–144) |
| | K | 5.4 | 4.1 | mmol/l | (3.2–4.8) |
| | Cl | 105 | 115 | mmol/l | (93–108) |
| | $HCO_3$ | 7 | 13 | mmol/l | (23–33) |
| | Urea | 6.2 | 4.2 | mmol/l | (3.0–8.0) |
| | Anion gap | 34 | 15 | mEq/l | (7–17) |
| | Glu. | 31 | 14 | mmol/l | (3.0–5.5) |

Untreated diabetic ketoacidosis usually presents with normo- or hypochloraemia, depending on plasma sodium levels, and an increased anion gap acidosis (see p. 92). Shortly after the commencement of treatment with saline infusion and insulin therapy a proportion of these patients will exhibit a hyperchloraemic (normal anion gap) metabolic acidosis. For a discussion of this response see Diabetes (p. 223).

### RESPIRATORY ALKALOSIS

Psychogenic overbreathing in a 42 year old woman.

| Plasma | Na | 135 | mmol/l | (132–144) | pH | 7.64 | (7.35–7.45) |
|---|---|---|---|---|---|---|---|
| | K | 3.3 | mmol/l | (3.2–4.8) | $P_{CO_2}$ | 2.0 kPa | (4.6–6.0) |
| | Cl | 112 | mmol/l | (98–108) | | | |
| | $HCO_3$ | 16 | mmol/l | (23–33) | | | |

In an attempt to compensate for the respiratory alkalosis, $HCO_3^-$ is excreted by the kidney. This $HCO_3^-$ loss is balanced by a loss of $Na^+$, leading to sodium depletion. In response to this there is an increased proximal renal tubular reabsorption of $Na^+$ and $Cl^-$ in the proportion of 1:1 (extracellular fluid normally has a ratio of $\sim 1.4:1$). This causes an increase in the plasma chloride concentration.

# HYPOCHLORAEMIA

Hypochloraemia may be associated with:

*Hyponatraemia*

*Metabolic alkalosis*

*Respiratory acidosis*

*Chloride loss*  Vomiting, chloride diarrhoea.

**Case examples/pathophysiology**

### HYPONATRAEMIA

Dilutional hyponatraemia due to inappropriate ADH secretion.

Plasma Na 115 mmol/l (132-144)
K    4.2 mmol/l (3.2-4.8)
Cl   78  mmol/l (98-108)
$HCO_3$  29  mmol/l (23-33)
Urea 3.0 mmol/l (3.0-8.0)

Haemodilution due to excessive secretion of ADH.

### METABOLIC ALKALOSIS

Mineralocorticoid excess due to ectopic ACTH syndrome in a 63 year old man with a carcinoma of the bronchus.

Plasma Na 144 mmol/l (132-144)
K    2.4 mmol/l (3.2-4.8)
Cl   86  mmol/l (98-108)
$HCO_3$ >40 mmol/l (23-33)

(1) ↑ mineralocorticoid activity → ↑ distal renal tubular $Na^+$ reabsorption + $K^+$ secretion → $K^+$ depletion and $Na^+$ retention
(2) $K^+$ depletion and mineralocorticoid excess → metabolic alkalosis (↑$[HCO_3^-]$)
(3) $Na^+$ retention → water retention → ↑ ECF
(4) ↑ ECF → ↓ proximal NaCl reabsorption (? 3rd factor) → ↑ NaCl excretion → hypochloraemia

## VOMITING

Pyloric stenosis in a 43 year old male.

Plasma Na 141 mmol/l (132–144)
K 3.1 mmol/l (3.2–4.8)
Cl 84 mmol/l (98–108)
$HCO_3$ 44 mmol/l (23–33)

Loss of $H^+$ and $Cl^-$ in vomitus results in metabolic alkalosis and hypochloraemia.

## RESPIRATORY ACIDOSIS

Chronic obstructive airway disease in a 73 year old man.

Plasma Na 135 mmol/l (132–144)   Blood pH 7.35 (7.35–7.45)
K 3.8 mmol/l (3.2–4.8)   $P{CO_2}$ 8.9 kPa (4.6–6.0)
Cl 88 mmol/l (98–108)
$HCO_3$ 37 mmol/l (23–33)

The acidosis caused by retention of $CO_2$ is compensated by renal retention and generation of $HCO_3^-$. For each mole of $HCO_3^-$ reabsorbed or generated a mole of $Na^+$ is also reabsorbed. Therefore:

(1) ↑ reabsorption of $NaHCO_3$ → ↑ reabsorption of $Na^+$ → ↑ retention of water ↑ ECV
(2) ↑ ECV → ↑ renal secretion of NaCl → $Cl^-$ loss → hypochloraemia

# 5 Anion gap

The analytes that are usually included in a plasma electrolyte profile are $Na^+$, $K^+$, $Cl^-$ and $HCO_3^-$. The difference between the sum of the cations and the sum of the anions is termed the anion gap.

Anion gap (mEq/l) = ($Na^+$ mEq/l + $K^+$ mEq/l) − ($Cl^-$ mEq/l + $HCO_3^-$ mEq/l)

A positive charge must always be balanced by the presence of a negative charge in the body fluids, so that the anion gap represents those unmeasured anions that, along with chloride and bicarbonate, balance the sum of measured cations (sodium and potassium). Note that the term anion gap is really a misnomer because there are an equal number of anions; the so-called gap arises because all the anion species present in a plasma sample are not measured. Note that the units for the anion gap are mEq, as it is the number of positive and negative charges present that is important. Approximate values are:

|  | Major cations | (mEq/l) | Major anions | (mEq/l) |
|---|---|---|---|---|
| Plasma | $Na^+$ | 140 | $Cl^-$ | 100 |
|  | $K^+$ | 4 | $HCO_3^-$ | 27 |
|  | *$Ca^{2+}$ | 4.5 | *$protein^{n-}$ | 15 |
|  | *$Mg^{2+}$ | 2.5 | *$PO_4^{2-}$ | 2 |
|  |  |  | *$SO_4^{2-}$ | 1 |
|  |  |  | *$Organic\ acids^{n-}$ | 5 |
|  | Total | 150 | Total | 150 |

* Unmeasured ions.

In the plasma the total cations should equal the total anions.

$$Na^+ + K^+ + Uc = Cl^- + HCO_3^- + Ua$$

where Uc = unmeasured cations, Ua = unmeasured anions. Therefore:

$$(Na^+ + K^+) - (Cl^- + HCO_3^-) = Ua - Uc = \text{anion gap}$$

Thus

↑ anion gap = ↑ Ua or ↓ Uc
↓ anion gap = ↓ Ua or ↑ Uc

## LABORATORY INVESTIGATION

### ANION GAP

If the anion gap is calculated from $(Na^+ + K^+) - (Cl^- + HCO_3^-)$ the total experimental error equals the square root of the sum of the squares of the individual analytical errors. Thus if the analytical errors of each of the above analytes is 1 mmol/l the error of the anion gap will be

$$\sqrt{1^2 + 1^2 + 1^2 + 1^2} \text{ mmol/l} = 2 \text{ mmol/l}$$

Therefore, there can be considerable error in this calculation and this should be taken into account when using the anion gap as a diagnostic tool.

The most useful application of the anion gap is that it allows metabolic acidosis to be divided into two categories.

*High anion gap acidosis*   (1) Renal failure, (2) lactic acidosis, (3) ketoacidosis, (4) ingestions, (a) paraldehyde, (b) methanol, (c) ethylene glycol, (d) salicylate (↑ production of organic acids, ketones, lactic acid).

*Normal anion gap acidosis*   (1) Renal tubular acidosis, (2) carbonic anhydrase inhibitors, (3) diarrhoea, (4) ureterosigmoidostomy, (5) aldosterone deficiency, (6) HCl, $NH_4Cl$, arginine HCl, lysine HCl therapy, (7) early uraemic acidosis, (8) treated diabetic ketoacidosis.

### PLASMA POTASSIUM

Usually acidosis is associated with hyperkalaemia. In situations where potassium loss is a part of the disease the acidosis may be associated with hypokalaemia, e.g. diarrhoea, renal tubular acidosis, carbonic anhydrase inhibitors (acetazolamide), ureterosigmoidostomy.

## INCREASED ANION GAP

**Causes**

*Increased unmeasured anions*   (1) Ketoacidosis, (2) lactic acidosis,

(3) renal failure, (4) ingestions (a) salicylate, (b) methanol/ethanol, (c) paraldehyde, (d) ethylene glycol.

*Decreased unmeasured cations*   $Ca^{2+}$, $Mg^{2+}$, etc.

*Laboratory error*

**Case examples/pathophysiology**

DECREASED UNMEASURED CATIONS

Theoretically it is possible to get an increased anion gap with hypocalcaemia and hypomagnesaemia. However, the magnitude of change in the level of cations that would be necessary to produce an abnormally raised anion gap are probably incompatible with life.

LABORATORY ERROR

Errors in the measurement of sodium (too high) or chloride and bicarbonate (too low) will produce an increased anion gap. In this context the anion gap is a useful quality control tool.

*Example*   A female aged 70 years who presented at the Accident and Emergency Department with right upper quadrant pain. She was not severely distressed.

| | | | |
|---|---|---|---|
| Plasma Na | 143 | mmol/l | (132–144) |
| K | 4.8 | mmol/l | (3.2–4.8) |
| Cl | 92 | mmol/l | (98–108) |
| $HCO_3$ | 15 | mmol/l | (23–33) |
| Urea | 9.3 | mmol/l | (3.0–8.0) |
| Creat. | 0.16 | mmol/l | (0.06–0.12) |
| Anion gap | 41 | mEq/l | (7–17) |

Apart from the abdominal pain this patient was well and had no signs or symptoms of a severe electrolyte disturbance. The anion gap is extremely high (levels above 30–35 mEq/l are unusual) and suggests an analytical error. In this case the error is most likely to be in either the bicarbonate or chloride estimation (both low), i.e. a normal or high chloride associated with a low bicarbonate or *vice versa* is compatible with a number of common electrolyte disturbances. As there was no clinical evidence for an electrolyte disturbance in this patient the laboratory was consulted. Further investigation revealed that the above results represented the values of a quality control specimen that had been incorrectly reported for those of the patient.

## Chapter 5    INCREASED UNMEASURED ANIONS: DIABETIC KETOACIDOSIS

Adult in diabetic coma.

| Plasma | Na | 135 | mmol/l | (132–144) | pH | 7.09 | (7.35–7.45) |
|---|---|---|---|---|---|---|---|
| | K | 5.7 | mmol/l | (3.2–4.8) | $P\text{co}_2$ | 4.4 kPa | (4.6–6.0) |
| | Cl | 101 | mmol/l | (98–108) | $P\text{o}_2$ | 14.8 kPa | (10.6–13.3) |
| | $HCO_3$ | 10 | mmol/l | (23–33) | | | |
| | Glu. | 31.0 | mmol/l | (3.0–5.5) | | | |
| Plasma ketones | | +ve | at 1/32 dilution (−ve) | | | | |
| Anion gap | | 30 | mEq/l | (7–17) | | | |

(1) ↓ insulin → ↑ production and ↓ oxidation of ketone bodies → ↑ plasma ketones (acetoacetate, hydroxybutyrate) → ↑ unmeasured anions
(2) ↑ keto-acids → acidosis

### INCREASED UNMEASURED ANIONS: URAEMIA

Chronic renal failure.

| Plasma | Na | 142 | mmol/l | (132–144) | pH | 7.22 | (7.35–7.45) |
|---|---|---|---|---|---|---|---|
| | K | 5.5 | mmol/l | (3.2–4.8) | $P\text{co}_2$ | 5.32 kPa | (4.6–6.0) |
| | Cl | 104 | mmol/l | (98–108) | | | |
| | $HCO_3$ | 15 | mmol/l | (23–33) | | | |
| | Urea | 63 | mmol/l | (3.0–8.0) | | | |
| | Creat. | 0.79 | mmol/l | (0.06–0.12) | | | |
| Anion gap | | 28 | mEq/l | (7–17) | | | |

(1) ↓ glomerular filtration rate → ↓ secretion of acid anions ($PO_4$, $SO_4$, organic acids) → ↑ unmeasured anions
(2) distal tubule disease → ↓ $H^+$ secretion → ↓ plasma pH

High anion gap occurs late in renal failure (severe ↓ in GFR). In early renal failure the acidosis is usually hyperchloraemic in nature (see p. 85, 160).

## DECREASED ANION GAP

**Causes**

*Decreased unmeasured anions* Uncommon, may occur with hypoalbuminaemia.
*Increased unmeasured cations* Increased $[Ca^{2+}]$ and $[Mg^{2+}]$. These would need to be increased to life-threatening levels in order to produce a decreased anion gap. Increased abnormal cation — IgG myeloma.

## Case examples/pathophysiology

LABORATORY ERROR

(1) Errors in the measurement of sodium (too low) or chloride and bicarbonate (too high) will give a decreased anion gap.
(2) Bromide intoxication — bromide ions in the plasma interfere with the measurement of chloride. The chloride concentration is overestimated and this results in a decreased or negative anion gap.

*Example*   A female aged 47 years with moderate renal failure due to analgesic nephropathy.

| | | | | |
|---|---|---|---|---|
| Plasma | Na | 143 | mmol/l | (132–144) |
| | K | 5.1 | mmol/l | (3.2–4.8) |
| | Cl | 112 | mmol/l | (98–108) |
| | $HCO_3$ | 35 | mmol/l | (23–33) |
| | Urea | 32.5 | mmol/l | (3.0–8.0) |
| | Creat. | 0.29 | mmol/l | (0.06–0.12) |
| Anion gap | | 1 | mEq/l | (7–17) |

In clinical practice the finding of low anion gaps are an uncommon occurrence. As this patient has renal failure the expected anion gap would be high or high normal, suggesting a possible error in the above results. The analyte that does not fit the clinical picture is the plasma bicarbonate concentration, i.e. patients with renal failure tend to have a metabolic acidosis (low $[HCO_3^-]$). The plasma bicarbonate in this patient, which had been incorrectly recorded, was 20 mmol/l. This corrected result gives an anion gap of 16 mEq/l and categorizes the patient as 'early uraemic' or hyperchloraemic acidosis.

DECREASED UNMEASURED ANIONS:
HYPOALBUMINAEMIA

Liver failure in a 49 year old male.

| | | | | |
|---|---|---|---|---|
| Plasma | Na | 134 | mmol/l | (132–144) |
| | K | 4.0 | mmol/l | (3.2–4.8) |
| | Cl | 109 | mmol/l | (98–108) |
| | $HCO_3$ | 24 | mmol/l | (23–33) |
| | Urea | 3.0 | mmol/l | (3.0–8.0) |
| Anion gap | | 5 | mEq/l | (7–17) |
| | Alb. | 19 | g/l | (30–50) |
| | AP | 1750 | U/l | (30–120) |

Chapter 5

| | ALT | 44 | U/l | (<35) |
|---|---|---|---|---|
| | Bili. | 130 | μmol/l | (<20) |

A plasma albumin concentration of 40 g/l contributes about 11 mEq of anion. A drop in plasma albumin to around 20 g/l will cause a drop in the unmeasured anions of around 5 mEq/l.

### INCREASED UNMEASURED CATIONS: IgG MYELOMA

| Plasma | Na | 134 | mmol/l | (132–144) |
|---|---|---|---|---|
| | K | 4.0 | mmol/l | (3.2–4.8) |
| | Cl | 107 | mmol/l | (98–108) |
| | $HCO_3$ | 25 | mmol/l | (23–33) |
| Anion gap | | 6 | mEq/l | (7–17) |
| | TP | 120 | g/l | (60–80) |
| | Alb. | 27 | g/l | (30–50) |

(1) IgG of myeloma possesses an overall positive charge and thus increases the amount of unmeasured cation.
(2) The increased viscosity of the plasma due to the high protein level may interfere with the analysis of sodium and potassium, i.e. a smaller aliquot than normal is aspirated into the flame photometer resulting in falsely low sodium and potassium levels.

# 6 Acid-base and bicarbonate

The oxidative processes that occur within cells yield some 20 000–25 000 mmol of carbonic acid ($H_2CO_3$) each day which, because of the equilibrium

$$H_2CO_3 \rightleftharpoons H_2O + CO_2$$

can be excreted as $CO_2$ by the lungs. The incomplete oxidation of fats, carbohydrates, sulphoproteins, phosphoproteins, etc. are responsible for a net body gain of 50–100 mmol of hydrogen ion ($H^+$) per day as non-volatile acids. Thus one major, and potentially harmful, side-effect of metabolism is that the cells and tissues continually tend towards a state of acidosis. The body therefore requires mechanisms that provide protection from, and then eliminate, the excess acids.

The laboratory tests that are available for routinely investigating the acid-base status of patients provide some measure of $H^+$ and of $CO_2$ at the very least, so in order to understand and interpret such tests it is helpful to consider the acid-base homeostatic mechanisms in terms of the metabolism of (1) $H^+$ and (2) $CO_2$.

Before these areas are discussed it is important to understand the following terms:

## Definitions

### ACID

A substance that dissociates in water to produce hydrogen ions ($H^+$). A strong acid almost completely dissociates to produce $H^+$:

Hydrochloric acid    $HCl \rightarrow H^+ + Cl^-$

whilst a weak acid shows poor dissociation and produces relatively few $H^+$:

Acetic acid    $CH_3COOH \rightleftharpoons H^+ + CH_3COO^-$

# Chapter 6

## ALKALI

A substance that dissociates in water to form hydroxyl ions ($OH^-$), e.g.

$$NaOH \rightarrow Na^+ + OH^-$$

## BASE

A substance that can accept $H^+$. In the above example the chloride ion ($Cl^-$) is a weak base because it barely combines with $H^+$ in water, while the acetate ion ($CH_3COO^-$) is acting as a strong conjugate base because it remains largely associated with $H^+$.

## BUFFER

A mixture of a weak acid and its conjugate base that attenuates a change in $H^+$ concentration when a strong acid or base is added to it. In effect the buffer does this by forming a weaker acid or base. Bicarbonate buffer is a mixture of carbonic acid ($H_2CO_3$) and bicarbonate ($HCO_3^-$):

Buffer + strong acid  $HCO_3^- + H^+ \rightleftharpoons H_2CO_3$
Buffer + strong alkali  $H_2CO_3 + OH^- \rightleftharpoons HCO_3^- + H_2O$

The carbonic acid/bicarbonate pair is the major buffer system of the extracellular fluid.

## pH

A measure of the hydrogen ion concentration [$H^+$], defined as the logarithm of the reciprocal of the $H^+$ concentration:

$$pH = \log \frac{1}{[H^+]} \text{ (or } = -\log [H^+])$$

## pH OF A BUFFER SYSTEM

This may be calculated from the Henderson-Hasselbalch equation, which relates the pH of a buffer to the concentration of the buffer acid and base present. Understanding this equation is very helpful to interpreting clinical acid-base measurements. In general, for the bicarbonate buffer system:

$$pH = pK + \log \frac{[\text{base}]}{[\text{acid}]}$$

where $K$ = overall dissociation equilibrium constant, $pK = \log^1/K$. For the bicarbonate system $pK = 6.1$.

For the bicarbonate buffer system in plasma the equation may be stated:

$$pH = 6.1 + \log \frac{[HCO_3^-]}{[H_2CO_3]}$$

The equation shows that the pH of a buffer solution is dependent on the *ratio* of the amounts of $HCO^-{}_3$ and $H_2CO_3$, not on their actual concentrations. Thus for plasma to remain at an ideal pH of 7.4

$$pH = 6.1 + 1.3 = 7.4$$

i.e. $\log \dfrac{[HCO_3^-]}{[H_2CO_3]}$ must be approximately equal to 1.3 (antilog 1.3 = 20). Thus for plasma:

$$\frac{[HCO_3^-]}{[H_2CO_3]} = \frac{20}{1}$$

when $pH = 7.4$.

An important implication of this equation is that the pH of a buffered solution (plasma) may be kept normal at the expense of abnormal concentrations of its buffer components.

*Simplification* In a fluid such as plasma $H_2CO_3$ is in equilibrium:

$$H_2CO_3 \rightleftharpoons CO_2 + H_2O$$

so that the concentration of $H_2CO_3$ is directly proportional to the partial pressure of $CO_2$ ($P_{CO_2}$), and may be related as follows:

$$[H_2CO_3] = P_{CO_2} \times 0.225$$

where $0.225 = CO_2$ solubility constant, $P_{CO_2}$ = partial pressure in kiloPascals (kPa). Thus the Henderson-Hasselbalch equation may be written as:

$$pH = pK + \log \frac{[HCO_3^-]}{P_{CO_2} \times 0.225}$$

(If $P_{CO_2}$ is in mm Hg, the solubility constant = 0.03.) When interpreting acid-base studies in patients it can be useful to memorize the equation simply as the relationship:

$$pH \propto \frac{[HCO_3^-]}{P_{CO_2}}$$

### RESPIRATORY COMPONENT

$P_{CO_2}$ is so-called because its level in the body is ultimately controlled by lung function.

### METABOLIC COMPONENT

Describes $HCO_3^-$ because its concentration reflects the non-respiratory contribution to the body's acid-base status.

### ACIDAEMIA

An abnormally raised [H$^+$] (normal range $\simeq$ 35–45 nmol/l) in the body (a decreased pH). Since pH depends on the ratio $\frac{[HCO_3^-]}{P_{CO_2}}$ acidosis can arise in two ways:

(1) a fall in [$HCO_3^-$] with a normal $P_{CO_2}$

$$pH \downarrow \propto \frac{\downarrow}{N} \quad \text{metabolic acidosis}$$

(2) a rise in $P_{CO_2}$ with a normal [$HCO_3^-$]

$$pH \downarrow \propto \frac{N}{\uparrow} \quad \text{respiratory acidosis}$$

### ALKALAEMIA

An abnormally low [H$^+$] in the body (an elevated pH). Alkalosis may arise in two ways:
(1) A rise in [$HCO_3^-$] with a normal $P_{CO_2}$

$$pH \uparrow \propto \frac{\uparrow}{N} \quad \text{metabolic alkalosis}$$

(2) a fall in $P_{CO_2}$ with a normal [$HCO_3^-$]

$$pH \uparrow \propto \frac{N}{\downarrow} \quad \text{respiratory alkalosis}$$

*Note* The suffix 'osis' refers to the underlying metabolic or respiratory disturbance while 'aemia' refers to the net effect of these disturbances on the blood $H^+$ concentration.

COMPENSATION

The Henderson-Hasselbalch equation for the bicarbonate buffer system shows that the plasma pH would be about 7.4 as long as the ratio of $[HCO_3^-]/[H_2CO_3]$ remains at 20/1. If a disease process altered the concentration of one of the components the ratio, and therefore the pH, can be returned to normal if the other component concentration is sufficiently altered in the same direction. Both component concentrations may be abnormal, but if the pH is restored to normal, then compensation is said to have occurred. (*Complete* compensation if the pH is restored to normal, *partial* if the pH has moved towards normal but remains abnormal.) For example, respiratory disease causes retention of $CO_2$, and therefore respiratory acidosis

$$pH \downarrow\downarrow \propto \frac{N}{\uparrow\uparrow}$$

A healthy kidney will respond over the next few days (2–3 days) by increasing the plasma level of $HCO_3^-$. If enough $HCO_3^-$ is retained to restore the ratio to normal then 'compensation' has occurred (by creating a metabolic alkalosis) and the pH will be normal, although at the expense of abnormal component concentrations, i.e.

$$pH\ N \propto \frac{\uparrow\uparrow}{\uparrow\uparrow}$$

If the increase in $[HCO_3^-]$ is insufficient to restore the ratio to 20/1 the pH may be improved but still abnormal

$$pH \downarrow \propto \frac{\uparrow}{\uparrow\uparrow}$$

In this case partial 'compensation' has occurred.

*Note* In practice a complete compensation rarely occurs, the pH returning 'towards' normal but rarely 'to' normal. In the above example if the pH fell to 7.0, then after 2–3 days the compensatory rise in $HCO_3^-$ would return the pH to $\sim 7.2–7.3$, but not to 7.4 (ideal pH). Only in prolonged respiratory alkalosis can the pH sometimes be returned to normal, but this depends on the ability of the kidney to excrete enough $HCO_3^-$.

*Chapter 6*  HYDROGEN ION METABOLISM

Each day the body produces, as end products of metabolism, 50–100 mmol of $H^+$ in the form of non-volatile acids, which can only be excreted by the kidney. When the $H^+$ ions leave the cells they are buffered by the extracellular buffers, particularly bicarbonate, which is consumed in the process.

$$H^+ + HCO_3^- \rightleftharpoons H_2CO_3 \rightleftharpoons H_2O + CO_2$$

This buffering mechanism is not perfect so there is a residual acidosis. However, in the short term this acidosis is minimized by the removal of $CO_2$ (lowering of $P\text{CO}_2$) from the extracellular fluid by the lungs so as to restore the $[HCO_3^-]/P\text{CO}_2$ ratio towards 20/1 (respiratory compensation). At this stage, although the blood pH is near normal, the original $H^+$ ions are still present and are represented by the decreased (consumed) $HCO_3^-$. The system is finally normalized when the excess $H^+$ are excreted by the kidney and the bicarbonate is regenerated.

Thus hydrogen ion metabolism should be considered in terms of: (1) $H^+$ production, (2) extracellular buffering of $H^+$ and respiratory compensation, (3) renal excretion of $H^+$.

## $H^+$ production

The 50–100 mmol ($\sim$ 0.7–1.4 mmol/kg) of $H^+$ produced daily originate from:
(1) Dietary sulphur-containing proteins → sulphuric acid
(2) Dietary phosphoproteins → phosphoric acid
(3) Incomplete oxidation of dietary carbohydrates and fats → lactic acid

## Extracellular buffering and respiratory compensation

### EXTRACELLULAR BUFFERING

Of the extracellular buffers (bicarbonate, phosphate, proteins) the bicarbonate system is the most important because it is: (1) present in large quantities and (2) the only system that can be easily measured with some accuracy.

As noted above the addition of acid to this system results in the consumption of bicarbonate, but leaves a residual $H^+$ activity ($\downarrow$ pH and $\downarrow [HCO_3^-]$). This problem of a decreased pH is overcome by respiratory compensation.

## RESPIRATORY COMPENSATION

A high extracellular $H^+$ concentration stimulates the respiratory centre, so that more $CO_2$ is blown off in the lungs, thus lowering the $P\text{co}_2$ of the ECF. This action returns the 'ratio' $HCO_3^-/P\text{co}_2$ back towards 20/1 and brings the pH towards normal, even though the $P\text{co}_2$ and $[HCO_3^-]$ are now abnormal.

(1) Addition of acid:

$$\text{pH} (\downarrow) \propto \frac{[HCO_3] (\downarrow)}{p\text{CO}_2 (N)}$$

(2) With respiratory compensation:

$$\text{pH} (N) \propto \frac{[HCO_3] (\downarrow)}{p\text{CO}_2 (\downarrow)}$$

Note that the pH is returned towards normal at the expense of abnormal concentrations of buffer components. In practice, although the respiratory centre responds promptly to a change in extracellular $[H^+]$, maximum compensation may not be reached for 12–24 hours.

## Renal excretion of $H^+$ and $HCO_3$ regeneration

The role of the kidney in $H^+$ metabolism is three-fold: (1) secretion of hydrogen ion, (2) conservation of filtered bicarbonate, (3) regeneration of consumed bicarbonate and $H^+$ excretion.

### SECRETION OF HYDROGEN ION

$H^+$ is said to be 'secreted' when it moves from the tubule cell into the tubular lumen. 'Excretion' occurs when some of this secreted $H^+$ is lost from the body in the urine. Each day 4000–5000 mmol of $H^+$ are secreted by the kidney but only a small portion of this (50–100 mmol) is actually excreted, i.e. that excess which has been produced in the form of non-volatile acids by metabolic processes (see Regeneration of Bicarbonate below). The major part ($\sim 4500$ mmol) of the $H^+$ secreted is concerned with bicarbonate conservation (see below) and during this process there is no net loss of $H^+$ from the body.

In the tubular cells hydrogen ion is generated from $CO_2$ and $H_2O$, promoted by the action of carbonic anhydrase ($H_2O + CO_2 \overset{CA}{\leftrightarrows} H_2CO_3 \rightarrow H^+ + HCO_3^-$) and secreted into the tubular lumen in exchange for sodium ions (see Fig. 6.1).

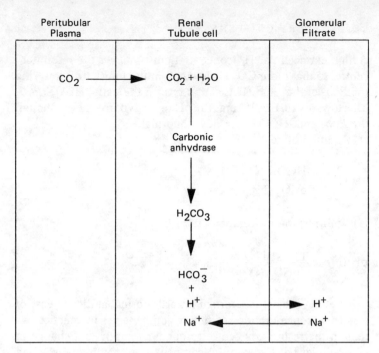

**Fig. 6.1** Generation and secretion of $H^+$ by renal tubule cells.

As the luminal hydrogen ion concentration increases, secretion of $H^+$ is suppressed because of the high gradient attained. In fact the highest luminal concentration of $H^+$ that can be reached is around 0.03 mmol/l (pH 4–4.5), so that if the luminal fluid was pure water and 1 l was passed a day only 0.03 mmol of hydrogen ion could be excreted, whereas the amount required to be excreted is of the order of 50–100 mmol.

This gradient problem is overcome by two processes.

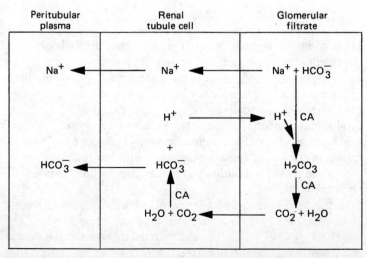

**Fig. 6.2** $H^+$ secretion and reabsorption of $HCO_3^-$ in the proximal renal tubule. CA, carbonic anhydrase.

*Proximal nephron* (Fig. 6.2)  Secreted hydrogen ion combines with filtered bicarbonate to form carbonic acid. This is converted to $CO_2$ and water under the influence of the carbonic anhydrase that is located in the luminal brush border of the tubule. The tubule cells then absorb the $CO_2$ along a concentration gradient (bicarbonate conservation).

*Distal nephron* (Fig. 6.3)  Secreted hydrogen ion is bound by the urinary buffers (ammonia and phosphate), thus preventing an increase in the $H^+$ concentration of the tubular fluid.

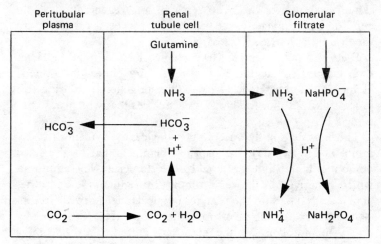

**Fig. 6.3**  $H^+$ secretion and $HCO_3^-$ generation by the distal nephron.

BICARBONATE CONSERVATION

Normally about 4500 mmol of bicarbonate is filtered by the glomeruli and reabsorbed by the tubules each day (85% in the proximal tubule and 15% distally). Reabsorption of bicarbonate from the tubular lumen requires the secretion of hydrogen ion (see Fig. 6.2).

The hydrogen ion, secreted in exchange for luminal sodium, reacts with luminal bicarbonate and, under the influence of carbonic anhydrase (cell brush border), produces $CO_2$; this is absorbed by the cell and enters the ongoing $H^+$-producing process. The reabsorbed sodium, and the bicarbonate produced within the cell, diffuses across the cell membrane and back into the blood.

Control of bicarbonate reabsorption

Proximal bicarbonate reabsorption is influenced by intravascular volume (IVV), potassium levels and plasma $P\text{co}_2$, i.e.

| Increased reabsorption | Decreased reabsorption |
|---|---|
| ↓ IVV | ↑ IVV |
| hypokalaemia | hyperkalaemia |
| ↑ $P\text{co}_2$ | ↓ $P\text{co}_2$ |

In the distal tubule aldosterone, which plays no part in the proximal process, increases hydrogen ion secretion and bicarbonate generation. Aldosterone deficiency has the reverse effect.

REGENERATION OF CONSUMED BICARBONATE
AND $H^+$ EXCRETION

The addition of hydrogen ions to the extracellular buffers results in consumption of bicarbonate. Therefore, to normalize this system excretion of the hydrogen ions and regeneration of the consumed bicarbonate has to occur. This process is carried out in the distal tubules (see Fig. 6.3).

Hydrogen ion is secreted in exchange for sodium. The $H^+$ is then trapped in the lumen by the urinary buffers (phosphate, ammonia) and excreted. The bicarbonate generated during this process diffuses into the blood and replaces that which was consumed.

The phosphate buffers account for the loss in the urine of 20–30 mmol of $H^+$ per day, the capacity being limited by the fixed amount of buffer that is filtered by the glomeruli. The ammonia, which is derived in the renal tubule from glutamine and other amino acids, takes up the remaining $H^+$. Under normal circumstances the excretion rate of ammonium ion ($NH_4^+$) is of the order of 20–40 mmol/day. If the body is required to excrete large amounts of $H^+$ (e.g. diabetic acidosis), the production of tubular ammonia and the excretion of ammonium ion can be increased ten-fold.

The important points to notice in this distal tubular process are:
(1) For each hydrogen ion secreted one bicarbonate molecule is generated and returned to the extracellular fluid.
(2) For hydrogen ion secretion to continue, its concentration in the lumen must be kept low. This is achieved by the binding action of the urinary buffers (during bicarbonate conservation the concentration is lowered by combination with luminal bicarbonate).

# CARBON DIOXIDE METABOLISM

Under basal conditions the normal adult generates around 25 000 mmol of $CO_2$ daily. This is transported in the blood to the lungs where it is excreted.

**Carbon dioxide transport**

Most of the $CO_2$ that arises as a result of aerobic metabolism diffuses from the cells into plasma and then into the red cells. In the erythrocytes a small proportion of the $CO_2$ combines with haemoglobin to form carbamino compounds, whilst the major part is converted, under the influence of carbonic anhydrase, to hydrogen and bicarbonate ions. The hydrogen ions are buffered by haemoglobin (reduced haemoglobin is less acid than oxy-haemoglobin), and the bicarbonate ions diffuse out into the plasma in exchange for chloride ions, the 'chloride shift' (see Fig. 6.4).

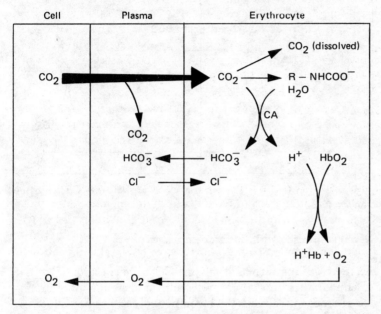

**Fig. 6.4** Removal of $CO_2$ from peripheral tissues. CA, carbonic anhydrase.

The quantitative distribution of $CO_2$ present in blood is approximately:
60% in the form of bicarbonate (generated by red cells but carried mostly in the plasma).
30% as carbamino compounds (virtually all attached to haemoglobin).
10% as dissolved $CO_2$ (including carbonic acid).

## Carbon dioxide excretion

When blood reaches the lungs oxygenation of haemoglobin causes release of the bound hydrogen ions; these combine with bicarbonate to produce $CO_2$, which can then be secreted into the alveoli. As the concentration of red cell bicarbonate decreases it is replenished by plasma bicarbonate which moves into the cell in exchange for chloride ions, i.e. the reactions involving $CO_2$ in the lung area are the reverse of those which occur at the cellular level (see Fig. 6.5).

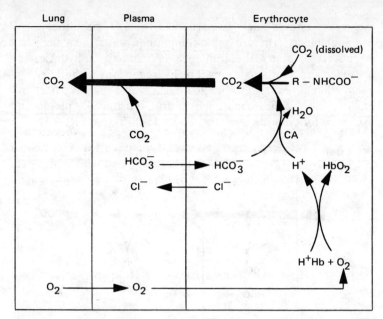

**Fig. 6.5** Transfer of $CO_2$ from the blood to the lung. CA, carbonic anhydrase.

### Renal response to plasma $P_{CO_2}$ changes

An increased (or decreased) plasma $CO_2$ concentration ($P_{CO_2}$), if sustained, will result in renal retention (or excretion) of bicarbonate in order to maintain the blood pH at near normal values.

In respiratory acidosis

$$pH (\downarrow) \propto \frac{[HCO_3^-] (N)}{P_{CO_2} (\uparrow)}$$

the increased $P_{CO_2}$ results in an increased proximal reabsorption of $HCO_3^-$, i.e. a metabolic alkalosis is created

$$pH (\sim N) \propto \frac{[HCO_3^-] (\uparrow)}{P_{CO_2} (\uparrow)}$$

in order to try and compensate for a respiratory acidosis. The reverse action (renal $HCO_3^-$ excretion and creation of a metabolic acidosis) occurs in respiratory alkalosis. This renal compensation of respiratory acid-base disorders is slow and up to 3-5 days may elapse before a steady state is reached.

## OTHER BUFFERING MECHANISMS

### Intracellular buffers

The intracellular buffers are protein, phosphate and, in the red cell, haemoglobin.

Cellular buffering of hydrogen ion has an important effect on plasma potassium. If extracellular hydrogen ion concentration is increased the $H^+$ moves into cells, electroneutrality being maintained by the reverse movement of potassium and sodium; this may result in severe hyperkalaemia. The changes in extracellular potassium concentration are more pronounced when acid-base disturbances are due to metabolic rather than respiratory causes. On the other hand, if the extracellular hydrogen ion concentration is lowered hydrogen ions are released from the cell in exchange for potassium, resulting in a fall in plasma potassium concentration. Conversely, cellular movements of potassium ions in hyper- and hypokalaemia have similar effects on hydrogen ion movements.

**Bone buffers**

Bone carbonate represents a large store of buffer that in severe acidosis can help to stabilize the plasma bicarbonate by absorbing hydrogen ion. This mechanism is probably responsible for the maintenance of the low, but stable, plasma bicarbonate in chronic renal failure where there is a positive hydrogen ion balance, i.e. bicarbonate concentration tends to remain in the range 12–18 mmol/l despite a hydrogen ion retention of around 10–20 mmol/day.

## LABORATORY INVESTIGATION

In the investigation of suspected acid-base disturbances two groups of tests are required; those that allow a biochemical diagnosis (acidosis, alkalosis, etc.) and those which give some indication of the aetiology (e.g. renal failure).

*Biochemical diagnosis*  (1) plasma bicarbonate, (2) blood 'gases'— pH, $P_{CO_2}$, $P_{O_2}$, (3) actual bicarbonate.

*Aetiological diagnosis*  (1) plasma electrolytes, (2) plasma urea and creatinine, (3) plasma glucose, (4) plasma lactate, (5) plasma (urine) ketones.

PLASMA BICARBONATE

A raised bicarbonate concentration represents a metabolic alkalosis, whilst a low level indicates a metabolic acidosis. However, the bicarbonate value in isolation is not particularly helpful in diagnosing acid-base disturbances because it gives no information about the hydrogen ion concentration (pH) or $P_{CO_2}$. Neither does it indicate whether the metabolic lesion is primary or secondary to a disorder of $P\text{co}_2$. However, if some clinical details are known about the patient (respiratory or metabolic disease) it is a useful parameter, particularly if considered in conjunction with other biochemical estimations (Figs 6.6 and 6.7).

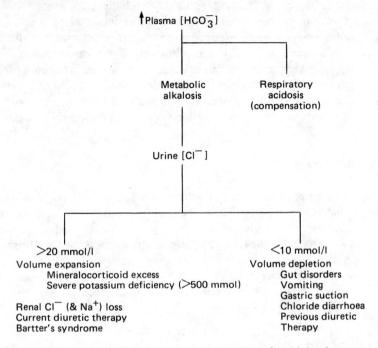

**Fig. 6.6** Scheme for the laboratory investigation of a high plasma bicarbonate.

Many laboratories report a plasma bicarbonate result along with their other electrolyte values. Generally, the electrolyte analysers measure the total $CO_2$ content of plasma ($HCO_3^-$, $H_2CO_3$, dissolved $CO_2$) and not just the bicarbonate ion. Therefore the values obtained with such instruments will be 1–2 mmol/l higher than the actual value of $[HCO_3^-]$. This however does not detract from its value as a measure of the metabolic component of acid-base disturbances.

### Blood gases

Results from blood gas analysers provide information on all aspects of the bicarbonate buffer system, as well as the oxygen status of the blood.

### pH

A scale for the measurement of hydrogen ion in which, as defined earlier

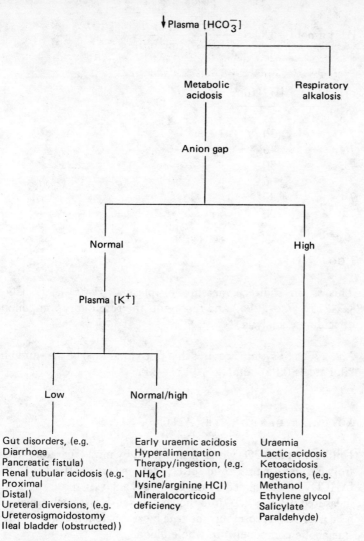

**Fig. 6.7** Scheme for the laboratory investigation of a low plasma bicarbonate.

$$pH = \log \frac{1}{[H^+]} \quad \text{or} \quad -\log [H^+]$$

Hydrogen ion concentration could just as easily be reported in the same way as other analytes; for example in our hospital the reference range for plasma $[H^+] = 35–45$ nmol/l. However, the traditional use of the pH scale means that users should be aware of how logarithmic scales can mislead the unwary. For example, a pH change from 7.45 to 7.15 is a fall of 0.3 units (antilog $0.3 = 2$) so that the pH change is indicating that the $[H^+]$ has doubled.

It is useful to remember the following data in order to check blood gas results for possible errors (e.g. inaccurate figures given over the telephone).

$$[H^+] \text{ nmol/l} = 180 \times \frac{P\text{CO}_2 \text{ (kPa)}}{[\text{HCO}_3^-] \text{ (mmol/l)}}$$

| pH  | $[H^+]$ (nmol/l) |
|-----|------------------|
| 7.6 | 25               |
| 7.5 | 30               |
| 7.4 | 40               |
| 7.3 | 50               |
| 7.2 | 60               |
| 7.1 | 80               |
| 7.0 | 100              |

$P\text{CO}_2$

This is an indirect measure of plasma $[H_2CO_3]$, ($[H_2CO_3]$ $= 0.225 \times P\text{CO}_2$ (kPa)) and represents the respiratory component of acid-base analysis.

*Note*   The SI system requires that $P\text{CO}_2$ (and $P\text{O}_2$) be measured in kiloPascals (kPa). mmHg $\times 0.133 =$ kPa.

### ACTUAL BICARBONATE ($\text{AHCO}_3^-$)

Unlike the total $CO_2$ which is measured by most laboratory 'electrolyte' instruments, and often designated as $[\text{HCO}_3^-]$, the $[\text{AHCO}_3^-]$ is calculated by the blood gas analyser substituting the measured pH and $P\text{CO}_2$ into the Henderson-Hasselbach equation. The $[\text{AHCO}_3^-]$ represents the true bicarbonate ion concentration (metabolic component) in whole blood.

There are two other values which also quantitate the metabolic component—standard bicarbonate and base excess—these are often calculated by modern blood gas analysers.

### STANDARD BICARBONATE

This is the plasma bicarbonate concentration in whole blood which in theory has been equilibrated at 37°C with a gas mixture having a $P\text{CO}_2$ of 5.32 kPa. It is a calculated parameter that reflects the acid-base status of the plasma if the lung function was normal. That is, in a patient with a mixed respiratory and metabolic acidosis the standard bicarbonate would reflect the metabolic abnormality.

## BASE EXCESS

A parameter that is calculated by some blood gas analysers after measurement of pH and $P_{CO_2}$. The base excess is an alternative way of expressing the metabolic component. It is best described by outlining how it might be measured.

A blood sample would be equilibrated with $P_{CO_2}$ at 5.32 kPa at 37°C. This mimics normal lung function and restores any abnormal respiratory component to normal, e.g.

$$\text{pH (N)} \propto \frac{[HCO_3^-]\,(\uparrow)}{P_{CO_2}\,(\uparrow)}$$

a sample from a patient with a compensated metabolic alkalosis.

Equilibrate with a normal $P_{CO_2}$.

$$\text{pH}\,(\uparrow) \propto \frac{[HCO_3^-]\,(\uparrow)}{P_{CO_2}\,(N)}$$

The sample would then be titrated with acid until the pH is restored to normal. The amount of acid needed to do this indicates the amount of excess base that was present in the sample. If 10 mmol/l of acid was added, then the base excess must have been 10 mmol/l, a reflection of the metabolic component.

If, as is more common, the sample was acidaemic then

$$\text{pH (N)} \propto \frac{[HCO_3^-]\,(\downarrow)}{P_{CO_2}\,(\downarrow)} = \text{compensated metabolic acidosis}$$

after equilibration with normal $CO_2$

$$\text{pH}\,(\downarrow) \propto \frac{[HCO_3^-]\,(\downarrow)}{P_{CO_2}\,(N)}$$

This sample would need titration with base to normalize the pH, so that the base 'excess' is negative. Thus if 10 mmol/l of base restored the pH to normal the base excess (BE) is $-10$ mmol/l. The reference range for BE is approximately $-3 - +3$ mmol/l. Negative values indicate acidosis and positive values suggest alkalosis.

In the interpretation of blood gas values the standard bicarbonate and base excess give similar information to that of actual bicarbonate and add nothing of practical value. They will not be used in the case examples that follow this section.

## $Po_2$

Although not a component of the acid-base system, the $Po_2$ indicates the patient's oxygenation status. A $Po_2$ estimation is of great value in respiratory disorders and, in the absence of clinical details, may give an indication of the aetiology of acid-base disturbance, e.g. a low $Po_2$ is often associated with a metabolic or respiratory acidosis (see p. 113, Interpretation of Blood Gas Results).

## PLASMA POTASSIUM

An acidosis is usually associated with a high plasma potassium concentration, whilst alkalosis may be associated with hypokalaemia; there are exceptions to this rule (see p. 64).

## PLASMA CHLORIDE AND ANION GAP

The anion gap (AG), representing the anions other than $Cl^-$ and $HCO_3^-$ which neutralize the positive charges of $Na^+$ and $K^+$, is useful for defining two types of metabolic acidosis.

$$AG\ mEq/l = [Na^+\ (mmol/l) + K^+\ (mmol/l)] - [Cl^-\ (mmol/l) + HCO_3^-\ (mmol/l)]$$

(1) Acid anions which are produced in excess amounts (diabetes mellitus), or which are not secreted because of glomerular disease (renal failure), remain in the plasma thus displacing bicarbonate ions. This produces the condition of normochloraemic acidosis or 'high anion gap acidosis'.

(2) In metabolic acidosis due to predominantly renal tubular disease acid anions, along with sodium, are excreted in the urine and replaced in the extracellular fluid by chloride ions absorbed from the gut or proximal tubule. This produces a 'hyperchloraemic' or 'normal anion gap' acidosis (high plasma chloride concentration and low bicarbonate). This condition will also occur with ingestion of substances which are metabolized to HCl ($NH_4Cl$, arginine HCl), i.e. the acid anion (chloride) replaces the bicarbonate ion that was consumed by the addition of hydrogen ion (HCl) to the buffer system. It will also occur in situations where $HCO_3^-$ is lost from the extracellular fluid, e.g. diarrhoea and proximal renal tubular acidosis.

## PLASMA UREA AND CREATININE

A metabolic acidosis associated with a high plasma urea (or

creatinine) occurs in renal failure due to intrinsic renal disease and in severe dehydration (prerenal uraemia).

### PLASMA GLUCOSE

The hyperglycaemia of severe diabetes mellitus is usually accompanied by metabolic acidosis, except in the less common condition of hyperosmolar coma.

### PLASMA (URINE) KETONE BODIES

Ketosis, mainly associated with diabetes mellitus and starvation, has a concomitant metabolic acidosis.

### PLASMA LACTIC ACID

Metabolic acidosis due to accumulation of lactic acid occurs in severe anoxic conditions, primary lactic acidosis and occasionally in diabetes mellitus.

### URINARY CHLORIDE

On the basis of spot urinary chloride concentrations the metabolic alkaloses can be divided into two types: those associated with (1) volume depletion, urine Cl < 10 mmol/l, (2) volume expansion, urine Cl > 20 mmol/l.

The inference from these findings is that volume depleted conditions (vomiting, previous diuretic therapy) can be treated with saline infusions alone, whereas those alkalotic conditions associated with a high urinary chloride (mineralocorticoid excess) will not respond to saline therapy.

## INTERPRETATION OF BLOOD GASES

The simplest approach to 'blood gas' interpretation is to consider the pH, [$HCO_3^-$] and $P\text{CO}_2$ results separately and then combine the information from each so as to arrive at a biochemical diagnosis (Figs 6.8, 6.9 and 6.10). It is helpful to recall the essential relationship of the Henderson-Hasselbalch equation,

$$\text{pH} \propto \frac{[HCO_3^-]}{P\text{CO}_2} \quad \text{(see p. 98).}$$

pH may be ↑, N or ↓. If ↑ = alkalaemia, ↓ = acidaemia, N = compensation (see below).

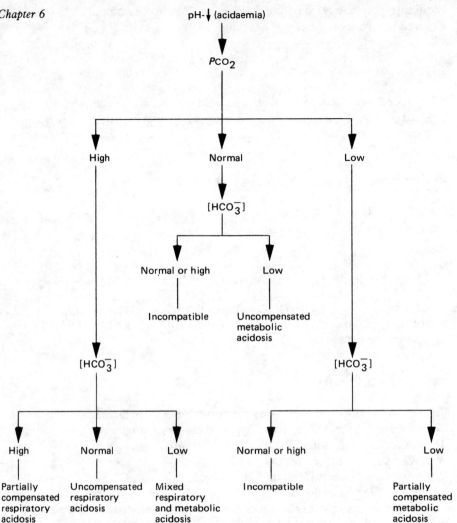

**Fig. 6.8** Scheme for determining biochemical diagnosis of acidaemia.

[HCO₃⁻] may be ↑, N or ↓. If ↑ = metabolic alkalosis, ↓ = metabolic acidosis.

$P$co₂ may be ↑, N or ↓. If ↑ = respiratory acidosis, ↓ = respiratory alkalosis.

Example 1

pH    = 7.09         (7.35–7.45)
HCO₃  = 6    mmol/l  (24–32)
$P$co₂  = 2.7  kPa     (4.7–6.0)

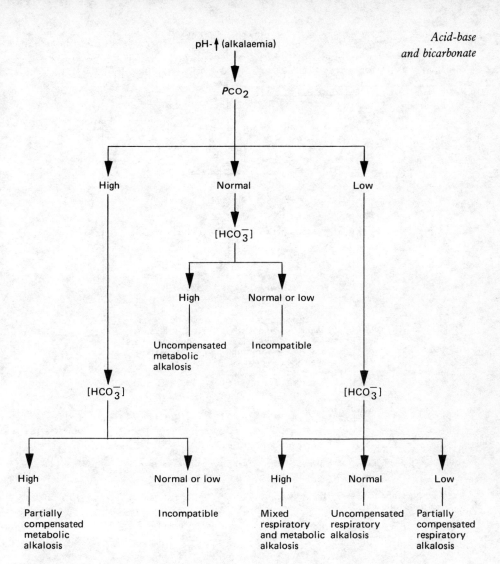

**Fig. 6.9** Scheme for determining the biochemical diagnosis of alkalaemia.

pH $= \downarrow =$ acidaemia, and could be due to $\frac{\downarrow}{N}$ or $\frac{N}{\uparrow}$, but [$HCO_3^-$] $= \downarrow =$ metabolic acidosis, $P_{CO_2} = \downarrow =$ respiratory alkalosis.

As the pH is low the primary lesion must be an acidosis. The only acidosis indicated by the other parameters is metabolic and, therefore, this case must be a primary metabolic acidosis. The low $P_{CO_2}$ indicates that a respiratory alkalosis is present and, therefore, the overall situation represents a partially compensated metabolic acidosis, e.g.

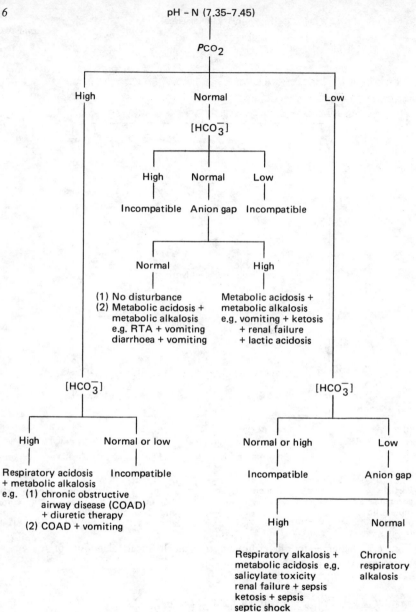

**Fig. 6.10** Scheme for determining the biochemical diagnosis of a patient with an acid-base disturbance but a normal blood pH.

$$\text{pH} \downarrow \propto \frac{\downarrow \downarrow}{\downarrow}$$

Example 2

pH       = 7.54            (7.35–7.45)
$[HCO_3^-]$ = 41   mmol/l  (24–32)
$P\text{co}_2$  = 6.7 kPa  (4.7–6.0)

pH = ↑ = alkalaemia, could be due to $\frac{↑}{N}$ or $\frac{N}{↓}$, but [HCO$_3^-$]
= ↑ = metabolic alkalosis, $P\text{CO}_2$ = ↑ = respiratory acidosis.

The high pH indicates a primary alkalosis. The only alkalosis indicated by the measured parameters is of the metabolic variety. Thus this is a partially compensated metabolic alkalosis, e.g.

$$↑ \propto \frac{↑↑}{↑}$$

## Example 3

pH      = 7.33        (7.35–7.45)
[HCO$_3^-$] = 34   mmol/l   (24–32)
$P\text{CO}_2$  = 8.7   kPa     (4.7–6.0)

pH = acidaemia, could be due to $\frac{N}{↑}$ or $\frac{↓}{N}$, but [HCO$_3^-$] = ↑
= metabolic alkalosis, $P\text{CO}_2$ = ↑ = respiratory acidosis.

This is a case of partially compensated respiratory acidosis, e.g.

$$↑ \propto \frac{↑}{↑↑}$$

## Example 4

pH      = 7.54        (7.35–7.45)
[HCO$_3^-$] = 25   mmol/l   (24–32)
$P\text{CO}_2$  = 3.99 kPa    (4.7–6.0)

pH = alkalaemia, could be due to $\frac{↑}{N}$ or $\frac{N}{↓}$, [HCO$_3^-$] = normal range, $P\text{CO}_2$ = respiratory alkalosis.

Thus this case must be an acute (uncompensated) respiratory alkalosis, e.g.

$$↑ \propto \frac{↑}{N}$$

## Compensation

If homeostatic mechanisms are normal the body attempts to correct an H$^+$ disturbance by creating the opposite situation, e.g.

| Primary disturbance | Normal response |
|---|---|
| Respiratory acidosis | Metabolic alkalosis |
| Respiratory alkalosis | Metabolic acidosis |
| Metabolic acidosis | Respiratory alkalosis |
| Metabolic alkalosis | Respiratory acidosis |

If the pH is returned to within the normal range, then 'compensation' is said to occur. If the pH is not normalized, the homeostatic response has not been fully successful, i.e. partial compensation. In practice the pH is rarely compensated except in chronic respiratory alkalosis and, therefore, a normal pH (with abnormal $P_{CO_2}$ and [$HCO_3^-$]) strongly suggests the presence of a mixed disturbance, e.g. a patient with chronic obstructive airways disease and on diuretics.

For any given primary disturbance to the plasma levels of $P_{CO_2}$ or [$HCO_3^-$] there is an approximate compensating response that can normally be expected. If the actual compensating response is markedly different from that expected, then the presence of a second acid-base disorder should be strongly suspected. The expected compensating response can be calculated using one of several published formulae that have been derived from patient data. The tables below are based on the formulae quoted by Narins and Emmett, modified for SI units (*Medicine* **59**, 161, 1980).

METABOLIC ACIDOSIS

| Primary disturbance of [$HCO_3^-$] mmol/l | Expected compensating response by $P_{CO_2}$ (kPa) |
|---|---|
| 20 | ~5.00 |
| 16 | ~4.25 |
| 12 | ~3.50 |
| 8 | ~2.70 |
| 4 | ~1.90 |
| lower limit of compensation = | ~1.30 kPa |

METABOLIC ALKALOSIS

| Primary disturbance of [$HCO_3^-$] mmol/l | Expected compensating response by $P_{CO_2}$ (kPa) |
|---|---|
| 32 | ~5.00 |
| 36 | ~5.50 |
| 40 | ~6.00 |
| 44 | ~6.50 |
| 48 | ~7.00 |
| Upper limit of compensation = | ~7.40 kPa |

*Note* (1) Expected compensation level may not be reached for 12-24 hours. (2) Respiratory response to metabolic alkalosis is variable and responses differing from the above do not necessarily mean the presence of an associated respiratory disorder.

## ACUTE RESPIRATORY ACIDOSIS

In the acute stage there is a moderate elevation of $[HCO_3^-]$ by around 2-4 mmol/l. This occurs in the first 10 minutes and is due to $HCO_3^-$ generated from the retained $CO_2$, i.e.

$$CO_2 + H_2O \rightarrow H_2CO_3 \rightarrow H^+ + HCO_3^-$$

Some of the excess $H^+$ is immediately buffered by intracellular and extracellular buffers. This initial increase in $HCO_3^-$ may raise the plasma $[HCO_3^-]$ to levels of 28-30 mmol/l but not higher.

## CHRONIC RESPIRATORY ACIDOSIS

Maximal renal compensation occurs over 2-4 days. In primary uncomplicated respiratory acidosis the plasma $[HCO_3^-]$ rarely rises above 45 mmol/l (limit of compensation).

| Primary disturbance of $P\text{co}_2$ (kPa) | Expected compensating response of $[HCO_3^-]$ mmol/l |
|---|---|
| 6.00 | ~27 |
| 7.00 | ~30 |
| 8.00 | ~33 |
| 9.00 | ~36 |
| 10.00 | ~39 |
| 11.00 | ~42 |
| 12.00 | ~45 |

In chronic respiratory acidosis patients with a $P\text{co}_2$ of less than 8 kPa may show complete compensation (pH ~7.4).

## ACUTE RESPIRATORY ALKALOSIS

Within 10 minutes of the acute onset of respiratory alkalosis the plasma $[HCO_3^-]$ will show a slight decrease (2-4 mmol/l), i.e. as the blood $CO_2$ level decreases (↑ respiration) $HCO_3^-$ moves into the erythrocytes from the plasma, combines with $H^+$ (released from haemoglobin) and forms carbonic acid, which breaks down to $CO_2$ and water. This acute decrease in plasma $[HCO_3^-]$ rarely reaches lower than about 18 mmol/l.

## CHRONIC RESPIRATORY ALKALOSIS

Maximal renal compensation occurs over the next 2-4 days. This is the only primary acid-base disorder that may exhibit complete compensation. However the plasma [$HCO_3^-$] rarely falls below 12 mmol/l (limit of compensation).

| Primary disturbance $P_{CO_2}$ (kPa) | Expected compensating response by [$HCO_3^-$] mmol/l |
|---|---|
| 4.0 | ~20 |
| 3.6 | ~18.5 |
| 3.2 | ~17 |
| 2.8 | ~15.5 |
| 2.4 | ~14 |

If the primary disturbance lasts several days a complete compensation (pH = ~7.4) may occur, probably depending on the ability of the kidney to excrete sufficient bicarbonate.

*Case example* 72 year old man with chronic obstructive airways disease who is taking diuretics for associated cardiac failure.

| Plasma | K | 2.6 | mmol/l | (3.2-4.8) |
| Blood | pH | 7.41 | | (7.35-7.45) |
| | $P_{CO_2}$ | 10.4 | kPa | (4.7-6.0) |
| | $HCO_3$ | 47 | mmol/l | (24-32) |

Since $N \propto \dfrac{\uparrow}{\uparrow}$ it is not possible to easily identify the primary disturbance using the above relationship as previously described. However, looking at the above tables it can be seen that a primary disturbance to $P_{CO_2}$ of 10.4 kPa should cause a compensatory response by [$HCO_3^-$] of approximately 39-42 mmol/l, whilst a primary disturbance to [$HCO_3^-$] of 47 mmol/l would lead to an expected compensating $P_{CO_2}$ of around 7.0 kPa. The actual response of both $P_{CO_2}$ and [$HCO_3^-$] greatly exceeds that predicted and, therefore, it can be said that the respiratory acidosis and the metabolic alkalosis are both primary disturbances (mixed) in this patient.

## LOW PLASMA BICARBONATE

Low plasma bicarbonate may be caused by:

Metabolic acidosis

Respiratory alkalosis (compensation)

## Metabolic acidosis

*Increased $H^+$ load*  (1) Ketoacidosis, (2) lactic acidosis, (3) ingestion  (a) salicylates, (b) ethylene glycol, (c) methanol/ethanol, (d) paraldehyde.

*Decreased secretion of $H^+$*  (1) Renal failure, (2) distal renal tubular acidosis, (3) mineralocorticoid deficiency, (4) failure of tubule response to aldosterone.

*Loss of bicarbonate*  (1) Gastrointestinal tract  (a) diarrhoea, (b) pancreatic fistula, (2) renal—proximal renal tubular acidosis.

## Case examples/pathophysiology

### DIABETIC KETOACIDOSIS

Known insulin dependent diabetic presenting in coma.

Plasma

| | | | | | | | |
|---|---|---|---|---|---|---|---|
| Na | 135 | mmol/l | (132–144) | Blood pH | 7.09 | | (7.35–7.45) |
| K | 5.7 | mmol/l | (3.2–4.8) | $H^+$ | 80 | nmol/l | (35–45) |
| Cl | 101 | mmol/l | (95–110) | $P_{CO_2}$ | 2.7 | kPa | (4.7–6.0) |
| $HCO_3$ | 10 | mmol/l | (23–33) | $P_{O_2}$ | 14.8 | kPa | (10.6–13.3) |
| Urea | 12 | mmol/l | (3.0–8.0) | A $HCO_3$ | 6 | mmol/l | (24–32) |
| Glu. | 30.5 | mmol/l | | | | | |
| Ketones | +ve | | | | | | |
| Anion gap | 30 | mEq/l | (8–17) | | | | |

(1) Increased production of non-volatile acids (acetoacetate, hydroxybutyrate) causes rapid consumption of buffer bicarbonate, and thus a decreased plasma bicarbonate.

$$\text{pH} (\downarrow\downarrow) \propto \frac{[HCO_3^-] (\downarrow\downarrow)}{P_{CO_2} (N)}$$

(2) The kidney is unable to rapidly  (a) excrete hydrogen ion and regenerate bicarbonate, (b) excrete the endogenous acid anions, so the low serum bicarbonate continues and the acid anions build up in the plasma, causing an increased anion gap.

(3) Respiratory compensation is attempted by increasing ventilation so as to lower the $P_{CO_2}$, but in this case the response is inadequate and the hydrogen ion concentration remains above normal (partial compensation).

$$\text{pH} (\downarrow) \propto \frac{[HCO_3^-] (\downarrow\downarrow)}{P_{CO_2} (\downarrow)}$$

(4) Hyperkalaemia is due to (a) increased release of $K^+$ from cellular protein which, in the absence of insulin, is now being catabolized to fuel gluconeogenesis, (b) acidaemia.

### RENAL FAILURE

Adult with chronic renal failure.

| Plasma | | | | Blood | | | |
|---|---|---|---|---|---|---|---|
| Na | 135 | mmol/l | (132–144) | pH | 7.28 | | (7.35–7.45) |
| K | 5.6 | mmol/l | (3.2–4.8) | $H^+$ | 52 | nmol/l | (35–45) |
| Cl | 96 | mmol/l | (93–108) | $P_{CO_2}$ | 3.5 | kPa | (4.7–6.0) |
| $HCO_3$ | 11 | mmol/l | (23–33) | $P_{O_2}$ | 12.2 | kPa | (10.6–13.3) |
| Urea | 75 | mmol/l | (3.0–8.0) | A $HCO_3$ | 12 | mmol/l | (24–32) |
| Creat. | 1.83 | mmol/l | (0.06–0.12) | | | | |
| Anion gap | 34 | mEq/l | (8–17) | | | | |

Pathophysiology

See Fig. 6.11.

Renal failure and diabetic ketoacidosis are examples of a high anion gap acidosis. Major causes of this condition are: (1) ketoacidosis, (2) lactic acidosis, (3) renal failure, (4) ingestions (see p. 121).

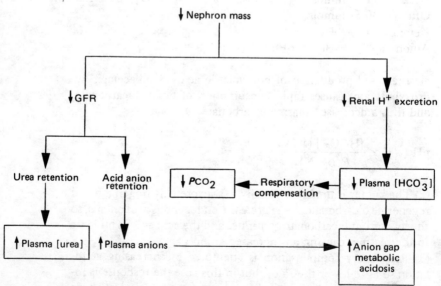

**Fig. 6.11** Mechanism of high anion gap metabolic acidosis in chronic renal failure.

# RENAL TUBULAR ACIDOSIS (RTA)

*Acid-base and bicarbonate*

Infant with chronic acidosis and failure to thrive (distal RTA).
Plasma

| | | | | | | | |
|---|---|---|---|---|---|---|---|
| Na | 139 | mmol/l | (132–144) | Blood pH | 7.24 | | (7.35–7.45) |
| K | 2.9 | mmol/l | (3.1–4.8) | $H^+$ | 58 | nmol/l | (35–45) |
| Cl | 116 | mmol/l | (93–108) | $P_{CO_2}$ | 3.2 | kPa | (4.7–6.0) |
| $HCO_3$ | 11 | mmol/l | (23–33) | $P_{O_2}$ | 10.7 | kPa | (10.6–13.3) |
| Urea | 5.3 | mmol/l | (3.0–8.0) | A $HCO_3$ | 10 | mmol/l | (24–32) |
| Anion gap | 15 | mEq/l | (8–17) | | | | |

Renal tubular acidosis (RTA) results from ineffective tubular secretion of hydrogen ion, and depending on the location of the defect it may result in $HCO_3^-$ wasting (proximal RTA, Type II) or failure to excrete the daily metabolic load of acid (distal RTA, Type I). In the proximal form the $HCO_3^-$ wasting is usually associated with other proximal tubular defects (phosphaturia, glycosuria, aminoaciduria). The distal $H^+$ secretory mechanism is intact and these patients are capable, during severe acidosis, of producing a urine of low pH ($<5.3$).

Subjects with the distal defect are able to reabsorb $HCO_3^-$ in the proximal tubule, but cannot under any circumstances acidify the urine, i.e. urine pH is always greater than 5.3, even during severe systemic acidosis.

Pathophysiology (see Fig. 6.12)

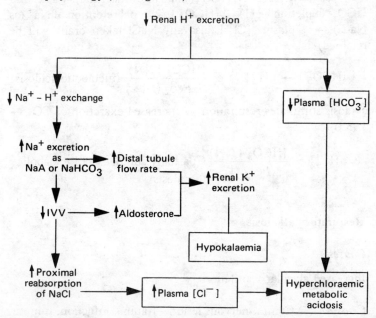

**Fig. 6.12** Mechanism of hyperchloraemic metabolic acidosis in renal tubular acidosis. NaA = sodium salt of acid anions (A).

This is an example of hyperchloraemic or normal anion gap acidosis. Major causes of this condition are:
(1) Gastrointestinal disorders—diarrhoea.
(2) Ureteral diversions—ureterosigmoidostomy.
(3) Renal tubular acidosis—proximal, distal.
(4) Hypoaldosteronism—mineralocorticoid deficiency, failure of tubular response.
(5) Ingestions/therapy—$NH_4Cl$, lysine HCl, arginine HCl.
(6) Early 'uraemic' acidosis.
(7) Treated phase of diabetic ketoacidosis.

LOSS OF BICARBONATE

A 22 year old man admitted with a history of several days diarrhoea.

| Plasma | | | | | | | |
|---|---|---|---|---|---|---|---|
| Na | 138 | mmol/l | (132–144) | Blood pH | 7.29 | | (7.35–7.45) |
| K | 3.1 | mmol/l | (3.2–4.8) | $H^+$ | 52 | nmol/l | (35–45) |
| Cl | 114 | mmol/l | (98–108) | $P_{CO_2}$ | 3.45 | kPa | (4.7–6.0) |
| $HCO_3$ | 11 | mmol/l | (23–33) | $P_{O_2}$ | 11.0 | kPa | (10.6–13.3) |
| Urea | 1.8 | mmol/l | (3.0–8.0) | A $HCO_3$ | 12 | mmol/l | (24–32) |
| Anion gap | 16 | mEq/l | (7–17) | | | | |

*Diarrhoea* → loss of $NaHCO_3$ (small intestinal fluid) → $Na^+$ and $HCO_3^-$ depletion → (1) ↓↓ $[HCO_3^-]$, (2) renal retention of $Na^+$ (as NaCl) → ↑ plasma $[Cl^-]$ (also any NaCl taken orally will be retained)

$$\downarrow\downarrow [HCO_3^-] \rightarrow pH (\downarrow\downarrow) \propto \frac{[HCO_3^-](\downarrow\downarrow)}{P_{CO_2}(N)} \text{ (metabolic acidosis)}$$

acidosis stimulates respiration → increased excretion of $CO_2$ → ↓ $P_{CO_2}$,

i.e. $pH (\downarrow) \propto \dfrac{[HCO_3^-](\downarrow\downarrow)}{P_{CO_2}(\downarrow)}$ (partially compensated metabolic acidosis)

**Respiratory alkalosis**

*Causes:*

Increased alveolar ventilation

*Central* (1) Central nervous lesion—trauma, infection, tumour,

(2) drugs—salicylates, (3) anxiety/hysteria, (4) pregnancy, (5) septicaemia, (6) liver failure.

*Pulmonary*  (1) Pneumonia, (2) asthma, (3) congestive cardiac failure, (4) embolism.

*Mechanical ventilation*

Patient with gram-negative septicaemia and hyperventilation.

Plasma
| | | | | | | | |
|---|---|---|---|---|---|---|---|
| Na | 144 | mmol/l | (132–144) | Blood pH | 7.44 | | (7.35–7.45) |
| K | 3.5 | mmol/l | (3.2–4.8) | $H^+$ | 36 | nmol/l | (35–45) |
| Cl | 111 | mmol/l | (98–108) | $P_{CO_2}$ | 4.0 | kPa | (4.7–6.0) |
| $HCO_3$ | 21 | mmol/l | (23–33) | $P_{O_2}$ | 14.0 | kPa | (10.6–13.2) |
| | | | | A $HCO_3$ | 20 | mmol/l | (24–32) |

Pathophysiology (see Fig. 6.13)

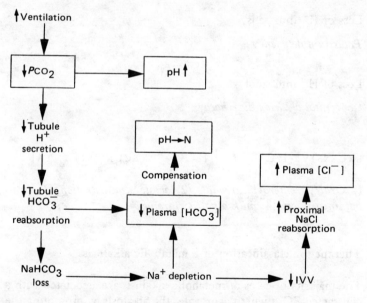

**Fig. 6.13**  Mechanism of a compensated respiratory alkalosis.

## HIGH PLASMA BICARBONATE

High plasma bicarbonate can be due to  (1) metabolic alkalosis, (2) respiratory acidosis (compensation).

*Chapter 6* **Metabolic alkalosis**

LOSS OF H$^+$ FROM EXTRACELLULAR FLUID

Loss in gastric juice

(1) *vomiting*, (2) *gastric drainage*.

Loss of H$^+$ in urine

*Mineralocorticoid excess* syndromes: (1) Cushing's syndrome, (2) primary aldosteronism, (3) secondary aldosteronism (a) renal artery stenosis, (b) malignant hypertension, (c) Bartter's syndrome, (d) diuretic therapy, (4) enzyme defects (a) 17 hydroxylase, (b) 11 hydroxylase, (5) Liddle's syndrome, (6) carbenoxolone therapy, (7) licorice ingestion.

*Severe potassium deficiency*

Loss of H$^+$ into cells
*Potassium deficiency*

Loss of H$^+$ into stool
*Congenital chloride diarrhoea*

EXCESSIVE $HCO_3^-$ (OR $HCO_3^-$ PRECURSOR) INTAKE

(1) *Oral/parenteral bicarbonate*, (2) *lactate administration*, (3) *citrate administration*, (4) *milk-alkali syndrome*.

**Therapeutic classification of a metabolic alkalosis**

The majority of cases of metabolic alkalosis are associated with a contracted ECV, which aggravates the alkalosis by promoting the renal retention of bicarbonate. Patients with this disorder will respond to IV infusions of saline. On the other hand patients with a metabolic alkalosis associated with a normal ECV will not respond to saline infusions; these patients are invariably severely potassium deficient and will respond to potassium therapy. These two groups can be differentiated on the basis of the urinary chloride concentration:

*Saline responsive* (urinary [Cl$^-$]<10 mmol/l   (1) vomiting, (2) chloride diarrhoea, (3) diuretic therapy (previous), (4) alkali ingestion.

*Saline unresponsive* (urinary [Cl$^-$]>20 mmol/l   (1) mineralocorticoid excess, (2) Bartter's syndrome, (3) severe potassium deficiency.

## Case examples/pathophysiology

VOMITING

Infant with projectile vomiting (pyloric stenosis).

Plasma

| | | | | Blood | | | |
|---|---|---|---|---|---|---|---|
| Na | 131 | mmol/l | (132–144) | | pH | 7.58 | (7.35–7.45) |
| K | 2.1 | mmol/l | (3.1–4.8) | | H$^+$ 26 | nmol/l | (35–45) |
| Cl | 76 | mmol/l | (93–108) | | $P\mathrm{co}_2$ 6.0 | kPa | (4.7–6.0) |
| HCO$_3$ | 40 | mmol/l | (23–33) | | $P\mathrm{o}_2$ 6.8 | kPa | (10.6–13.3) |
| Creat. | 0.52 | mmol/l | (0.06–0.12) | | A HCO$_3$ 41 | mmol/l | (24–32) |

Urine

| Na | 32 | mmol/l |
|---|---|---|
| K | 25 | mmol/l |
| Cl | 10 | mmol/l |

Pathophysiology (see Fig. 6.14)

Figure 6.14 indicates the pathophysiology during the active phase of vomiting. During the recovery phase (no further generation of bicarbonate) the contracted extracellular volume and state of potassium deficiency will increase proximal reabsorption of bicarbonate and sodium. Therefore, now instead of the high urinary [Na$^+$] and low [Cl$^-$] of the active phase, both ions are decreased. During vomiting urinary chloride levels are a better indicator of volume depletion than urinary sodium levels.

PRIMARY ALDOSTERONISM

Adult with moderate hypertension (see Fig. 6.15).

Plasma

| | | | | Blood | | | |
|---|---|---|---|---|---|---|---|
| Na | 144 | mmol/l | (132–144) | | pH | 7.56 | (7.35–7.45) |
| K | 1.7 | mmol/l | (3.1–4.8) | | H$^+$ 27 | nmol/l | (35–45) |
| Cl | 85 | mmol/l | (93–108) | | $P\mathrm{co}_2$ 5.8 | kPa | (4.7–6.0) |
| HCO$_3$ | 40 | mmol/l | (23–33) | | $P\mathrm{o}_2$ 9.0 | kPa | (10.6–13.3) |
| Urea | 3.4 | mmol/l | (3.0–8.0) | | A HCO$_3$ 38 | mmol/l | (24–32) |
| Creat. | 0.08 | mmol/l | (0.06–0.12) | | | | |

Urine

| Na | 71 | mmol/l |
|---|---|---|
| K | 22 | mmol/l |
| Cl | 84 | mmol/l |

# Chapter 6

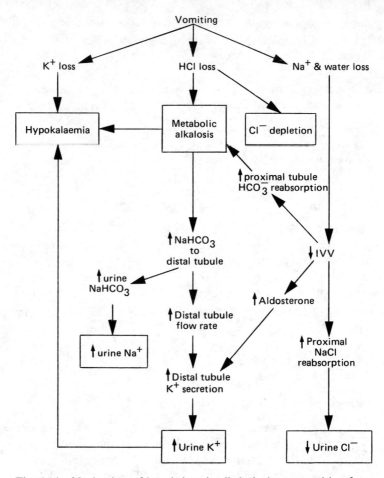

**Fig. 6.14** Mechanism of hypokalaemic alkalosis due to vomiting from above the pylorus.

The increased extracellular volume, due to the increased distal sodium reabsorption caused by the excess aldosterone, results in decreased proximal tubular sodium reabsorption. Thus the patient comes back into sodium balance. This is known as the 'escape phenomenon'.

These patients will have a spot urinary chloride concentration greater than 20 mmol/l and their alkalosis will not respond to saline infusions—'saline resistant alkalosis'.

INGESTION OF ALKALI

Excessive ingestion of magnesium carbonate and sodium bicarbonate over a period of months for a gastric ulcer.

*Acid-base and bicarbonate*

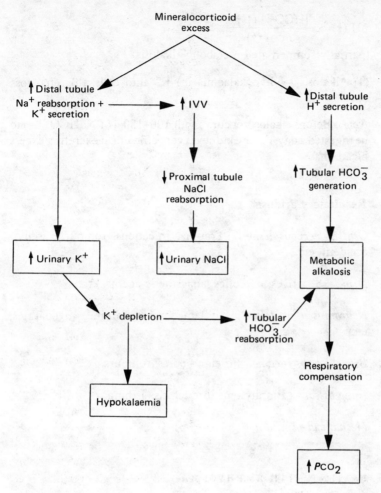

**Fig. 6.15** Mechanism of metabolic alkalosis in mineralocorticoid excess.

Plasma
| | | | | | | | | |
|---|---|---|---|---|---|---|---|---|
| Na | 145 | mmol/l | (132–144) | Blood pH | 7.57 | | (7.35–7.45) |
| K | 2.9 | mmol/l | (3.2–4.8) | $H^+$ | 27 | nmol/l | (35–45) |
| Cl | 91 | mmol/l | (98–108) | $P\text{co}_2$ | 5.3 | kPa | (4.7–6.0) |
| $HCO_3$ | 37 | mmol/l | (23–33) | $P\text{o}_2$ | 6.1 | kPa | (10.6–13.3) |
| Urea | 9.5 | mmol/l | (3.0–8.0) | A $HCO_3$ | 35 | mmol/l | (24–32) |

(1) ↑ intake of $NaHCO_3$ and $MgCO_3$ → ↑ $[HCO_3^-]$ ($CO_3^{2-}$ is converted to $HCO_3^-$)

(2) ↑↑ $[HCO_3^-]$ → ↑ pH    pH (↑↑) ∝ $\dfrac{[HCO_3^-]\,(\uparrow\uparrow)}{P\text{co}_2\,(N)}$

(3) ↑ pH → ↓ pulmonary ventilation → ↑ $P\text{co}_2$ →

$$pH\,(\uparrow) \propto \frac{[HCO_3^-]\,(\uparrow\uparrow)}{P{CO_2}\,(\uparrow)}$$

(partially compensated metabolic alkalosis)

(4) Alkalosis → hypokalaemia (a) $K^+$ into cells, (b) renal loss of $K^+$

*Note* Before alkalosis occurs about 100–150 g of $NaHCO_3$ has to be ingested a day, i.e. the kidney is very efficient at excreting excess bicarbonate.

## Respiratory acidosis

*Central nervous lesions* Trauma, infections, tumour, vascular accident, etc.

*Drugs* Sedatives, narcotics, anaesthetic agents, etc.

*Neuromuscular defects* Guillain-Barré syndrome, myopathies, etc.

*Bronchial obstruction* Foreign bodies, tumours.

*Lung disease* Emphysema, bronchitis, asthma.

*Mechanical ventilation*

PATIENT WITH EMPHYSEMA

| Plasma | | | | | | | |
|---|---|---|---|---|---|---|---|
| Na | 135 | mmol/l | (132–144) | Blood pH | 7.36 | | (7.35–7.45) |
| K | 3.8 | mmol/l | (3.2–4.8) | $H^+$ | 43 | nmol/l | (35–45) |
| Cl | 88 | mmol/l | (95–110) | $P{CO_2}$ | 8.4 | kPa | (4.7–6.0) |
| $HCO_3$ | 37 | mmol/l | (23–33) | $P{O_2}$ | 7.0 | kPa | (10.6–13.3) |
| | | | | A $HCO_3$ | 35 | mmol/l | (24–32) |

Pathophysiology (see Fig. 6.16)

## MIXED RESPIRATORY AND METABOLIC ACIDOSIS

### Causes

(1) Respiratory arrest with severe anoxia, (2) respiratory arrest superimposed on a previous metabolic acidosis.

*Acid-base and bicarbonate*

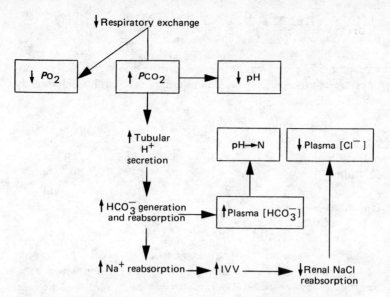

**Fig. 6.16** Mechanism of respiratory acidosis with partial compensation.

PATIENT WITH CARDIO-PULMONARY ARREST

| | | | |
|---|---|---|---|
| Blood pH | 7.01 | | (7.35–7.45) |
| $H^+$ | 98 | nmol/l | (35–45) |
| $P\text{co}_2$ | 8.8 | kPa | (4.7–6.0) |
| $P\text{o}_2$ | 6.8 | kPa | (10.6–13.3) |
| A $HCO_3$ | 16 | mmol/l | (24–32) |

Pathophysiology (see Fig. 6.17)

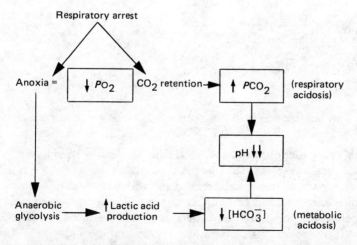

**Fig. 6.17** Mechanism of mixed respiratory and metabolic acidosis in respiratory arrest.

*Chapter 6*  # MIXED RESPIRATORY AND METABOLIC ALKALOSIS

Male aged 70 years, with respiratory failure due to chronic obstructive airway disease. He was admitted to ICU and mechanically ventilated.

|  |  | On admission | 5 h of ventilation |  |
|---|---|---|---|---|
| Blood | pH | 7.19 | 7.66 | (7.35–7.45) |
|  | $H^+$ (nmol/l) | 64 | 22 | (35–45) |
|  | $P_{CO_2}$ (kPa) | 12.9 | 4.2 | (4.6–6.0) |
|  | $P_{O_2}$ (kPa) | 7.6 | 9.9 | (10.6–13.3) |
|  | A $HCO_3$ (mmol/l) | 36 | 35 | (24–32) |

On admission this patient had a partially compensated respiratory acidosis, i.e. ↑ $P_{CO_2}$ and ↑ [$HCO_3^-$]. Ventilation was overenthusiastic and produced a respiratory alkalosis ($P_{CO_2}$ < 4.6 kPa) by removing too much $CO_2$. The patient's plasma [$HCO_3^-$], originally increased to compensate the respiratory acidosis, remains high because the renal compensatory mechanism takes 2–4 days to respond.

## MIXED RESPIRATORY ACIDOSIS AND METABOLIC ALKALOSIS

Male aged 82 years, with chronic obstructive airway disease and congestive cardiac failure, who has been on long term thiazide therapy.

| Plasma |  |  |  |  |  |  |  |  |
|---|---|---|---|---|---|---|---|---|
| Na | 131 | mmol/l | (132–144) | Blood pH | 7.43 |  | (7.35–7.45) |
| K | 2.2 | mmol/l | (3.2–4.8) | $P_{CO_2}$ | 9.9 | kPa | (4.6–6.0) |
| Cl | 72 | mmol/l | (98–108) | $P_{O_2}$ | 8.1 | kPa | (10.6–13.3) |
| $HCO_3$ | 47 | mmol/l | (23–33) | A $HCO_3$ | 48 | mmol/l | (24–32) |
| Urea | 15.3 | mmol/l | (3.0–8.0) |  |  |  |  |

This patient has a metabolic alkalosis (↑ $HCO_3^-$) and a respiratory acidosis (↑ $P_{CO_2}$). As the pH is normal, it is not possible to tell which is the primary disturbance and which is the compensating response. Thus this case could represent: (1) a compensated metabolic alkalosis, (2) a compensated respiratory acidosis.

If this patient had a pure primary respiratory acidosis the following would be expected after compensation (see p. 119):

(1) Plasma $HCO_3$ less than 45 mmol/l (limit of compensation). The expected level would be approximately 39 mmol/l.

(2) pH would be less than 7.40—overcompensation of primary disorders does not occur.

If the patient had a pure metabolic alkalosis the $P_{CO_2}$ would be expected to be about 7.0 kPa. In this case the $P_{CO_2}$ is too high for it to be a compensatory response to a pure metabolic alkalosis with a plasma $[HCO_3^-]$ of 48 mmol/l.

Thus this patient's plasma $[HCO_3^-]$ is too high for the degree of respiratory acidosis, and the blood $P_{CO_2}$ is too high for the degree of metabolic alkalosis. He therefore has a primary respiratory acidosis (due to airway disease) and a primary metabolic alkalosis (due to potassium deficiency consequent to thiazide diuretic therapy).

## MIXED RESPIRATORY ALKALOSIS AND METABOLIC ACIDOSIS

A 45 year old female who attempted suicide by consuming large amounts of aspirin (approximately 100 g, 12 hours prior to admission). On presentation she was comatose and had peripheral cyanosis.

| Plasma | | | | | | | |
|---|---|---|---|---|---|---|---|
| Na | 142 | mmol/l | (132–144) | Blood pH | 7.38 | | (7.35–7.45) |
| K | 5.5 | mmol/l | (3.2–4.8) | $P_{CO_2}$ | 3.5 | kPa | (4.6–6.0) |
| Cl | 98 | mmol/l | (98–108) | $P_{O_2}$ | 8.0 | kPa | (10.6–13.3) |
| $HCO_3$ | 19 | mmol/l | (23–33) | A $HCO_3$ | 15 | mmol/l | (24–32) |
| Urea | 12.0 | mmol/l | (3.0–8.0) | | | | |
| Anion gap | 30 | mEq/l | (7–17) | | | | |
| Plasma salicylate | 5 | mmol/l | (toxic level > 2 mmol/l) | | | | |

Salicylate acts on the adult central nervous system to stimulate respiration ($\downarrow P_{CO_2}$). Salicylate also alters certain metabolic pathways causing overproduction of organic acids, including lactic acid ($\downarrow [HCO_3^-] + \uparrow$ anion gap). A perusal of the above blood gas parameters suggests that these results may represent a mixed acid-base disturbance, i.e.

(1) Complete compensation (pH 7.38). This is unusual except in chronic respiratory alkalosis.

(2) If this was a primary metabolic acidosis the $P_{CO_2}$ would be expected to be around 4 kPa, i.e. above value is less than expected.

(3) If it was a primary respiratory alkalosis the expected $[HCO_3^-]$ would be around 18 mmol/l and overcompensation would not occur (above pH is on the acid side of normal).

In children aspirin intoxication commonly causes a pure metabolic acidosis rather than the above type of mixed disturbance.

# 7 Mineralocorticoids

The mineralocorticoids are $C_{21}$ steroids that are produced by the adrenal cortex. Their predominant action is to influence the uptake and distribution of sodium and potassium. The most important of these steroids is aldosterone, but an intermediate in its synthetic pathway, 11-deoxycorticosterone (DOC), also has significant mineralocorticoid activity.

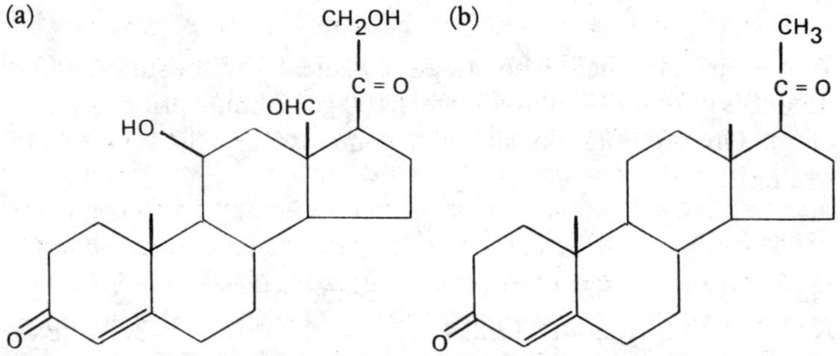

**Fig. 7.1** (a) Aldosterone. (b) Progesterone.

## ALDOSTERONE SYNTHESIS

The synthesis of aldosterone occurs in the zona glomerulosa region of the adrenal cortex.

Progesterone is hydroxylated at the $C_{21}$ position to form 11-deoxycorticosterone, which is then hydroxylated at $C_{11}$ to produce corticosterone. The conversion of the $C_{18}$ methyl group of corticosterone to an aldehyde group leads to the formation of aldosterone (Figs 7.1 and 7.2). The precise method of this conversion is unclear but it is thought that the enzyme corticosterone methyl oxidase Type I (CMO I) converts corticosterone to a labile intermediate product which either decomposes to 18-hydroxycorticosterone, or serves as a substrate for the enzyme corticosterone methyl oxidase Type II (CMO II) which then converts it to aldosterone.

## ALDOSTERONE METABOLISM

Under normal circumstances 50–250 µg (0.14–0.7 µmol) of aldo-

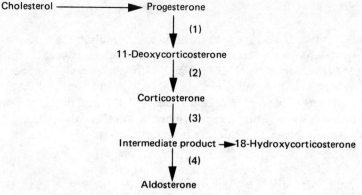

**Fig. 7.2** Pathway for the biosynthesis of aldosterone. (1) $C_{21}$ hydroxylase, (2) $C_{11}$ hydroxylase, (3) Corticosterone methyl oxidase type I (CMO I), (4) CMO II.

sterone are produced daily. It circulates in the plasma weakly bound to albumin and more tightly bound to transcortin and a specific binding protein, aldosterone-binding globulin.

The principal site of catabolism is the liver, where over 90% of aldosterone is cleared from the blood during a single passage. In the liver the aldosterone is inactivated by reduction to the tetrahydro derivative, which is then conjugated to glucuronic acid ($C_3$ position) and excreted in the urine. More than 50% of aldosterone is excreted as the glucuronide. Another 10–20% of aldosterone is conjugated at the $C_{18}$ position to form a glucuronide which, if allowed to stand at pH 1, dissociates into free aldosterone. Measurements of 'urinary aldosterone' usually refer to this metabolite rather than aldosterone itself which is excreted in minute quantities.

The normal plasma level of aldosterone is of the order of 0.03–0.55 nmol/l; the 24 hour excretion rate of the acid labile conjugate is around 5–55 nmol/day.

## Control of secretion

There are three major control mechanisms for aldosterone release: (1) ACTH (corticotropin), (2) potassium, (3) renin-angiotensin system.

### ACTH

In the normal subject aldosterone has a circadian rhythm similar to that of cortisol and ACTH. However, only pharmacological doses of ACTH will increase aldosterone secretion and the abnormal levels of ACTH seen in disease states (e.g. Cushing's syndrome, pan-hypopituitarism) do not affect the plasma aldosterone levels. Thus ACTH plays only a minor role in aldosterone release.

# POTASSIUM

Potassium loading (e.g. infusions) stimulates aldosterone secretion, whilst depletion of body potassium is associated with decreased secretion. This regulatory action by potassium appears to occur at the zona glomerulosa cell and acts independently of the renin-angiotensin system. It is unclear whether it is the potassium concentration within the zona glomerulosa cell, or the circulating potassium level, that affects aldosterone output.

### RENIN-ANGIOTENSIN SYSTEM

Renin, an enzyme synthesized in the kidney by the juxtaglomerular apparatus, regulates aldosterone secretion through the action of angiotensin II (Fig. 7.3).

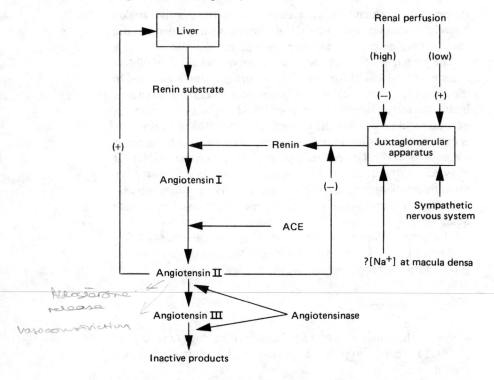

**Fig. 7.3** Control of renin-angiotensin system. ACE = angiotensin converting enzyme.

Renin is synthesized in and released from the juxtaglomerular apparatus, a group of cells comprising the juxtaglomerular cells of the afferent glomerular arteriole and the macula densa cells of the distal tubule. Renin converts renin substrate (angiotensinogen), a glycoprotein synthesized by the liver, to a decapeptide, angiotensin I. This substance is converted, mainly in the lung, by angiotensin converting enzyme (ACE) to an octapeptide, angiotensin II.

Angiotensin II has four main actions: (1) arteriolar vasoconstriction, (2) aldosterone release, (3) inhibition of renin release (short feed back loop), (4) stimulation of liver synthesis of renin substrate.

Angiotensin II is rapidly destroyed by a number of peptidases (angiotensinase) that are present in plasma and tissues. Recent studies have suggested that a heptapeptide (angiotensin III), the immediate product of angiotensinase action, may mediate the aldosterone effect as well as angiotensin II.

Control of renin secretion

Renin release, and thus indirectly aldosterone secretion, is controlled by several factors: (1) renal perfusion pressure, (2) sodium concentration at the macula densa, (3) sympathetic nervous system, (4) angiotensin II.

*Renal perfusion pressure* A fall in renal perfusion (blood) pressure, which decreases the pressure in the afferent arterioles, evokes renin release, and thus shock, haemorrhage, dehydration, etc. will increase renin secretion. Increased secretion also occurs when the subject moves from the supine to the upright position (decreased venous return and cardiac output due to gravitational pooling of blood in the extremities). Renin release is reduced by a raised renal perfusion pressure, e.g. ECF volume expansion. These pressure changes are thought to be detected by the juxtaglomerular cells of the afferent arterioles.

*Sodium concentration at the macula densa* It was originally thought that a decreased sodium concentration in the fluid perfusing the distal renal tubule increases renin secretion, however it has recently been suggested that the macula densa responds to the rate of chloride absorption in the ascending loop of Henle, i.e.

↑ reabsorption (high NaCl load) → ↓ renin; ↓ reabsorption (low NaCl load) → ↑ renin

*Sympathetic nervous system* The sympathetic nervous system activates renin release through β-receptors. This mechanism is thought to be important for the increased renin release associated with postural changes, and with minor decreases in the extracellular fluid volume. Infusions of catecholamines will also increase renin release.

*Angiotensin II and III* These two substances inhibit renin release and thus constitute a short negative feedback.

Chapter 7   Plasma volume and aldosterone

From the above discussion it can be seen that there is an intimate relationship between plasma volume, renin, and aldosterone. For example, if plasma volume falls (dehydration) the following sequence of events occurs:

(1) ↓ plasma volume → ↑ renin → ↑ angiotensin II → (a) vasoconstriction, (b) ↑ aldosterone
(2) ↑ aldosterone → ↑ renal $Na^+$ reabsorption (see below) → ↑ extracellular $Na^+$
(3) ↑ extracellular $[Na^+]$ → ↑ osmolality → (a) thirst, (b) ↑ ADH release
(4) ↑ ADH + thirst → +ve water balance → ↑ plasma volume
(5) ↑ plasma volume → ↓ renin secretion

**Action of aldosterone**

The overall effect of aldosterone is that of sodium conservation and potassium loss by specifically stimulating $Na^+$ uptake and $K^+$ secretion by the epithelial cells of the body, particularly: distal renal tubule cells, mucosal cells of the gut, sweat gland epithelium, salivary gland epithelium.

The most important site of action is the distal convoluted tubule, where $Na^+$ is reabsorbed from the lumen and $K^+$ and $H^+$ secreted. Although there is a loose relationship between $Na^+$ reabsorption and $K^+$ excretion, this does not occur on a 1:1 basis ($Na^+$-$K^+$ exchange). The experimental evidence for this is as follows:
(1) Amount of reabsorbed $Na^+$ is often greater than the combined secretion of $K^+$ and $H^+$.
(2) Actinomycin D, an inhibitor of RNA synthesis, prevents the reabsorption of sodium but does not affect the secretion of potassium.
(3) Micropuncture studies have shown that sodium is mainly reabsorbed in the early part of the distal convoluted tubule, whereas potassium is secreted in the late part of the tubule.

**Mineralocorticoid excess**

The features of mineralocorticoid excess are: (1) sodium retention and extracellular volume excess, (2) potassium depletion and hypokalaemia, (3) metabolic alkalosis.

The plasma sodium level is usually not in the 'hypernatraemia range', due to concomitant water retention, but it is often at the upper limit of normal and always greater than 140 mmol/l. The patient may be hypertensive (volume expansion), but not oedematous because after a period sodium balance is regained. This is termed the 'mineralocorticoid escape phenomenon' and is possibly due to the action of the so-called 'third factor' or natriuretic hormone (see Chapter 2).

Hypokalaemia is common, but may be masked if sodium is withheld from the diet, i.e. decreased distal tubular secretion due to ↓ tubular flow rate. The metabolic alkalosis is due to a combination of mineralocorticoid excess and potassium deficiency (see Chapter 3).

**Mineralocorticoid deficiency**

The biochemical features of deficiency are: (1) sodium depletion, hyponatraemia and volume depletion, (2) hyperkalaemia, (3) mild metabolic acidosis.

These patients do not become completely depleted of sodium because when the extracellular volume falls the renal proximal tubular reabsorption of sodium is stimulated (? 'third factor').

If the deficiency is of long standing a mild normal anion gap metabolic acidosis may occur. This is sometimes referred to as Type IV renal tubular acidosis and may be due to potassium excess (potassium depresses renal $NH_3$ production which is necessary for the efficient excretion of $H^+$ by the kidney).

# LABORATORY INVESTIGATION

The complete investigation of disorders of mineralocorticoid metabolism would ideally involve the measurements of the following analytes: (1) electrolytes, (2) aldosterone, (3) precursors and metabolites of aldosterone, (4) plasma renin activity (PRA).

However, the estimation of electrolytes, PRA and aldosterone is all that is necessary to confirm a clinical suspicion of primary aldosteronism. Although most routine laboratories do not offer PRA and aldosterone assays these are often available in regional centres.

**Electrolytes**

PLASMA

Mineralocorticoid excess or deficiency is often first suspected on the basis of the plasma potassium levels. In the deficiency syndromes (e.g. Addison's) hyperkalaemia is invariably associated with hyponatraemia and a minimal acidosis. In the excess syndromes the hypokalaemia is usually associated with a metabolic alkalosis. The plasma sodium levels are variable, depending on the type of disorder; they may vary from hypernatraemia (e.g. primary hyperaldosteronism) to hyponatraemia (e.g. secondary hyperaldosteronism associated with sodium deficiency). Hypokalaemia does not occur in all cases of primary hyperaldosteronism (20–50% have plasma potassium levels greater than 3.5 mmol/l) but the potassium level in all cases is invariably below 4.0 mmol/l.

It is important to remember that hypertension with associated hypokalaemic alkalosis has many causes other than primary hyperaldosteronism (see p. 144).

URINE

The measurement of urinary electrolytes in the investigation of suspected mineralocorticoid disorders is generally unhelpful, unless the exact dietary intake of sodium and potassium is known. 'Spot' urinary potassium levels may occasionally be helpful in determining the cause of hypokalaemia, i.e. hypokalaemia associated with a high 'spot' urinary potassium ($> \sim 40$ mmol/l) suggests renal potassium loss, which is one of the features of mineralocorticoid excess.

### Aldosterone

Both plasma and urinary aldosterone concentrations can be measured by radioimmunoassay techniques. However, as with plasma renin activity, these assays are usually carried out by either regional laboratories or by those that have a research interest in this area and are able to supply expertise in the interpretation of the results.

Plasma levels of aldosterone are increased in both primary (Conn's syndrome) and secondary hyperaldosteronism but in the latter case the plasma renin activity is high, whilst in the primary disease the activity is suppressed.

Urinary levels usually reflect plasma levels but give a better estimation of the aldosterone secretion rate. Some adenomas may only secrete aldosterone intermittently and, in this event, the 24 hour urinary excretion rate is likely to be of more use than a single plasma level.

### Plasma renin activity (PRA)

The assay of plasma renin is based on the controlled incubation of the plasma with its own renin substrate or added substrate and estimation, by radioimmunoassay, of the generated angiotensin. The method requires some technical expertise and is not yet widely available.

Low levels of renin activity are found in both primary hyperaldosteronism, and in up to 20% of cases of essential hypertension (low renin hypertension, LRH). High levels of PRA are present in hypertension associated with renal disease, and in malignant hypertension. Renin activity is high in all cases of secondary aldosteronism.

## Diagnosis of primary hyperaldosteronism

The investigation/diagnosis of primary hyperaldosteronism (Conn's syndrome), which is usually due to an adrenal adenoma or occasionally a carcinoma, is conveniently approached in three phases: (1) screening, (2) diagnosis of primary aldosteronism, (3) differentiation of adenoma from idiopathic adrenal hyperplasia.

### SCREENING

The prevalence of primary hyperaldosteronism is of the order of 0.5–1.0% of the hypertensive population. The full investigation of these patients is costly and time consuming, so it is necessary to screen for those patients most likely to have the disorder. Although it is recognized that not all cases of Conn's syndrome have hypokalaemia, most authorities agree that a plasma potassium level is the most cost-effective method for selecting patients. The criteria for further investigation should include the following: (1) hypertension, (2) on no medication (especially diuretics), (3) on an unrestricted salt diet (50–100 mmol/day), (4) hypokalaemia (plasma potassium < 3.5 mmol/l), (5) urinary potassium > 40 mmol/l.

*Note* Some specialists investigate all hypertensives who have a plasma potassium less than 4.0 mmol/l, others use the criteria of a plasma potassium less than 3.5 mmol/l associated with a urinary potassium greater than 40 mmol/l.

### DIAGNOSIS

The next step is to measure the PRA to identify those patients with suppressed plasma renin activity. Those with suppressed PRA fall into one of two groups: (1) primary hyperaldosteronism, (2) low renin hypertension. These can be differentiated on the basis of inappropriate secretion of aldosterone in the following way.
 Blood is taken for a plasma aldosterone before and after salt loading (e.g. 2 litres of normal saline IV over 4 hours, or a high salt diet (150–250 mmol/day for a week), or a high salt diet plus a mineralocorticoid). In those cases with primary aldosteronism, the plasma aldosterone level will be initially high and will not suppress below the upper limit of the normal range after salt loading.

Approximately one third of patients with primary hyperaldosteronism have bilateral idiopathic adrenal hyperplasia. It is important to distinguish hyperplasia from the adenomas as the latter can be surgically treated, whereas surgical intervention in bilateral hyperplasia (adrenalectomy) may not resolve the associated hypertension.

(1) Adrenal venous sampling

An adenoma is producing the excess aldosterone, therefore the opposite healthy adrenal will be suppressed and the plasma level of aldosterone draining from it will be low. In hyperplasia both adrenals will secrete high levels of aldosterone.

(2) Aldosterone response to ambulation

Plasma aldosterone levels taken before and after 4 hours of ambulation has been suggested as the ideal method of differentiating adenoma and hyperplasia. In adenoma the plasma aldosterone shows a fall after ambulation, whereas in hyperplasia the aldosterone remains the same or increases.

*Note* Some authorities have recently suggested that idiopathic adrenal hyperplasia (IAH) is not a variant of primary hyperaldosteronism (Conn's syndrome), but is in fact a variant of low renin essential hypertension (see pp. 140, 506). They suggest that IAH represents low renin hypertension with a higher than usual circulating level of aldosterone. This is an attractive hypothesis as it fits in well with the biochemical abnormalities that are found and the results of treatment, i.e. removal of the adrenal glands in IAH does not cure the hypertension.

## MINERALOCORTICOID EXCESS

**Causes**

Aldosterone

*Primary*   (1) Adenoma, (2) idiopathic adrenal hyperplasia.

*Secondary*   (1) Hypertensive (a) malignant hypertension, (b) renovascular hypertension, (c) oestrogen therapy, (d) renin-secreting tumour, (2) oedematous states (a) cirrhosis, (b) nephrotic syndrome, (c) cardiac failure, (3) hypovolaemia (a) shock/blood loss, (b) dehydration, (c) diuretic therapy, (4) Bartter's syndrome.

### Mineralocorticoids other than aldosterone

*Cushing's syndrome* (1) Hyperplasia, (2) adenoma/carcinoma, (3) ectopic ACTH (see p. 493).

*Adrenal enzyme defects* (1) 11β-hydroxylase, (2) 17α-hydroxylase.

*Drugs* (1) Carbenoxolone, (2) licorice.

Renal tubule defect

*Liddle's syndrome*

## Case examples/pathophysiology

PRIMARY ALDOSTERONISM

A 47 year old female with weakness and hypertension (180/110).

| | | | | |
|---|---|---|---|---|
| Plasma | Na | 143 | mmol/l | (132–144) |
| | K | 2.7 | mmol/l | (3.2–4.8) |
| | Cl | 98 | mmol/l | (98–108) |
| | $HCO_3$ | 37 | mmol/l | (23–33) |
| | Urea | 6.6 | mmol/l | (3.0–8.0) |
| Urine | K | 56 | mmol/l | |

*Endocrine studies*
PRA             0.04    ng/ml/h  (0.1–0.4)

| | Basal | Post salt load | |
|---|---|---|---|
| Plasma aldosterone: | 3.6 | 3.3 nmol/l | (0.03–0.55) |

A tumour of 2.5 mm diameter was found in the left adrenal. Postoperatively the blood pressure and plasma aldosterone levels reverted to normal.

Pathophysiology

↑ aldosterone → ↑ distal tubular $Na^+$ reabsorption and $K^+$ secretion → (1) renal $K^+$ loss → hypokalaemia → metabolic alkalosis, (2) $Na^+$ retention → ↑ extracellular volume → (a) hypertension, (b) ↓ renin secretion

Causes of hypertension associated with hypokalaemic alkalosis

(1) Essential hypertension treated with diuretics (common), (2) primary aldosteronism, (3) hypertension plus secondary aldosteronism (a) malignant hypertension, (b) renovascular hypertension, (c) oestrogen therapy, (d) renin-secreting tumour, (4) Cushing's syndrome, (5) adrenal enzyme defects ($C_{17}$ and $C_{11}$ hydroxylase), (6) carbenoxolone/licorice ingestion, (7) Liddle's syndrome.

## SECONDARY ALDOSTERONISM

Secondary aldosteronism describes an increased secretion of aldosterone that is due, unlike primary hyperaldosteronism, to a stimulus arising from outside the adrenal cortex activating the renin-angiotensin system. There may be associated hypokalaemia. Unlike primary hyperaldosteronism the PRA is increased.

## HYPERTENSIVE STATES

### Malignant hypertension

This condition is associated with a rapid increase in hypertension, which leads to papilloedema, hypertensive encephalopathy and nephrosclerosis. The increased renin activity probably results from a decreased local perfusion of the kidney due to intrarenal vascular disease.

### Renovascular hypertension

Stenosis (narrowing) of a major renal artery will result in a decreased blood perfusion of that kidney. This may cause increased renin production, and subsequent aldosterone excess, in an attempt to raise blood volume and therefore perfusion pressure. Increased sodium reabsorption causes extracellular expansion and hypertension. However, not all cases of renovascular hypertension can be explained along these lines.

### Oestrogen therapy

A small proportion of women on oestrogen-containing oral contraceptives develop hypertension that is associated with aldosterone excess. It is thought that in these patients there is increased synthesis of renin substrate by the liver (oestrogen effect), and that this results in increased production of angiotensin and aldosterone.

## Renin-secreting tumours

Several cases of renin-secreting haemangiopericytomas (tumours of the juxtaglomerular cells) have been reported in the last few years. These subjects have had hyperaldosteronism and hypertension.

### OEDEMATOUS STATES

Secondary aldosteronism may occur in the oedematous states that are associated with cardiac failure, nephrosis and cirrhosis. It is probable that the elevated aldosterone is due to increased renin secretion subsequent to an inadequate or ineffective blood volume which occurs in these diseases.

i.e. $\downarrow$ IVV (1) $\rightarrow$ $\uparrow$ renin $\rightarrow$ $\uparrow$ ALDO $\rightarrow$ $Na^+$ retention, (2) $\rightarrow$ $\uparrow$ ADH $\rightarrow$ $H_2O$ retention

### HYPOVOLAEMIA

A decreased blood volume due to any cause (haemorrhage, shock, dehydration etc.) is a potent stimulus to renin secretion.

### BARTTER'S SYNDROME

An 11 year old male presented with growth failure, enuresis and normotension.

| | | | | |
|---|---|---|---|---|
| Plasma | Na | 136 | mmol/l | (132–144) |
| | K | 2.1 | mmol/l | (3.2–4.8) |
| | Cl | 94 | mmol/l | (98–108) |
| | $HCO_3$ | 29 | mmol/l | (23–33) |
| | Urea | 6.0 | mmol/l | (3.0–8.0) |
| Urine | K | 44 | mmol/l | |

*Endocrine studies*
PRA 2.8 ng/ml/h (0.1–0.4)
Plasma aldosterone 1.2 nmol/l (0.03–0.55)

A renal biopsy revealed juxtaglomerular hyperplasia and the patient was considered to have Bartter's syndrome.

The nature of the defect in this syndrome is poorly understood. Several possibilities have been put forward, including increased prostaglandin synthesis and subnormal vasopressor response to angiotensin. The most plausible suggestion is that there is a defective tubular reabsorption of sodium (or chloride) leading to chronic sodium depletion, hypovolaemia and stimulation of renin secretion. Prolonged stimulation of the juxtaglomerular apparatus would eventually lead to hyperplasia of this body. The hypokalaemia is the result of excess aldosterone.

## MINERALOCORTICOIDS OTHER THAN ALDOSTERONE

Cushing's syndrome (see also p. 487)

Mild hypokalaemic alkalosis and hypertension is commonly associated with Cushing's syndrome. The mineralocorticoid effect in this disorder reflects the large amounts of circulating cortisol (cortisol has a weak mineralocorticoid action) and increased secretion of 11-deoxycortisol that arises as a result of inappropriate ACTH stimulation.

Adrenal enzyme defects

*$C_{17}$-hydroxylase*   Deficiency of this enzyme reduces production of cortisol by depressing the conversion of progesterone to 17-hydroxyprogesterone. The raised ACTH activity (diminished negative feedback by cortisol) causes increased production of progesterone, which is then shunted into the mineralocorticoid pathway. There is increased 11-deoxycorticosterone and corticosterone production. Aldosterone production is often suppressed; this may be secondary to hypokalaemia, i.e. mineralocorticoid action of 11-deoxycorticosterone. Clinically the patients have hypogonadism and hypertension.

*$C_{11}$-hydroxylase*   The impaired conversion of 11-deoxycorticosterone to corticosterone (Fig. 7.2) results in the accumulation of 11-deoxycorticosterone. Clinically these patients have virilization and hypertension.

Drugs

*Carbenoxolone/licorice*   Both of these substances contain glycyrrhizinic acid derivatives, which have an aldosterone-like action.

Renal tubule defect

*Liddle's syndrome (pseudohyperaldosteronism)*   This rare familial disease is characterized by hypertension and hypokalaemia, and is remarkable in that there is biochemical evidence of mineralocorticoid excess in the presence of low circulating levels of aldosterone and other mineralocorticoids. The exact nature of the defect is not known. The patients do not respond to spironolactone (a diuretic that blocks the action of aldosterone at the renal tubule) but respond to triamterene, which causes natriuresis and potassium retention in the absence of mineralocorticoids. A primary renal transport defect has been proposed.

## MINERALOCORTICOID DEFICIENCY

### Causes

*Primary adrenal failure* (Addison's)    (1) acute, (2) chronic.

*Adrenal enzyme defects*   (1) 3β-ol-dehydrogenase, (2) $C_{21}$ hydroxylase, (3) corticosterone methyl oxidase.

*Renal tubular dysfunction*   (1) pseudohypoaldosteronism, (2) tubule disease.

*Drugs*   (1) heparin infusions, (2) spironolactone therapy.

*Hyporeninaemic hypoaldosteronism*   (1) diabetes mellitus, (2) tubulointerstitial disease.

#### PRIMARY ADRENAL FAILURE

See Addison's disease (p. 481).

#### ADRENAL ENZYME DEFECTS

3β-ol-dehydrogenase

In this rare defect there is decreased synthesis of both cortisol and mineralocorticoids due to decreased conversion of pregnenolone to progesterone. Should the infants survive they present with severe salt wasting.

### $C_{21}$-hydroxylase

This is the commonest form of congenital adrenal hyperplasia. There is suppressed formation of both cortisol and aldosterone. Deficient mineralocorticoid activity results in salt wasting and hyperkalaemia. The lack of cortisol stimulates ACTH production, which in turn increases adrenal androgen synthesis and results in virilization (see p. 485 for case example).

### Corticosterone methyl oxidase

Deficiency of this enzyme system results in a syndrome of isolated aldosterone deficiency associated with salt wasting and hyperkalaemia.

## RENAL TUBULAR DYSFUNCTION

### Pseudohypoaldosteronism

In this rare disorder biochemical mineralocorticoid deficiency (salt wasting, hyperkalaemia) is associated with high circulating levels of aldosterone. It appears to be due to renal tubular resistance to the action of mineralocorticoids.

### Renal tubule disease

Tubular resistance to the action of aldosterone resulting in hyperkalaemia may occur in such diseases as systemic lupus erythematosis, amyloidosis and interstitial nephritis. A similar situation may occur in the transplanted kidney.

## DRUGS

### Heparin

Prolonged heparin therapy may result in severe hyperkalaemia due to deficiency of aldosterone. Heparin has been shown to inhibit the adrenocortical synthesis of aldosterone.

### Spironolactone

This drug is a competitive inhibitor of aldosterone at the cell receptor level, and therefore has its major effect on the distal renal tubule.

# HYPORENINAEMIC HYPOALDOSTERONISM

This recently described syndrome occurs in middle-aged to elderly patients and is characterized by: (1) hyperkalaemia (25% symptomatic), (2) low circulating levels of renin and aldosterone.

Salt wasting is uncommon because, although the aldosterone levels are reduced enough to cause hyperkalaemia, they are still sufficient to prevent salt wasting. Hyperchloraemic acidosis occurs in 50% of cases. About two thirds of the cases have chronic renal failure although renal function is insufficiently compromised (GFR>20 ml/m) to cause hyperkalaemia. Over half of the patients have insulin dependent diabetes.

The cause of this syndrome is unclear but the possibilities include the following:
(1) Hyporeninism due to renal pathology.
(2) Primary defect in aldosterone biosynthesis. The resulting hyperkalaemia suppressing renin production.
(3) Two primary defects (kidney and adrenal).
(4) Cellular potassium deficiency due to insulinopenia (suppression of aldosterone synthesis) and hyperkalaemic suppression of renin.

# 8 Urea and creatinine

## UREA

### Production

Dietary and endogenous amino acids provide an important energy source, but their catabolism by transamination and oxidative deamination produces large amounts of ammonia. The level of ammonia, and therefore its toxic effect, is normally kept very low by the efficient conversion of the ammonia to urea in the liver (urea cycle). The rate of urea production varies with the amount of dietary protein ingested; a daily protein intake of 100 g yields about 25-35 g of urea. In severe liver disease the production rate may be sufficiently depressed as to result in a low plasma level of urea.

### Excretion

*Kidney* Renal excretion depends on (1) renal blood flow, (2) glomerular filtration rate (GFR), (3) rate of urine flow. 40-60% of the urea filtered by the glomerulus is reabsorbed in the collecting ducts. The rate of reabsorption varies inversely with the urine flow rate.

*Gut secretion and sweat* is not a significant route.

## CREATININE

### Production

Phosphocreatine, a high energy compound present in muscle, is utilized in maintaining the level of ATP in working muscle. A proportion of the phosphocreatine spontaneously loses its phosphate acid and cyclizes to form creatinine. Since the amount of phosphocreatine is dependent on the mass of working muscle, the quantity of creatinine produced each day is fairly constant for a given individual. Small amounts of creatine occur in the diet, mainly in foods such as roast meats.

**Excretion**

*Kidney* The rate of creatinine excretion varies with renal blood flow and GFR, but unlike urea there is no tubular reabsorption. A small amount of creatinine is secreted by the tubules.

*Other body secretions* are unimportant as excretory routes.

## LABORATORY INVESTIGATION

Investigations that are helpful in determining the cause of high plasma urea and creatinine concentrations are:

*Plasma* (1) electrolytes, (2) urea, creatinine, (3) osmolality.

*Urine* (1) electrolytes, (2) urea, creatinine, (3) osmolality.

*Derived indices* (1) $\frac{urine}{plasma}$ urea, creatinine, osmolality, (2) $FE_{Na}$ (fractional excretion of sodium), (3) clearance—urea, creatinine.

### PLASMA SODIUM

See Chapter 2.

### PLASMA POTASSIUM

Acute renal failure is invariably associated with hyperkalaemia, but in chronic renal failure (CRF) hyperkalaemia usually does not occur until the GFR falls to below 20 ml/min ($\sim 0.3$ ml/s). Hyperkalaemia in the presence of mild renal insufficiency may be due to: (1) distal tubular disorders, (2) mineralocorticoid deficiency, (3) other extrarenal factors.

### PLASMA BICARBONATE

In CRF an acidosis (low $[HCO_3^-]$), like hyperkalaemia, usually does not occur until there is a very significant fall in GFR. In early chronic renal insufficiency the acidosis may be hyperchloraemic (normal anion gap) in nature (early uraemic acidosis). This suggests that tubular dysfunction is relatively greater than glomerular dysfunction. In severe renal insufficiency with poor glomerular function, the acidosis is of the high anion gap type (see pp. 90, 160).

PLASMA UREA

Urea levels vary widely with diet, rate of protein metabolism, liver production and GFR. A high protein intake (diet, haemorrhage into gut) in the absence of renal insufficiency may produce slightly elevated plasma urea levels (up to 10 mmol/l). Concentrations >10 mmol/l invariably mean a decreased GFR due either to intrinsic renal disease or to a decreased renal blood flow (shock, dehydration, etc.). As a rule of thumb it can be said that plasma urea concentrations greater than 20 mmol/l are due to intrinsic renal disease whilst levels up to 20 mmol/l result from either intrinsic renal disease or decreased renal blood flow in the absence of renal disease (prerenal uraemia). In renal insufficiency the plasma urea level does not begin to rise until the GFR falls by about 50% (Fig. 8.1).

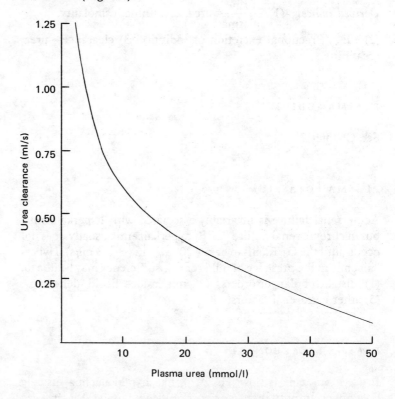

Fig. 8.1 Plasma urea levels in renal insufficiency.

## PLASMA CREATININE

It is generally accepted that the plasma creatinine concentration is usually a better indicator of GFR than urea because the former is less affected by diet, but this is true only when the creatinine level is significantly raised. Mild changes in the plasma creatinine level may be misleading because:
(1) Slight increases may occur after a meal of roast meats.
(2) 5–20% of the measured chromogen formed during the laboratory analysis of plasma creatinine may be due to substances other than creatinine. Such considerations are important when interpreting results from patients with mild renal impairment and a small muscle mass.

## PLASMA OSMOLALITY

By itself this measurement is of little diagnostic help as a renal function test, but can be of more value when used as a ratio with the urine osmolality (see below).

## URINE SODIUM

Sodium concentrations on spot urines may be helpful in the differential diagnosis of oliguria.

Prerenal uraemia (PRU)     urine $[Na^+] = <10$ mmol/l
Acute tubular necrosis (ATN)    urine $[Na^+] = >20$ mmol/l

Better diagnostic discrimination may be obtained if the $FE_{Na^+}$ is applied to this situation (see p. 33).

PRU     $FE_{Na^+} = <1\%$
ATN     $FE_{Na^+} = >1\%$

*Note* The $FE_{Na^+}$ is also increased (>1%) in: (1) obstructive nephropathy, (2) diuretic therapy, (3) osmotic diuresis, (4) chronic renal failure, (5) vomiting (acute phase).

## URINE OSMOLALITY

The most useful measurement is the urine:plasma ratio. This can be helpful in differentiating the causes of oliguria.

PRU    urine:plasma osmolality $= >1.3$ (2–3)
ATN    urine:plasma osmolality $= <1.3$

In chronic renal failure the ratio approximates 1:1 because of the inability of the diseased kidney to either concentrate or dilute urine.

### URINARY UREA AND CREATININE

In isolation these estimations are of little diagnostic use but indices derived from them may be helpful in certain circumstances.

Urine/plasma ratio

This ratio (of urea or creatinine) provides a rough index of the ability of the kidney to concentrate urine. Either the urea or creatinine ratio can assist in differentiating PRU from ATN.

PRU ratio $= >14$ (usually $>20$)
ATN ratio $= <14$

As with the estimation of urinary sodium concentration, and the urine:plasma osmolality ratio, the urea or creatinine ratios do not always indicate the correct pathophysiology and should be used only in conjunction with other clinical and diagnostic evidence.

Renal clearance

Clearance tests measure the amount of plasma (or blood) that is theoretically completely cleared of the considered substance in a given time (minutes or seconds).
In the case of creatinine it is calculated as follows:

$$\text{creatinine clearance (ml/s)} = \frac{\text{urinary [creatinine]}}{\text{plasma [creatinine]}} \times \frac{\text{urine volume (ml)}}{\text{collection time (s)}}$$

Either urea clearance or creatinine clearance can be used to estimate the GFR, but of the two the creatinine clearance is the more accurate indicator. Both have advantages and disadvantages.

Creatinine clearance

*Advantages* (1) Minimal fluctuations of plasma concentration over 24 h, and therefore can utilize a long urinary collection period (increased accuracy of collection), (2) results correlate well with those determined by inulin clearance.

*Disadvantages* Creatinine (urine and plasma) is technically difficult to measure accurately, particularly at low levels. Most methods overestimate the true value, due to the presence of non-creatinine chromogens.

Urea clearance

*Advantages* Urine and plasma levels are technically easy to measure in the laboratory.

*Disadvantages* (1) Wide plasma fluctuations over 24 hours and therefore have to use a short collection period (poor accuracy), (2) 40–60% of filtered urea is reabsorbed in the collecting ducts, and therefore it gives a poor estimation of GFR.

# RENAL FAILURE

The major functions of the kidney are:
(1) Regulation of water, electrolyte and osmolal balance.
(2) Regulation of acid-base status.
(3) Removal of waste products of metabolism and certain toxic substances.
(4) Conversion of vitamin D to its active metabolite.
(5) Production of erythropoietin.

Renal failure is said to occur when the kidneys are unable to maintain a normal internal environment. The characteristic early feature of renal failure is a high or rising level of plasma urea. This has been loosely termed uraemia (uraemia = urine in the blood), and is due to a decreased GFR. A decreased GFR resulting in an increased plasma urea may be due to:
(1) Decreased renal blood flow (dehydration, shock, etc.)—prerenal uraemia (PRU), prerenal failure, functional renal impairment.
(2) Decreased number of functioning nephrons—intrinsic renal disease.
(3) Obstruction to urine outflow—obstructive nephropathy.

## Chronic renal failure

In this condition renal failure develops over a long period of time (months, years), and is associated with both glomerular and tubular dysfunction.

## Acute renal failure

Failure develops over a short period of time (hours, days) and is usually associated with oliguria (urine output < 400 ml/day). The aetiology may be prerenal, renal or postrenal.

PRERENAL

The increased plasma urea is due to a decreased GFR as a consequence of a decreased renal blood flow (blood loss, shock, congestive cardiac failure, dehydration) and can be reversed when the predisposing causes have been corrected. If it is not treated promptly this condition may progress to acute tubular necrosis.

RENAL

This condition is irreversible in the short term but may be reversed in the long term (7–10 days) if the patient can be kept alive. In these patients the finding of areas of tubular necrosis in the kidneys has led to the term acute tubular necrosis (ATN). However, not all patients demonstrate this pathology and it is has been suggested that the term vasomotor nephropathy be used. Other terms used for this condition are acute renal failure, crush syndrome and haemoglobinuric nephropathy.

POSTRENAL

Acute bilateral ureteric obstruction or acute obstruction of the bladder outlet will also cause an oliguria associated with a rising plasma urea.

## INCREASED PLASMA UREA AND CREATININE

Increased plasma concentrations of both these analytes may be due to:

Decreased renal excretion

*Prerenal uraemia* (reduced RBF) (1) Dehydration, (2) shock/blood loss, (3) congestive cardiac failure.

*Renal uraemia* (intrinsic renal disease) (1) Renal failure, (2) acute/chronic.

*Postrenal uraemia* (urinary tract obstruction) (1) Bilateral ureteric obstruction, (2) bladder neck obstruction, (3) urethral obstruction.

Increased production of urea

*Increased protein catabolism* (1) Heavy protein meal, (2) haemorrhage into gut.

Increased production of creatinine

*Increased dietary intake* Roast meats.

*Large muscle mass* Acromegaly, gigantism.

**Laboratory diagnosis of a high plasma urea concentration**

See Fig. 8.2.

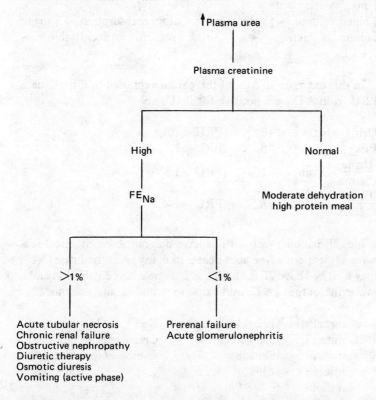

Fig. 8.2 Scheme for the laboratory diagnosis of a high plasma urea concentration.

Chapter 8    **Case examples/pathophysiology**

PRERENAL URAEMIA (functional renal insufficiency)

Male, 50 years, with septicaemia, dehydration and oliguria.

Plasma
| | | | | |
|---|---|---|---|---|
| Na | 145 | mmol/l | (132–144) | |
| K | 3.5 | mmol/l | (3.2–4.8) | |
| Cl | 104 | mmol/l | (98–108) | |
| $HCO_3$ | 22 | mmol/l | (23–33) | |
| Urea | 26.0 | mmol/l | (3.0–8.0) | |
| Creat. | 0.20 | mmol/l | (0.06–0.12) | |
| Osmo. | 305 | mmol/kg | (275–295) | |

| Urine | | |
|---|---|---|
| Na | <10 | mmol/l |
| Creat. | 7.7 | mmol/l |
| Osmo. | 529 | mmol/kg |
| $FE_{Na}$ | <0.18% | |
| $\frac{Urine}{Plasma}$ Osmo. | 1.7 | |
| $\frac{Urine}{Plasma}$ Creat. | 38.5 | |

If renal function is intact dehydration is associated with maximal conservation of sodium and water by the kidney, i.e.

↓ blood volume
+                    } → ↑ ADH → ↑ renal water reabsorption
↑ plasma osmolality       ↑ urine osmolality (urine > plasma)

↓ blood volume → ↑ renal tubular sodium reabsorption → ↓ urine sodium excretion → ↓ urine sodium concentration → ↓↓ $FE_{Na}$

In this example all four of the parameters used to differentiate PRU from ATN are positive for PRU, i.e.

| | | | |
|---|---|---|---|
| Urine Na | | <10 | (PRU<10) |
| $FE_{Na}$ | | <0.18 | (PRU<1) |
| $\frac{Urine}{Plasma}$ | Osmo. | 1.7 | (PRU>1.3) |
| $\frac{Urine}{Plasma}$ | Creat. | 38.5 | (PRU>14) |

Not all patients with a PRU have diagnostic biochemical features as clear cut as the above case. In many cases the urine $[Na^+]$ may well be above 20 mmol/l, e.g. patient with dehydration and a low urine output (<12 ml/h) due to vomiting and diarrhoea.

Plasma
| | | | | |
|---|---|---|---|---|
| Na | 138 | mmol/l | (132–144) | |
| K | 3.9 | mmol/l | (3.2–4.8) | |
| Cl | 98 | mmol/l | (98–108) | |
| $HCO_3$ | 23 | mmol/l | (23–33) | |
| Urea | 16 | mmol/l | (3.0–8.0) | |
| Creat. | 0.21 | mmol/l | (0.06–0.12) | |
| Osmo. | 297 | mmol/kg | (275–295) | |

| Urine | | |
|---|---|---|
| Na | 51 | mmol/l |
| Creat. | 10.7 | mmol/l |
| Osmo. | 560 | mmol/kg |
| $FE_{NA}$ | 0.72% | |
| $\frac{Urine}{Plasma}$ Osmo. | 1.88 | |
| $\frac{Urine}{Plasma}$ Creat. | 51 | |

In this case the urine [Na$^+$] suggests ATN, but the other parameters indicate PRU. The high urine [Na$^+$] is a reflection of the concentrating ability of the kidney and not tubular dysfunction (i.e. FE$_{Na}$ = <1%).

*Urea and creatinine*

CHRONIC RENAL FAILURE

Patient with analgesic nephropathy.

Plasma
| | | | | | | | |
|---|---|---|---|---|---|---|---|
| Na | 133 | mmol/l | (132–144) | Urine Na | 26 | mmol/l | |
| K | 5.1 | mmol/l | (3.2–4.8) | Creat. | 6.6 | mmol/l | |
| Cl | 101 | mmol/l | (98–108) | Osmo. | 330 | mmol/kg | |
| HCO$_3$ | 16 | mmol/l | (23–33) | FE$_{Na}$ | | 3.3% | |
| Urea | 33.8 | mmol/l | (3.0–8.0) | Creatinine | | | |
| Creat. | 1.13 | mmol/l | (0.06–0.12) | clearance | | 0.1 | ml/s (1.5–2.0) |
| Ca | 1.69 | mmol/l | (2.15–2.55) | | | | |
| PO$_4$ | 2.81 | mmol/l | (0.6–1.25) | | | | |
| Osmo. | 315 | mmol/kg | (275–295) | | | | |
| Anion gap | 21 | mEq/l | (7–17) | | | | |

Chronic renal failure describes dysfunction that develops over a long period of time (months to years). The biochemical abnormalities found in this disorder can be explained in terms of the destruction of renal parenchyma and the resultant decrease in functional nephron mass.

Sodium

Providing there are no complications (vomiting, diarrhoea, etc.) the patient with chronic renal failure stays in sodium balance (i.e. input equals output). However, as there is a decrease in the number of working nephrons each has to excrete more sodium than it would under normal circumstances in order to maintain sodium balance. Therefore, the fractional excretion of sodium (FE$_{Na+}$) will be abnormally high. Tubular dysfunction and osmotic diuresis due to a high plasma urea will promote sodium excretion.

Potassium

In mild and moderate renal failure potassium balance is maintained by increased excretion per nephron (increased fractional excretion), and elimination via the large bowel. As the failure progresses hyperkalaemia will occur due to: (1) decreased number of nephrons, (2) acidosis.

Hyperkalaemia usually does not become evident until the GFR falls below 0.3 ml/s (plasma creatinine level of approximately 0.3–0.4 mmol/l).

### Bicarbonate

The metabolic acidosis ($\downarrow [HCO_3^-]$) of renal failure is due to an inability of the renal tubules to secrete $H^+$ because of the decrease in the number of nephrons. Some authorities suggest that the basic biochemical abnormality lies in the deficient production of $NH_3$ (decreased number of nephrons) which is necessary to trap the secreted $H^+$ in the tubular lumen. A modest amount of tubular bicarbonate wasting also contributes to the acidosis.

Retention of unmeasured acid anions ($PO_4^{2-}$, $SO_4^{2-}$, etc.) due to the decreased GFR causes an increase in the plasma anion gap. In some cases of early renal failure the tubular disorder ($H^+$ elimination) is proportionally greater than the decrease in GFR (anion elimination), resulting in $H^+$ retention associated with adequate anion excretion. This results in a normal anion gap (hyperchloraemic) acidosis, commonly called 'early uraemic acidosis'.

A feature of chronic renal failure is that, although there is a positive $H^+$ balance of some 10-20 mmol/day, the acidosis is relatively mild (plasma $[HCO_3^-]$ of 12-18 mmol/l), and nonprogressive in nature. This is probably due to absorption of the excess $H^+$ by bone buffers.

### Urea and creatinine

Increased levels reflect the decreased GFR. Plasma levels of these analytes tend not to rise until the GFR has fallen by 50-60% of the normal level (creatinine clearance of approximately 0.5-0.7 ml/s).

### Calcium and phosphate

Plasma phosphate levels begin to rise when the GFR has fallen to approximately 0.25-0.3 ml/s. Prior to this there is increased secretion of phosphate by the remaining nephrons, due to increased PTH secretion as a consequence of hypocalcaemia.

The hypocalcaemia of renal failure is due to a number of causes, including the following mechanisms:
(1) Destruction of renal parenchyma resulting in decreased conversion of 25-hydroxycholecalciferol to 1,25-dihydroxycholecalciferol.
(2) Phosphate retention, causing sequestration of calcium as calcium phosphate.
(3) Retained toxic substances which may prevent the normal action of PTH on bone.

### Osmolality

The high plasma osmolality in the above case reflects the high plasma urea level. This is not a hypertonic state because in the body the urea is evenly distributed in the water on both sides of the cell membranes. Thus fluid shifts do not occur between the intracellular and extracellular compartments and ADH secretion is not increased. The urine osmolality tends to be similar to that of the plasma ($\pm 50$ mmol/kg) because of the inability of the kidney to either concentrate or dilute the urine. This is due to tubular dysfunction and osmotic (urea) diuresis.

ACUTE RENAL FAILURE (acute tubular necrosis)

Patient with severe crush injuries and oliguria.

Plasma
| | | | | | | | |
|---|---|---|---|---|---|---|---|
| Na | 135 | mmol/l | (132–144) | Urine | Na | 55 | mmol/l |
| K | 6.3 | mmol/l | (3.1–4.8) | | Creat. | 5.3 | mmol/l |
| Cl | 96 | mmol/l | (98–108) | | Osmo. | 360 | mmol/kg |
| $HCO_3$ | 15 | mmol/l | (23–33) | Urine ⎱ | Creat. | 6.7 | |
| Urea | 55.0 | mmol/l | (3.0–8.0) | Plasma ⎰ | Osmo. | 1.07 | |
| Creat. | 0.79 | mmol/l | (0.06–0.12) | | | | |
| Osmo. | 335 | mmol/kg | (275–295) | $FE_{Na}$ | | 6% | |

Acute renal failure occurs over a short period (hours or days) and is usually associated with oliguria (urine output <400 ml/day). Although the causes of acute renal failure (acute tubular necrosis) are well defined, the nature of the pathogenesis is unclear and subject to controversy. There are three main theories of the causes of the oliguria. (1) tubular obstruction due to cellular debris, (2) tubular back flow (glomerular filtrate reabsorption) through damaged tubular epithelium, (3) alterations in renal haemodynamics, resulting in ischaemia and decreased GFR.

Whatever the pathogenesis there are two main features of ATN, those of decreased GFR and tubular dysfunction, which can explain the biochemical abnormalities.

Urea and creatinine

Increased plasma levels reflect decreased GFR.

Potassium

Hyperkalaemia is due to (1) acidosis, (2) decreased tubular secretion (a) tubule damage, (b) decreased tubular urine flow, (3) potassium release from damaged cells in crush injury.

Bicarbonate

The low plasma bicarbonate level is a consequence of $H^+$ retention due to (1) tubular damage, (2) decreased tubular flow (inadequate $Na^+$ available for exchange with $H^+$).

### Osmolality

The increased plasma osmolality reflects urea retention. Tubular damage results in an inability of the kidney to concentrate and dilute urine and, therefore, any urine that is excreted remains iso-osmotic with the plasma.

### Sodium

Tubular damage results in defective sodium reabsorption, and thus there is an increased sodium excretion per nephron. This will result in an increased urinary concentration ($>20$ mmol/l) and a high fractional excretion of sodium ($>1\%$).

After the oliguric phase (7–10 days), if the patient survives, there is a period of increasing diuresis (diuretic phase) as more nephrons become functional and the GFR returns towards normal. This diuresis is due to (1) high blood urea → osmotic diuresis, (2) tubular dysfunction, (3) release of accumulated water and electrolytes.

During this phase sodium and potassium is lost in the urine and depletion of these two ions may occur, resulting in hyponatraemia and hypokalaemia.

#### POSTRENAL URAEMIA

Prostatic hypertrophy and chronic urinary retention in a male aged 60 years.

| Plasma | | | | | | | |
|---|---|---|---|---|---|---|---|
| Na | 140 | mmol/l | (132–144) | Urine Na | | 74 | mmol/l |
| K | 5.0 | mmol/l | (3.1–4.8) | Creat. | | 3.6 | mmol/l |
| Cl | 107 | mmol/l | (98–108) | Osmo. | | 305 | mmol/kg |
| $HCO_3$ | 21 | mmol/l | (23–33) | $\dfrac{\text{Urine}}{\text{Plasma}}$ Osmo. | | 0.94 | |
| Urea | 37.2 | mmol/l | (3.0–8.0) | | | | |
| Creat. | 0.64 | mmol/l | (0.06–0.12) | $FE_{Na}$ | | 9.4% | |
| Osmo. | 323 | mmol/kg | (275–295) | | | | |

During urinary obstruction the back pressure of urine interferes with glomerular filtration and tubular function.

(1) ↓ GFR → ↑ plasma [urea] and [creatinine]
(2) tubular dysfunction → (a) retention of $H^+$ (acidosis), (b) retention of $K^+$ (hyperkalaemia), (c) interference with sodium reabsorption (increased urinary [$Na^+$] and $FE_{Na}$), (d) inability to concentrate urine

With relief of obstruction the ability to reabsorb sodium is often slow to return to normal and renal sodium loss associated with hyponatraemia is a common finding.

## INCREASED UREA PRODUCTION

Haemorrhage into the gut in a 46 year old man. Admission sample.

Plasma
| | | | | | | | |
|---|---|---|---|---|---|---|---|
| Na | 138 | mmol/l | (132–144) | Urine Na | 28 | mmol/l | |
| K | 3.8 | mmol/l | (3.1–4.8) | Creat. | 6 | mmol/l | |
| Cl | 100 | mmol/l | (98–108) | Osmo. | 470 | mmol/kg | |
| $HCO_3$ | 25 | mmol/l | (23–33) | Urine/Plasma Osmo. | 1.6 | | |
| Urea | 9.5 | mmol/l | (3.0–8.0) | | | | |
| Creat. | 0.10 | mmol/l | (0.06–0.12) | $FE_{Na}$ | 0.3% | | |
| Osmo. | 292 | mmol/kg | (275–295) | | | | |

In the presence of good renal function, increased plasma urea levels after a large protein meal are the exception rather than the rule because the increased amounts of urea are rapidly cleared by the kidneys. Thus a high plasma urea level that is associated with a gut haemorrhage (protein load in gut) usually indicates an associated decreased GFR. This will occur in two circumstances.
(1) Associated renal insufficiency due to intrinsic renal disease.
(2) Haemorrhage severe enough to decrease the renal blood flow and GFR in an otherwise normal kidney.

## DECREASED PLASMA UREA AND CREATININE

### Urea

CAUSES

*Decreased synthesis*   Liver disease.

*Decreased precursors*

*Decreased protein intake*   Diet, vomiting, IV feeding.

*Increased protein synthesis*   Infancy, pregnancy.

*Haemodilution*   Overhydration, pregnancy.

*Increased excretion* (increased GFR)   Nephrotic syndrome, pregnancy.

A decreased plasma urea is clinically significant in the following circumstances:

*Liver disease*   Persistent low levels indicate a severe degree of dysfunction.

*IV feeding*   Low levels may indicate an inadequate supply of aminoacid nitrogen.

*Overhydration*   In the absence of liver disease and a low protein intake a low plasma urea concentration in adults may indicate a mild to severe degree of overhydration. This is especially so if there are other indications of haemodilution (low haemoglobin, low plasma protein levels).

## Creatinine

Low plasma creatinine concentrations are not clinically important. However, it must be remembered that the plasma level is a function of muscle mass and that people with a small muscle mass tend to have low normal levels. This is particularly so in elderly females, infants and children. These patients may have a significant increase in their plasma level (renal failure, etc.) but still have a value within the adult reference range.

# 9 Calcium, phosphate and magnesium

## CALCIUM METABOLISM

**Body distribution**

| | |
|---|---|
| Total calcium | 25–35 mol |
| Skeleton | >99% |
| Soft tissues | 0.5% |
| Extracellular fluid | 0.1% |

The total calcium content of the body depends on the balance between intestinal absorption of calcium from the diet and losses via the faeces and urine.

**Gut uptake and loss**

Twenty to 40% of the average daily dietary intake of calcium (∼ 25 mmol) is absorbed by the gut. The actual amount taken up depends on:
(1) Amount of ionized calcium ($Ca^{2+}$) available. $Ca^{2+}$ levels in the intestinal lumen are reduced by dietary substances that form calcium complexes, e.g. phytates, oxalates. Acid pH increases the ionization of calcium, whilst alkalinity promotes complex formation and diminished absorption.
(2) Presence of the active metabolite of Vitamin D (1,25-$(OH)_2D_3$).
   Calcium is secreted into the gut as a normal constituent of bile and intestinal fluids. Faecal output of calcium could exceed intestinal absorption in situations where the diet contains high levels of phytates or other sequestrating substances. Under normal circumstances the faeces are not an important excretion route for calcium.

**Kidney**

The kidney filters about 250 mmol of $Ca^{2+}$ each day, some 98% of which is reabsorbed by the tubules. The major portion of this filtered $Ca^{2+}$ is taken up by the proximal tubules, whilst a fine adjustment to the amount reabsorbed occurs in the distal tubules under the influence of PTH (increased PTH → increased $Ca^{2+}$ uptake).

TUBULAR HANDLING OF CALCIUM

| Tubule section | Reabsorbed % |
| --- | --- |
| Proximal convoluted | ~60 |
| Thick portion of loop of Henle | ~20 |
| Distal convoluted | ~10 (↑ absorption by PTH and thiazides) |
| Collecting duct | ~5 |

**Plasma calcium**

Calcium is present in three forms in plasma, each of which is in equilibrium with the others.

| | % of total plasma calcium |
| --- | --- |
| Ionized ($Ca^{2+}$) | 40–50 |
| Bound to proteins | 40–50 |
| Complexed (citrates) | 5–10 |

The ionized species ($Ca^{2+}$) is the physiologically active fraction that directly affects such properties as neuromuscular excitability and the release of parathyroid hormone (PTH) from the parathyroid glands. The inactive protein-bound calcium is attached mainly to albumin, with a small amount being carried by globulins and lipoproteins. The distribution of calcium between the various forms is affected by plasma pH;

pH ↑ : ↓ $Ca^{2+}$
pH ↓ : ↑ $Ca^{2+}$

In addition to the effects of gut uptake and renal excretion already discussed, the concentration of total calcium in plasma is also dependent on the following

PLASMA ALBUMIN

Alteration to the level of circulating albumin will change the plasma concentration of both the bound calcium and the total calcium. However, the concentration of $Ca^{2+}$ will be kept to within normal limits by various factors, including PTH, renal clearance and re-equilibration with protein and complexes. A fall (or rise) in plasma albumin of 10 g/l is usually associated with a fall (or rise) in *total* calcium of 0.20–0.25 mmol/l.

BONE FLUX

Bone undergoes constant remodelling, some 5 mmol of calcium being deposited and resorbed each day. This balance can, for example, be disturbed by malignant disease, when excessive calcium may be released from bone into the plasma, due either to the erosive effects of secondary deposits, or the PTH-like activity of substances produced by the tumour.

CALCIUM PRECIPITATION

Transient hypocalcaemia can arise if $Ca^{2+}$ is rapidly removed from plasma. The hyperphosphataemia of renal failure, for example, can cause the solubility constant of calcium phosphate to be exceeded, and the complex to be deposited in soft tissues. Similarly, the elevated levels of fatty acids resulting from the action of lipases released during acute pancreatitis will also precipitate calcium as complexes.

# HORMONES AND NORMAL CALCIUM METABOLISM (Fig. 9.3)

## Parathyroid hormone (PTH)

PTH is a polypeptide of 84 amino acid residues (mol. wt. 9300 Daltons), that is formed within the parathyroid cells as pre-pro-PTH (115 amino acids). Pre-pro-PTH is then converted to pro-PTH (90 amino acids), from which six amino acids are removed prior to secretion as PTH. In the circulation there is further cleavage, mainly in the liver, between residues 33–34 or 36–37 to form two fragments (C-terminal and N-terminal). Only the N-terminal fragment is biologically active.

Degradation occurs mainly in the kidney. The C-terminal fragment is filtered by the glomerulus, reabsorbed, and degraded by the renal tubules. The N-terminal fragment is probably degraded after attaching to the renal receptors and exerting its biological effect. The C-terminal fragment has a longer plasma half-life than the N-terminal fragment.

The plasma calcium concentration is the principal regulator of PTH secretion, by a simple negative feedback mechanism. Magnesium has a similar effect, but is thought to be unimportant unless plasma magnesium levels are markedly abnormal. PTH secretion is also subject to negative feedback control by two vitamin D metabolites, 1,25-$(OH)_2D_3$ and 24,25-$(OH)_2D_3$.

### Bone

Stimulates osteoclast activity, the increased bone resorption causing an increase in plasma $Ca^{2+}$ and $PO_4$. The active metabolite of vitamin $D_3$ (1,25-$(OH)_2D_3$), plays a permissive role for this effect.

### Kidney

Increases the distal nephron reabsorption of $Ca^{2+}$ and decreases that of $PO_4$, in the proximal tubule. (For simplicity $PO_4$ is used throughout as the symbol for phosphate, regardless of its actual composition and ionic charge in any given physiological situation.)

### Vitamin D

Increases the renal formation of the active form of vitamin D (see below), so that the indirect effect of PTH is to increase the intestinal and renal uptake of $Ca^{2+}$ and $PO_4$.

The overall effect of PTH is to increase plasma $[Ca^{2+}]$ and to lower plasma $[PO_4]$.

## Vitamin D ($D_3$, cholecalciferol)

Vitamin D is a steroid that requires certain metabolic modifications before it can express its biological activity.

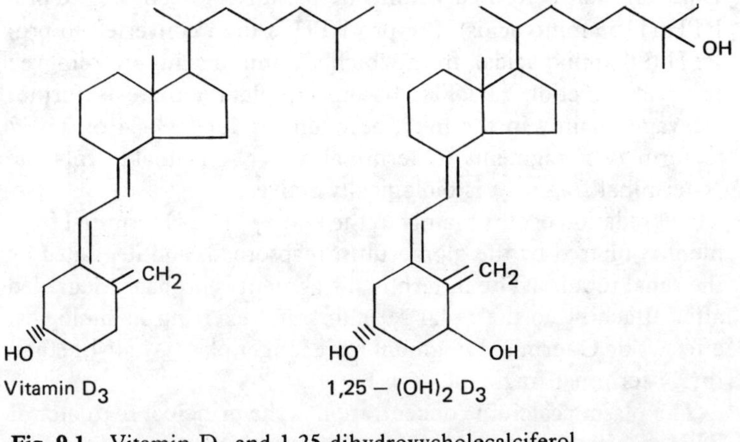

**Fig. 9.1** Vitamin $D_3$ and 1,25-dihydroxycholecalciferol.

*Calcium, phosphate and magnesium*

The majority of the body's vitamin D requirement is met by the conversion of 7-dehydrocholesterol, a sterol component of the epidermal layer of the skin, to vitamin $D_3$ (cholecalciferol) by ultraviolet light. With a prolonged absence of exposure to sunlight the diet provides the sole source of vitamin D. Vitamin $D_3$ is taken to the liver where it undergoes enzymatic hydroxylation to give 25-hydroxycholecalciferol (25-OHD$_3$). This intermediate is further hydroxylated by an enzyme system present in the kidney cortex to form 1,25-dihydroxycholecalciferol (1,25-$(OH)_2D_3$), the physiologically active metabolite of vitamin D (Fig. 9.1). The renal formation of 1,25-$(OH)_2D_3$ is promoted by PTH. It is not known if $Ca^{2+}$ has a direct negative feedback effect on this conversion step.

$\downarrow$ plasma $[Ca^{2+}] \rightarrow \uparrow$ PTH

25-OHD$_3$ $\xrightarrow{1\alpha\text{-hydroxylase}}$ 1,25-$(OH)_2D_3 \rightarrow \uparrow$ plasma $[Ca^{2+}]$

There is some evidence that during periods of increased calcium requirement (growth, pregnancy, lactation) the relevant hormones (i.e. GH, HPL, PRL) also stimulate the formation of 1,25-$(OH)_2D_3$.

If there is adequate 1,25-$(OH)_2D_3$ then it promotes the conversion of 25-OHD$_3$, via a separate renal enzyme system, to a relatively inactive metabolite, 24,25-$(OH)_2D_3$. Both 1,25- and 24,25-$(OH)_2D_3$ inhibit PTH secretion (Fig. 9.2).

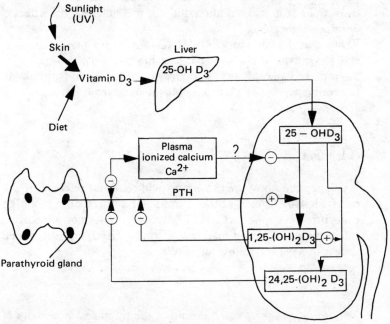

**Fig. 9.2** The formation of 1,25-dihydroxycholecalciferol.

The conversion of vitamin D to an active metabolite that acts on tissues other than its site of origin defines $1,25\text{-}(OH)_2D_3$ as a steroid hormone and vitamin D as a prohormone.

ACTIONS OF $1,25\text{-}(OH)_2D_3$

### Small intestine

$1,25\text{-}(OH)_2D_3$ increases the absorption of dietary calcium and phosphorus, by stimulating the synthesis of calcium-binding protein (CaBP). CaBP is located within the intestinal cells and is thought to facilitate the uptake of calcium across the brush border from the intestinal lumen. The mechanism by which $1,25\text{-}(OH)_2D_3$ promotes phosphate uptake is not known, other than it differs from that of calcium.

### Bone

Vitamin D, as the $1,25\text{-}(OH)_2D_3$ metabolite, mobilizes calcium and phosphorus from bone, a process that requires the presence of PTH. Vitamin D is also required for normal bone mineralization to occur.

### Kidney

There is evidence that $1,25\text{-}(OH)_2D_3$ causes enhancement of calcium and phosphate retention, but it is probable that this effect is of minor importance.

The overall effects of $1,25\text{-}(OH)_2D_3$ are to raise plasma calcium and phosphate levels by: (1) increasing intestinal absorption of calcium and phosphate, (2) increasing calcium and phosphate resorption from bone, (3) promoting the renal retention of calcium and phosphate.

## Calcitonin

A polypeptide hormone (32 amino acid residues) released by the parafollicular (C) cells of the thyroid gland in response to a high concentration of plasma $Ca^{2+}$. The role of this hormone in normal calcium metabolism in man is still controversial, but it has been shown to have the following effects:

### Bone

Opposes bone resorption by limiting the number and activity of osteoclasts, and by lowering the calcium efflux from bone to the ECF. Calcitonin also has a hypophosphataemic effect.

Kidney

Although pharmacological doses increase renal clearance of calcium and phosphate, it is not clear whether this is a physiological role of calcitonin.

Plasma calcitonin

Hypercalcaemia stimulates calcitonin release, whilst hypocalcaemia has an inhibitory effect. Plasma calcitonin levels are higher in males than females and are also raised during periods of increased calcium demand such as growth, pregnancy and lactation.

Role of calcitonin

Although calcitonin strongly inhibits osteoclastic bone resorption it is not clear whether skeletal preservation is the major physiological function of calcitonin. Removal of calcitonin (e.g. thyroidectomy) does not apparently disturb calcium homeostasis, nor does high plasma levels of calcitonin (e.g. medullary thyroid carcinoma), and therefore the role of calcitonin remains unclear.

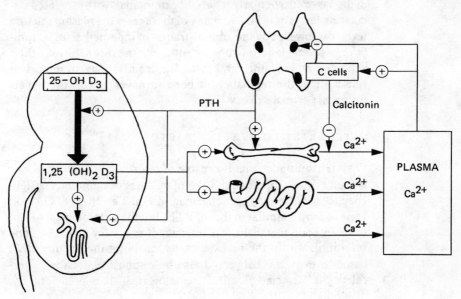

Fig. 9.3 Major hormone control mechanisms of plasma $Ca^{2+}$ homeostasis.

## THYROID HORMONES

Thyroid hormones influence the rate of calcium removal from bone. Marginal hypercalcaemia is evident in about 20% of cases of thyrotoxicosis, whilst osteoporosis may be associated with prolonged hyperthyroidism. In these cases treatment of the thyroid disease may result in hypocalcaemia when the calcium returns to the depleted bone.

### ADRENOCORTICAL STEROIDS

The occasional association of hypercalcaemia with Addison's disease is probably due to the following.
(1) Haemoconcentration following the loss of salt and water.
(2) Increased renal reabsorption of calcium. There is an association between sodium and calcium reabsorption in the proximal tubule and so the hypovolaemia which increases sodium uptake probably also promotes that of calcium.

### PROSTAGLANDINS

The prostaglandins, especially $E_1$ and $E_2$, are potent stimulators of bone resorption *in vitro*, whilst intravenous infusions of $PGE_2$ can cause hypercalcaemia in animals. There have been many reports in the literature of the association of hypercalcaemia of malignancy (particularly renal cell carcinoma), with increased circulating levels of $PGE_2$. In many of these cases the plasma calcium level was lowered by the administration of indomethacin, an inhibitor of prostaglandin synthesis. Although the mechanism of their action on bone is unclear, prostaglandins may play an important role both in the regulation of bone turnover and in the hypercalcaemia of malignancy.

### OSTEOCLAST ACTIVATING FACTOR (OAF)

OAF is a substance that is capable of resorbing bone *in vitro*, and is found in the supernatant fluid of cultures of mononuclear leucocytes stimulated by phytohaemaglutinin. The effect of OAF on bone *in vitro* is similar to that of PTH, but the physiological role of OAF in bone metabolism is unclear. It may play a role in bone resorption within the marrow cavity, enabling the haemopoietic tissue to expand. OAF could also be responsible for the hypercalcaemia associated with haematological malignancies, e.g. myeloma.

## LABORATORY INVESTIGATION OF CALCIUM METABOLISM

In the investigation of the causes of hyper- and hypocalcaemia the most useful initial biochemical tests are:

*Plasma* (1) Calcium, (2) albumin, (3) urea, (4) phosphate, (5) alkaline phosphatase.

Other tests that may be useful in certain situations, or have been used in the past, are:

*Plasma* (1) PTH, (2) bicarbonate, (3) steroid suppression.

*Urine* (1) Calcium excretion, (2) phosphate excretion, (3) hydroxyproline, (4) cyclic AMP excretion.

## Plasma calcium and albumin

The concentration of total calcium is directly affected by that of albumin and does not necessarily reflect the level of plasma $Ca^{2+}$. Although $Ca^{2+}$ would be the assay of choice the methods available are still unreliable and not widely available. This problem can be partly alleviated by measuring calcium and albumin, and then making allowances for any variation of the latter from an arbitrary norm. Various rule of thumb corrections are used, an example of which follows:

For each 1 g/l a patient's plasma albumin is below 40 g/l (mean value of the laboratory reference range), ADD to the total calcium value a correction factor of 0.02 mmol/l. (For each 1 g/l the albumin is above 40 g/l subtract the correction factor.) For example

Plasma calcium = 1.89 mmol/l (ref. range 2.15–2.55)
albumin = 22 g/l (for inpatients 30–50)
40 − 22 = 18 g/l
18 × 0.02 = 0.36 mmol/l
1.89 + 0.36 = 2.25 mmol/l (corrected value)

The corrected calcium value is compared with the reference range for total calcium, and if normal it suggests that the $Ca^{2+}$ level is also normal. The corrected value is not a measure of $Ca^{2+}$, but is thought to be a reflection of it in most cases.

Other important considerations in the measurement of plasma calcium are:
(1) Collect blood samples without a tourniquet, otherwise local haemoconcentration will elevate the calcium value.
(2) Plasma calcium levels rise slightly after meals, and therefore where possible, fasting samples should be obtained.
(3) Patient values should be compared with the appropriate reference range. Because of fluid redistribution the reference range for calcium in ambulant subjects is higher than for recumbent patients, as it also is for albumin, e.g.

|  | Outpatient | Inpatient |
|---|---|---|
| Plasma calcium (mmol/l) | 2.15–2.55 | 2.05–2.55 |
| albumin (g/l) | 37–52 | 30–50 |

(4) Because slightly raised plasma calcium values may have diagnostic significance, patients found to have a marginally elevated calcium (2.55–2.65 mmol/l using the above reference range) should have the test repeated, preferably on three consecutive days, to ensure that analytical or collection errors are not responsible for the hypercalcaemia.

PLASMA PHOSPHATE

Although plasma phosphate is elevated in classic cases of hypoparathyroidism and reduced in those of hyperparathyroidism, the value of phosphate as a diagnostic test in these diseases is limited because:
(1) Hypophosphataemia with hypercalcaemia may occur in those cases of malignancy in which ectopic PTH is produced.
(2) Chronic hypercalcaemia due to any cause has been shown to increase renal excretion of phosphate.
(3) Some studies have shown that more than 50% of subsequently proven primary hyperparathyroid patients have had normal plasma phosphate levels at initial diagnosis. Since the prevalence of this disease is about 0.1%, it is probable that the widespread use of biochemical screening has led to early diagnosis, whilst the phosphate is low but still normal.
(4) Hypercalcaemia, of any cause, may result in renal failure and consequent hyperphosphataemia.

PLASMA ALKALINE PHOSPHATASE

A raised alkaline phosphatase activity, in conjunction with a plasma calcium abnormality, is suggestive of:
(1) Increased osteoblast activity, e.g. primary hyperparathyroidism, malignancy of bone (primary, secondary), secondary hyperparathyroidism.
(2) Liver involvement due to metastases from a malignancy, which is also causing a disturbance in calcium metabolism.
However it is important to note that
(1) At detection over 50% of cases of primary hyperparathyroidism have normal plasma levels of alkaline phosphatase activity.
(2) There are numerous other causes of a raised alkaline phosphatase, including growth, pregnancy and liver disease.

## PLASMA UREA

The initial laboratory investigation of plasma calcium abnormalities should include an assessment of renal function, since renal failure is associated with hypocalcaemia and hyperphosphataemia (cf. hypoparathyroidism). Similarly, prolonged hypercalcaemia may result in renal damage. A urea or creatinine measurement is adequate.

## PLASMA PTH

The value of measuring PTH in the diagnosis of calcium abnormalities is established in theory, but less so in practice. This is largely due to the technical difficulties of the available assays. These problems mainly revolve around whether the antibody used in an assay is directed against the whole of, or only part of, the PTH molecule and whether the N-terminal fragment (active), as opposed to the C-terminal fragment (inactive), is measured. In addition, most available assays are insufficiently sensitive to measure low levels of PTH. With the reservation that both an assay and its associated reference range may be unreliable, the following points emerge:

(1) Primary hyperparathyroidism may be associated with an increased or a normal PTH level. The latter situation is of course an inappropriate one, since hypercalcaemia should suppress PTH secretion. The C-terminal assay appears to give better diagnostic discrimination than the N-terminal one; this may be due to the former fragment having a longer plasma half-life.

(2) Renal failure is associated with increased PTH levels (secondary, tertiary hyperparathyroidism). The raised PTH level is due to increased secretion ($\downarrow Ca^{2+}$), and also to reduced degradation, particularly of the C-terminal fragment which is degraded by the renal tubules after filtration by the glomeruli.

(3) The hypercalcaemia of malignancy is associated with increased circulating levels of PTH in a small number of cases. This may be due to: (a) Ectopic production of PTH by the malignant tissue. At the present time the evidence for ectopic production of PTH is poorly documented and it appears that if this condition exists it accounts for only a small proportion of the hypercalcaemia that is associated with malignancy. (b) Coexistence of renal failure. The C-terminal PTH fragment requires normal renal function for degradation. The association of renal insufficiency and malignant disease is common, particularly as they both tend to occur in the older age groups. (c) Coexistence of primary hyperparathyroidism. The prevalence of primary hyperparathyroidism in the community is of the order of 1/1000, so it is not surprising that this condition and malignant disease may both occur in the same patient.

(4) PTH assays on blood specimens from catheterization of neck veins are of value in localizing parathyroid adenomas.

(The PTH standards used in the following case examples for estimating plasma PTH/C-terminus were calibrated against the WHO First International Reference Preparation 71/324.)

PLASMA BICARBONATE

PTH decreases both the renal secretion of $H^+$ and the regeneration of $HCO_3^-$, so that primary hyperparathyroidism tends to be accompanied by a renal tubular type of acidosis (hyperchloraemia, low $HCO_3^-$), whereas hypercalcaemia due to other causes is associated with the reverse situation. However, these acid-base changes, when they occur, are usually minimal and are of little diagnostic value.

STEROID SUPPRESSION TEST

This test is of value in determining whether a hypercalcaemia of unknown aetiology is due to excess PTH secretion by the parathyroids or not.

*Outline protocol*
(1) Basal levels of plasma calcium, albumin, urea measured daily for 3 days.
(2) 40 mg hydrocortisone taken orally every 8 hours for 10 days.
(3) Plasma calcium, albumin, urea measured on days 6, 8, 10.
(4) Steroid dose tailed off.

*Interpretation*
(1) The steroid may cause water retention, and therefore calcium values should be carefully adjusted according to any alteration in the albumin level during the test.
(2) The plasma calcium concentration will show little or no change throughout the test if the hypercalcaemia is caused by an inappropriately high (autonomous) secretion of PTH by the parathyroids (primary hyperparathyroidism).
(3) The plasma calcium level often falls to normal values if the hypercalcaemia is not due to excess PTH from the parathyroids. It has been reported that the calcium normalizes even if ectopic PTH is the causative factor.
(4) Failure to suppress calcium may occur in severe cases of malignancy or hyperthyroidism, but in these situations the cause of the hypercalcaemia is usually clinically obvious and the test unnecessary. The test may be difficult to interpret in cases of marginal hypercalcaemia and may decline in value as good PTH assays become more widely available.

## URINARY CALCIUM EXCRETION

Adults with normal renal function excrete less than 7.5 mmol/24 h of calcium. A plasma calcium raised by any cause will result in hypercalciuria and, therefore, the estimation of urinary calcium excretion has generally been considered to be of little value in the differential diagnosis of hypercalcaemia. However, a condition termed familial hypocalciuric hypercalcaemia is associated with a low renal calcium excretion rate (<5 mmol/24 h) and therefore such measurements would be of value in detecting this rare cause of hypercalcaemia.

## URINARY PHOSPHATE EXCRETION

Although the effect of PTH on the renal excretion of $PO_4$ might be expected to render the estimation of urinary phosphate a useful test this has not been borne out in practice, possibly because hypercalcaemia *per se* tends to increase renal excretion of $PO_4$. Dietary effects and wide reference ranges have minimized any value in either the measurement of urinary $PO_4$ or the various indices based on urinary $PO_4$ levels (e.g. tubular reabsorption of phosphate).

## URINARY HYDROXYPROLINE

The plasma and urinary levels of the two amino acid constituents of collagen, hydroxyproline and hydroxylysine are increased when collagen turnover is increased. The need for special dietary preparation of the patient (no gelatin) and analytical problems have limited the value of these tests in investigating disorders of calcium metabolism.

## URINARY CYCLIC AMP

At the cellular level PTH expresses its action via cyclic AMP, the excretion of which is increased in hypercalcaemia due to raised PTH (hyperparathyroidism or ectopic production). Although literature reports have indicated that this test has some value it has not yet found widespread acceptance and so it is difficult to assess whether this test is little used because it is of limited diagnostic value or because few laboratories offer it.

# HYPERCALCAEMIA

The causes of hypercalcaemia may be conveniently classified as follows:

*Increased intake/absorption* (1) Vitamin D excess, (2) sarcoidosis—increased sensitivity to vitamin D, (3) milk alkali syndrome—high Ca and alkali intake, (4) hyperalimentation—imbalance of intravenous therapy.

*Increased plasma albumin* (1) Dehydration, (2) prolonged application of tourniquet during venepuncture.

*Increased bone resorption* (1) Malignancy, (2) hyperparathyroidism—primary, tertiary, (3) renal failure, (4) thyrotoxicosis, (5) immobilization.

*Increased renal resorption* (1) Thiazide diuretics, (2) familial hypocalciuric hypercalcaemia.

The commonest causes of hypercalcaemia are: (1) malignancy (carcinoma, lymphoma, multiple myeloma, leukaemia), (2) primary hyperparathyroidism, (3) iatrogenic, (4) vitamin D intoxication.

Elucidation of the cause of hypercalcaemia can usually be made on clinical grounds, without resort to sophisticated biochemical investigation. However, a major problem that occasionally arises is the differentiation between hypercalcaemia due to primary hyperparathyroidism and that due to 'occult' malignancy. The following general observations can be helpful.

|  | Malignancy | Hyperparathyroidism |
|---|---|---|
| Degree of hypercalcaemia usually | >3.0 mmol/l | <3.0 mmol/l |
| Duration | weeks/months | months/years |
| Rate of increase of plasma calcium | rapid | slow |
| Renal calculi | unusual | common |
| Steroid suppression test | suppression | no response |
| Plasma PTH | ↓ | ↑ |

**Case examples/pathophysiology**

VITAMIN D EXCESS

A 59 year old lady complaining of polyuria and nocturia. She had a long history of taking a self-prescribed multivitamin preparation.

| | | | |
|---|---|---|---|
| Plasma Ca | 3.21 | mmol/l | (2.15–2.55) |
| PO$_4$ | 2.01 | mmol/l | (0.60–1.25) |
| Alb. | 43 | g/l | (37–52) |
| AP | 97 | U/l | (30–120) |
| Urea | 12 | mmol/l | (3.0–8.0) |

Pathophysiology

(1) ↑ vit. D → ↑ gut absorption of $Ca^{2+}$ → ↑ plasma [Ca] (ionized + total)
(2) ↑ plasma $[Ca^{2+}]$ → ↓ PTH → ↓ renal $PO_4$ excretion → ↑ plasma $[PO_4]$

Prolonged hypercalcaemia may be associated with a high plasma urea due to: (1) renal insufficiency due to deposition of calcium around the tubules (nephrocalcinosis), (2) dehydration—hypercalcaemia interferes with the action of ADH and therefore the renal concentrating mechanism, producing polyuria.

SARCOIDOSIS

Rare disease of unknown aetiology, in which hypercalcaemia is caused by an increased sensitivity to vitamin D. The biochemical picture is similar to that of vitamin D excess.

MILK ALKALI SYNDROME

This syndrome (now rare) is characterized by hypercalcaemia, alkalosis and renal failure. It is due to excessive intake of milk (vitamin D, calcium) and alkali ($NaHCO_3$, $MgCO_3$), usually in patients with ulcers. If prolonged, the hypercalcaemia results in renal failure.

HYPERALIMENTATION

Poor control of intravenous therapy may result in either hypercalcaemia or hypocalcaemia. This is particularly likely to occur in premature infants in whom the calcium homeostatic mechanisms are usually immature.

DEHYDRATION

A 46 year old man admitted with small bowel obstruction following a 24 hour history of vomiting.

| | | | |
|---|---|---|---|
| Plasma Ca | 2.69 | mmol/l | (2.05–2.55) |
| $PO_4$ | 1.05 | mmol/l | (0.60–1.25) |
| Alb. | 54 | g/l | (30–50) |
| AP | 100 | U/l | (30–120) |
| Urea | 20 | mmol/l | (3.0–8.0) |

A raised plasma albumin is an unusual finding and usually indicates either an analytical error or haemoconcentration. In this case, the loss of fluid in the vomitus explains the elevated albumin and therefore allowance has to be made for this in interpreting the calcium result. The corrected calcium value is 2.41 mmol/l, suggesting that the $[Ca^{2+}]$ is also normal. On rehydration the plasma calcium should be rechecked.

USE OF TOURNIQUET

Prolonged application of a tourniquet during blood collection may result in haemoconcentration (fluid shifts out of intravascular space), with an increased plasma albumin and minimal hypercalcaemia. Blood for calcium analysis should be taken, if possible, without the use of a tourniquet.

MALIGNANCY

A 61 year old man with weight loss. Heavy cigarette smoker since aged 15. Relevant biochemical results at presentation shown below. Oat cell carcinoma of lung subsequently diagnosed.

| Plasma | Ca | 3.89 | mmol/l | (2.15–2.55) |
|---|---|---|---|---|
| | $PO_4$ | 0.55 | mmol/l | (0.60–1.25) |
| | Alb. | 36 | g/l | (37–52) |
| | AP | 235 | U/l | (30–120) |
| | Urea | 4.7 | mmol/l | (3.0–8.0) |

The high calcium value (>3.0) would strongly support an initial diagnosis of malignancy. Note the borderline low result for $PO_4$, and elevated alkaline phosphatase.

Pathophysiology

Various possible mechanisms include:
(1) Erosion of bone by secondaries with release of $Ca^{2+}$ and $PO_4$.
(2) Ectopic production of PTH by the tumour.
(3) Production of a PTH-like substance by the tumour.
(4) Production of prostaglandins by the tumour.
(5) Production of a vitamin D-like substance by the tumour (existence not proven).
(6) Hypercalcaemia due to other causes, e.g. primary hyperparathyroidism.

The biochemical picture in this case is similar to hyperparathyroidism and thus the mechanism could be due to ectopic production of PTH or secretion of a PTH-like substance by the tumour.

↑ PTH → (1) ↑ bone resorption → ↑ plasma [Ca], (2) ↑ renal $PO_4$ excretion → ↓ plasma $[PO_4]$, (3) ↑ osteoblast activity → ↑ plasma [AP]

Many cases of malignancy-associated hypercalcaemia have a low or low normal phosphate that is suggestive of ectopic production of PTH. However, it must be remembered that chronic hypercalcaemia *per se* in the presence of normal renal function will result in increased secretion of phosphate, and perhaps hypophosphataemia, and that ectopic PTH production is probably a rare occurrence. Definite implication of ectopic PTH production would require the demonstration of increased PTH activity in the plasma and in the tumour tissue. In cases with biochemical features similar to the above (↑ Ca, ↓ $PO_4$) it is probably better to use the term 'humoral hypercalcaemia of malignancy' rather than ectopic PTH production.

High plasma levels of alkaline phosphatase in malignancy suggest bone disease (PTH action, secondaries) or liver involvement by metastases.

MALIGNANCY

A 56 year old lady found to have a 4 cm diameter lump in her left breast.

| Plasma | Ca | 3.89 | mmol/l | (2.15–2.55) |
|---|---|---|---|---|
| | $PO_4$ | 1.55 | mmol/l | (0.60–1.25) |
| | Alb. | 38 | g/l | (37–52) |
| | AP | 306 | U/l | (30–120) |
| | Urea | 12.5 | mmol/l | (3.0–8.0) |

Again the severe hypercalcaemia is typical of that found in malignancy, in this case a carcinoma of the breast. The hyperphosphataemia points against ectopic PTH-like substance production (humoral hypercalcaemia of malignancy) as a cause of the hypercalcaemia, and suggests the cause as one of the following: (1) secondary tumour deposits in bone, causing resorption of $Ca^{2+}$ and $PO_4$, (2) tumour production of vitamin D-like substances (existence not proven).

However the hyperphosphataemia could also be caused by renal insufficiency, in which case a low $[PO_4]$ due to an ectopic PTH is being masked.

MULTIPLE MYELOMA

A 59 year old lady with T4 crush fracture and anaemia. Paraprotein found on serum electrophoresis.

| | | | |
|---|---|---|---|
| Plasma Ca | 3.84 | mmol/l | (2.15–2.55) |
| $PO_4$ | 1.41 | mmol/l | (0.60–1.25) |
| TP | 90 | g/l | (60–80) |
| Alb. | 32 | g/l | (37–52) |
| AP | 92 | U/l | (30–120) |
| Urea | 14.0 | mmol/l | (3.0–8.0) |

In this case of multiple myeloma the marked hypercalcaemia and slight hyperphosphataemia are probably due to local bone resorption by the tumour cells, which elaborate an osteoclast activating factor (OAF). Multiple myeloma are usually associated with normal levels of alkaline phosphatase activity, but this is not always the case, particularly if the liver becomes involved in the disease process. Other types of haematological neoplasm associated with hypercalcaemia are Hodgkin's disease, non-Hodgkin's lymphoma and leukaemia.

PRIMARY HYPERPARATHYROIDISM

A 37 year old man with a history of renal calculi composed largely of calcium and phosphorus. A parathyroid adenoma was removed on 27 August.

| Date | 26/8 | 28/8 | | |
|---|---|---|---|---|
| Plasma Ca | 2.78 | 1.86 | mmol/l | (2.15–2.55) |
| $PO_4$ | 0.53 | 0.65 | mmol/l | (0.60–1.25) |
| Alb. | 43 | 42 | g/l | (30–50) |
| AP | 195 | 180 | U/l | (30–120) |
| Urea | 5.0 | 6.5 | mmol/l | (3.0–8.0) |

This is the classical clinical presentation and biochemistry of primary hyperparathyroidism. Hypercalcaemia due to other causes may be associated with renal calculi but they are not usually of a recurrent nature.

Pathophysiology

↑ PTH: ↑ calcium resorption → ↑ plasma [Ca]
      ↑ renal $PO_4$ excretion → ↓ plasma [$PO_4$]
      ↑ osteoblast activity → ↑ plasma [AP]

Prolonged exposure to excessive PTH (from the adenoma) results in suppression of PTH release from the normal parathyroid tissue. When the adenoma is removed there is a sudden cut-off of PTH and of bone resorption. However, deposition of calcium in bone continues for some time, and this often results in transient postoperative hypocalcaemia. The hypocalcaemia persists (usually 1 or 2 days, but may continue for several months) until the normal parathyroid tissue recovers its function.

## PRIMARY HYPERPARATHYROIDISM

A 42 year old man complaining of depression, constipation and vague aches and pains.

| Plasma | Ca | 2.83 | mmol/l | (2.15–2.55) |
|---|---|---|---|---|
| | $PO_4$ | 0.82 | mmol/l | (0.60–1.25) |
| | Alb. | 45 | g/l | (37–52) |
| | AP | 80 | U/l | (30–120) |
| | Urea | 4.5 | mmol/l | (3.0–8.0) |
| | PTH (C term) | 9 | µg/l | (<6) |

The prevalence of primary hyperparathyroidism is about 1/1000 of the general population. With the widespread use of biochemical screening, and consequent early detection, the commonest presentation of this disorder is hypercalcaemia associated with vague symptoms and therefore, as in this case, the plasma $[PO_4]$ and [AP] are often found to be normal. A normal $[PO_4]$ may also be due to an associated renal insufficiency, and therefore renal function should always be examined. An occasional diagnostic problem is the differentiation between primary hyperparathyroidism and hypercalcaemia-associated malignancy in which the tumour is occult. In these cases the steroid suppression test, plasma PTH assay and bone radiology can be of additional diagnostic help.

## RENAL FAILURE

A 47 year old man being treated with intermittent haemodialysis for chronic renal failure.

| Plasma | Ca | 2.77 | mmol/l | (2.05–2.55) |
|---|---|---|---|---|
| | $PO_4$ | 0.90 | mmol/l | (0.60–1.25) |
| | Alb. | 46 | g/l | (30–50) |
| | AP | 155 | U/l | (30–120) |
| | Urea | 16.2 | mmol/l | (3.0–8.0) |
| | PTH (C term) | 15 | µg/l | (<6) |

Hypocalcaemia gradually develops as renal failure progresses (see Hypocalcaemia case). The continually low plasma $[Ca^{2+}]$ causes the output of abnormally large amounts of PTH in order to try and maintain a normal $[Ca^{2+}]$. The parathyroid glands become hyperplastic as they maintain the sustained high output of PTH (secondary hyperparathyroidism). At this stage the plasma [Ca] is usually within normal limits. Eventually the parathyroid glands lose the ability to respond to plasma $[Ca^{2+}]$, and then secrete the high levels of PTH autonomously (tertiary hyperparathyroidism). If the renal failure is unrelieved the low or low normal [Ca] will continue (see Hypocalcaemia case), but treatment (dialysis, renal transplant) may result in hypercalcaemia. If this patient had a renal transplant it is likely that the $[PO_4]$ would then be low, because the normal kidney function would allow the full phosphaturic effect of the elevated PTH to be expressed. Parathyroidectomy may have to be considered in the latter situation. Very high levels of C-terminal PTH may be found in patients with chronic renal failure. This is due to increased secretion (secondary hyperparathyroidism) and decreased degradation ($\downarrow$ GFR and renal tubule destruction).

IMMOBILIZATION

An 18 year old man hospitalized for seven weeks with a fractured femur.

| Plasma | Ca | 2.72 | mmol/l | (2.05-2.55) |
| --- | --- | --- | --- | --- |
| | $PO_4$ | 1.12 | mol/l | (0.60-1.25) |
| | Alb. | 43 | g/l | (30-50) |
| | AP | 160 | U/l | (30-120) |
| | Urea | 5.7 | mmol/l | (3.0-8.0) |

Complete immobilization in bed for a lengthy period may cause hypercalcaemia in a proportion of subjects in whom there is a reasonably rapid bone turnover, particularly children and young adults in traction for fractures, and in patients with Paget's disease. In patients with fractures the plasma calcium level may rise above 3.0 mmol/l. It is advisable to recheck the plasma calcium following mobilization. Hypercalcaemia occurred in some astronauts during the early space flights, and it is thought that lack of mobility was a major contributing factor.

THIAZIDE DIURETICS

A 52 year old woman on long term thiazide therapy for fluid retention secondary to congestive cardiac failure.

| Plasma | Ca | 2.66 | mmol/l | (2.15-2.55) |
| --- | --- | --- | --- | --- |
| | $PO_4$ | 1.05 | mmol/l | (0.60-1.25) |
| | Alb. | 39 | g/l | (37-52) |
| | AP | 55 | U/l | (30-120) |
| | Urea | 8.5 | mmol/l | (3.0-8.0) |

Prolonged thiazide treatment may occasionally be associated with hypercalcaemia. The mechanism is unclear, but may involve

(1) extracellular volume depletion, resulting in increased proximal tubule reabsorption of calcium, (2) increased sensitivity of the distal tubule to the action of PTH, (3) increased PTH secretion.

The hypercalcaemia is usually mild (<2.7 mmol/l) and may return to normal if thiazides are withdrawn. There is strong evidence that thiazide therapy will only produce hypercalcaemia if there is pre-existing metabolic bone disease (e.g. hyperparathyroidism). Therefore, all such cases should be thoroughly investigated for other causes of hypercalcaemia.

FAMILIAL HYPOCALCIURIC HYPERCALCAEMIA

This rare condition, usually confused with primary hyperparathyroidism, is inherited as an autosomal dominant and usually presents with hypercalcaemia before the age of 10 years. The aetiology is unclear, but the hypercalcaemia appears to be due to increased renal resorption of calcium. Diagnosis is based on the finding of hypercalcaemia associated with a low renal excretion rate of calcium (<5 mmol/d). Many of the reported cases of this disease have been diagnosed after parathyroidectomy for suspected primary hyperparathyroidism failed to resolve the hypercalcaemia. It is therefore good practice to estimate the 24 hour urinary calcium excretion rate on all suspected cases of primary hyperparathyroidism, particularly if they are in the young age group or if other members of the family also have hypercalcaemia.

# HYPOCALCAEMIA

The causes of hypocalcaemia may be classified as follows:

*Decreased intake*  (1) Inadequate intravenous nutrition, (2) vitamin D deficiency (a) poor nutrition, (b) malabsorption, (c) renal disease, (d) anticonvulsant therapy.

*Decreased plasma albumin*  (1) Malnutrition, (2) cirrhosis of the liver, (3) nephrotic syndrome, (4) protein-losing enteropathy, (5) burns.

*Decreased flux from bone*  (1) PTH deficiency (a) congenital, (b) idiopathic, (c) surgical removal of glands, (d) surgical removal of adenoma, (e) infarction of adenoma, (2) resistance to PTH action (a) uraemia, (b) Mg deficiency, (c) pseudohypoparathyroidism.

*Extraskeletal sequestration* (1) Acute pancreatitis, (2) hyperphosphataemia.

*Neonatal hypocalcaemia* (1) Prematurity (a) magnesium deficiency, immature parathyroid gland, immature vitamin D metabolism, (2) poor feeding (a) oral, (b) intravenous.

The most common causes of hypocalcaemia in clinical practice are: (1) hypoalbuminaemia, (2) chronic renal failure, (3) acute pancreatitis, (4) prematurity, (5) inadequate nutrition (IV or oral).

**Case examples/pathophysiology**

VITAMIN D DEFICIENCY

A 79 year old widow living alone. Diet mainly comprised toast and black tea for a long period prior to welfare intervention.

| Plasma | Ca | 1.94 | mmol/l | (2.15–2.55) |
|---|---|---|---|---|
| | $PO_4$ | 0.50 | mmol/l | (0.60–1.25) |
| | Alb. | 39 | g/l | (37–52) |
| | AP | 180 | U/l | (30–120) |
| | Urea | 6.0 | mmol/l | (3.0–8.0) |

Pathophysiology

(1) ↓ vit D → ↓ $Ca^{2+}$ absorption → ↓ plasma $[Ca^{2+}]$
(2) ↓ plasma $[Ca^{2+}]$ → ↑ PTH (secondary hyperparathyroidism)
(3) ↑ PTH → ↓ renal $PO_4$ resorption → ↓ plasma $[PO_4]$
(4) ↑ osteoblast activity → ↑ AP activity

The classical features of Vitamin D deficiency are therefore: (1) hypocalcaemia, (2) hypophosphataemia, (3) high alkaline phosphatase activity.

Early in the disease the raised alkaline phosphatase may be the only abnormality evident, whilst the high PTH maintains normocalcaemia. Other causes of vitamin D deficiency include: (1) malabsorption—vitamin D is a fat soluble vitamin, (2) long-term anticonvulsant therapy—phenylhydantoin may induce enzymes that catabolize vitamin D or block the action of 1,25-$(OH)_2D_3$ on the intestinal mucosa, (3) renal disease.

DECREASED PLASMA ALBUMIN

A 47 year old man with alcohol-induced cirrhosis of the liver.

| | | | |
|---|---|---|---|
| Plasma Ca | 1.84 | mmol/l | (2.05–2.55) |
| PO$_4$ | 0.71 | mmol/l | (0.60–1.25) |
| Alb. | 22 | g/l | (30–50) |
| AP | 700 | U/l | (30–120) |
| Urea | 2.5 | mmol/l | (3.0–8.0) |

Correction of the total calcium measurement for the severe hypoalbuminaemia yields a value of 2.20 mmol/l. This suggests, but does not guarantee, that the [Ca$^{2+}$] is within normal limits. The liver is involved in the conversion of D$_3$ to 25-OHD$_3$, and this may be reduced in advanced liver disease. Also, alcoholics often have a poor diet. These factors may cause [Ca$^{2+}$] to be lower than that suggested by the correction factor. Careful clinical evaluation of calcium status is important in these cases.

HYPOPARATHYROIDISM

A 60 year old lady presenting with carpopedal spasm. Previously undergone a total thyroidectomy for carcinoma of the thyroid gland.

| | | | |
|---|---|---|---|
| Plasma Ca | 1.70 | mmol/l | (2.15–2.55) |
| PO$_4$ | 2.18 | mmol/l | (0.60–1.25) |
| Alb. | 41 | g/l | (37–52) |
| AP | 90 | U/l | (30–120) |
| Urea | 6.0 | mmol/l | (3.0–8.0) |

A major potential complication of total thyroidectomy is the accidental removal of the parathyroid glands, or damage to their blood supply (transient hypocalcaemia).

↓ PTH → ↓ calcium removal from bone → ↓ plasma [Ca]
↓ PTH → ↑ renal PO$_4$ resorption → ↑ plasma [PO$_4$]

In hypoparathyroidism the plasma calcium level may fall as low as 1.2–1.3 mmol/l.

Hypocalcaemia occurring after thyroidectomy may be due to:
(1) Reversal of negative bone and calcium balance. Thyrotoxicosis is associated with increased bone resorption (promoted by thyroxine) and occasionally hypercalcaemia, which will result in the suppression of parathyroid gland activity. After thyroidectomy calcium is redeposited in bone, whilst recovery of parathyroid gland activity may be delayed. This is often manifested by a transient hypocalcaemia which is maximal on the first postoperative day.
(2) Hypoparathyroidism (a) removal of parathyroid glands, (b) injury to parathyroid glands at operation. Most of these patients

will eventually recover, but in some the condition may be permanent. The condition is probably due to ischaemia which may produce transient or permanent parathyroid gland dysfunction. In the latter case hypocalcaemia may not develop until one or many years after the operation.

RENAL FAILURE

A 27 year old man with severe bilateral polycystic renal disease.

| Plasma Ca | 1.86 | mmol/l | (2.15–2.55) |
|---|---|---|---|
| PO$_4$ | 2.14 | mmol/l | (0.60–1.25) |
| Alb. | 40 | g/l | (37–52) |
| AP | 152 | U/l | (30–120) |
| Urea | 35 | mmol/l | (3.0–8.0) |
| PTH (C term) | 12 | µg/l | (<6) |

Hypocalcaemia in renal failure may be due to:
(1) Hyperphosphataemia due to low GFR leads to (a) *in vivo* precipitation of calcium, (b) reduced calcium uptake from gut due to ↑[PO$_4$] of gut secretion.
(2) Reduced conversion of 25-(OH)D$_3$ to 1,25-(OH)$_2$D$_3$ by damaged kidneys and ↑[PO$_4$].
(3) Resistance of bone to PTH caused by unidentified, retained toxic substances.

The hypocalcaemia stimulates a high and continuous secretion of PTH (secondary hyperparathyroidism), whilst the renal insufficiency retards the degradation of the PTH C-terminal fragment. This results in very high plasma levels of plasma PTH (C-terminal assay). The elevated plasma alkaline phosphatase activity reflects metabolic bone disease due to 1,25-(OH)$_2$D$_3$ deficiency and increased PTH activity.

The hypocalcaemia of chronic renal failure is usually mild and not associated with symptoms (tetany). If the plasma calcium concentration is less than ~1.8 mmol/l other causes of hypocalcaemia should be considered.

MAGNESIUM DEFICIENCY

A 7 day old infant with convulsions. Hypocalcaemia corrected with calcium infusion, but a further convulsion occurred the next day. The pre-infusion test results are shown below.

| Plasma Ca | 1.66 | mmol/l | (1.95–2.40) |
|---|---|---|---|
| PO$_4$ | 1.91 | mmol/l | (1.20–2.80) |
| Alb. | 37 | g/l | (30–50) |

| | | | |
|---|---|---|---|
| AP | 180 | U/l | (70–250) |
| Urea | 2.5 | mmol/l | (2.5–5.0) |
| Mg | 0.5 | mmol/l | (0.7–1.0) |

The association of hypocalcaemia and hypomagnesaemia is unclear, but may be due to one or more of the following mechanisms: (1) skeletal resistance to PTH caused by low magnesium, (2) decreased PTH secretion due to reduced magnesium, (3) increased calcium flux into the bone to replace magnesium.

## PSEUDOHYPOPARATHYROIDISM

A very rare disease in which there is end organ (bone, kidney) resistance to the action of PTH. The plasma [Ca] is low, [PO$_4$] high and there is an elevated PTH. Clinically the typical case shows mental retardation, small stature and short fourth metacarpals. The biochemical features are the same as for hypoparathyroidism ($\downarrow$ Ca, $\uparrow$ PO$_4$).

## ACUTE PANCREATITIS

A 51 year old woman with a long history of alcoholism, now admitted to hospital with an acute central abdominal pain radiating through to the back.

| Plasma | Ca | 1.88 | mmol/l | (2.15–2.55) |
|---|---|---|---|---|
| | PO$_4$ | 1.22 | mmol/l | (0.60–1.25) |
| | Alb. | 47 | g/l | (37–52) |
| | AP | 75 | U/l | (30–120) |
| | Urea | 8.0 | mmol/l | (3.0–8.0) |
| | Amy. | 12 000 | U/l | (<300) |

In acute pancreatitis the hypocalcaemia is transient, rarely causes symptoms and usually occurs within 36 hours of the onset of the episode. The causes of the hypocalcaemia are poorly understood, but the following mechanisms have been suggested:
(1) Damaged pancreas → $\uparrow$ lipases → $\uparrow$ free fatty acids → calcium soaps formed.
(2) Glucagon released from the injured pancreas. Experimental infusions of glucagon lower plasma [Ca] by an unknown mechanism (?increased bone deposition).
(3) Hypomagnesaemia. Commonly occurs in chronic alcoholics.
(4) Toxic substances released by the damaged pancreas causing skeletal resistance to PTH.
(5) Toxic substances inhibiting PTH secretion.

Many reports in the literature have suggested that there is a causal relationship between primary hyperparathyroidism and acute pancreatitis because of the high incidence of these two conditions occurring in the same patient. Some authorities further suggest that if hypocalcaemia does not occur during the course of acute pancreatitis the patient is likely to have an associated primary hyperparathyroidism. However, more recent studies have not validated these two claims.

# PHOSPHATE

## PHOSPHATE METABOLISM

### Body distribution

| | |
|---|---|
| Total phosphate | $\sim 25$ mol |
| Skeleton | 85% |
| Soft tissues | 15% |
| Extracellular fluid | <1% |

### Gut uptake

90% of the daily dietary intake of 25–35 mmol is absorbed. This uptake is stimulated by both PTH and $1,25\text{-}(OH)_2D_3$.

### Kidney

About 85% of the phosphate filtered by the glomeruli is reabsorbed by the proximal renal tubules. The rate of reabsorption is increased by vitamin D and growth hormone and decreased by PTH.

## LABORATORY INVESTIGATION OF PHOSPHATE METABOLISM

#### PLASMA PHOSPHATE

Adult levels ($\sim 0.6\text{-}1.2$ mmol/l) are lower than those found during childhood ($\sim 1.3\text{-}2.8$ mmol/l) and are reached by 15–17 years of age.

There is often a slight fall in $[PO_4]$ after a meal rich in carbohydrate.

## PLASMA CALCIUM

If a plasma calcium level is abnormal it can give a clue to the cause of an associated hyper- or hypophosphataemia, e.g.

$\uparrow$ [Ca] + $\downarrow$ [PO$_4$] — primary hyperparathyroidism, malignancy (humoral hypercalcaemia)
$\uparrow$ [Ca] + $\uparrow$ [PO$_4$] — malignancy (1° or 2° tumour deposits in bone), postdialysis in renal failure
$\downarrow$ [Ca] + $\uparrow$ [PO$_4$] — hypoparathyroidism, renal failure (untreated)
$\downarrow$ [Ca] + $\downarrow$ [PO$_4$] — vitamin D deficiency

## PLASMA UREA

The reduced GFR due to renal failure causes hyperphosphataemia. However, in this situation the plasma phosphate level does not begin to rise until the GFR has fallen to below 0.3–0.15 ml/s, i.e. as the renal failure progresses and hypocalcaemia occurs there is an increased secretion of PTH, which prevents PO$_4$ resorption in the remaining normal nephrons.

## ACID-BASE

Acidosis may be associated with a slight hyperphosphataemia. Alkalosis causes phosphate to enter cells with a resultant hypophosphataemia; this reflects increased glycolysis which is associated with intracellular alkalosis.

## PLASMA GLUCOSE

Diabetes mellitus and its treatment are often associated with severe disturbances to phosphate metabolism. In insulin deficiency PO$_4$ moves out of the cells to the ECF and is excreted in the urine, resulting in phosphate depletion. During insulin therapy for diabetic ketoacidosis hypophosphataemia occurs due to increased cellular uptake (inorganic PO$_4$ is required for the production of phosphorylated carbohydrates).

## LIVER FUNCTION TESTS

Hypophosphataemia may be seen in liver failure, but the mechanism is unclear.

Chapter 9

# HYPOPHOSPHATAEMIA

Hypophosphataemia may be caused by the following.

*Decreased intake*   (1) Starvation, poor IV nutrition, (2) malabsorption, (3) vomiting, (4) antacid therapy.

*Increased cell uptake*   (1) High dietary carbohydrate, (2) treatment of diabetic acidosis, (3) alkalosis, (4) liver disease.

*Increased excretion*   (1) Diuretics, (2) hypomagnesaemia, (3) renal $PO_4$ leak, (4) excess PTH action, (5) dialysis.

**Case examples/pathophysiology**

VOMITING

Adult with vomiting during previous 48 hours due to pyloric stenosis.

| Plasma | Na | 138 | mmol/l | (132–144) |
|---|---|---|---|---|
| | K | 2.9 | mmol/l | (3.2–4.8) |
| | Cl | 86 | mmol/l | (98–108) |
| | $HCO_3$ | 37 | mmol/l | (23–33) |
| | Urea | 7.5 | mmol/l | (3.0–8.0) |
| | Ca | 2.27 | mmol/l | (2.15–2.55) |
| | $PO_4$ | 0.50 | mmol/l | (0.60–1.25) |
| | Alb. | 35 | g/l | (30–50) |
| | AP | 86 | U/l | (30–120) |

Pathophysiology

(1) Electrolytes (see p. 75, 128).
(2) Vomiting → loss of acid → alkalosis ($HCO_3$ ↑) → movement of plasma $PO_4$ into cells → ↓ plasma $[PO_4]$. Decreased dietary intake of $PO_4$ will also contribute to the hypophosphataemia.

ANTACID THERAPY

Both aluminium carbonate and magnesium carbonate can produce hypophosphataemia by binding $PO_4$ in the gut, thereby reducing absorption. Any associated metabolic alkalosis will aggravate the hypophosphataemia.

## HYPERALIMENTATION

*Calcium, phosphate and magnesium*

A 52 year old man on total parenteral nutrition following bowel surgery.

| Plasma | Ca | 2.20 | mmol/l | (2.15–2.55) |
|---|---|---|---|---|
| | $PO_4$ | 0.45 | mmol/l | (0.60–1.25) |
| | Alb. | 36 | g/l | (30–50) |
| | AP | 94 | U/l | (30–120) |

Pathophysiology

(1) Inadequate phosphate in the intravenous feed.
(2) High carbohydrate input. Cellular uptake of glucose is associated with a fall in plasma $[PO_4]$, probably due to the intra-cellular formation of glucose-6-phosphate.

## DIABETIC KETOACIDOSIS

A 34 year old woman hospitalized with diabetic ketoacidosis and treated with insulin.

| | | On admission | 8 h post insulin | | |
|---|---|---|---|---|---|
| Plasma | Ca | 2.38 | 2.04 | mmol/l | (2.15–2.55) |
| | $PO_4$ | 3.50 | 0.20 | mmol/l | (0.60–1.25) |
| | Alb. | 43 | 30 | g/l | (30–50) |
| | AP | 110 | 90 | U/l | (30–120) |
| | Glu. | 77 | 9.0 | mmol/l | (3.0–5.0) |

Pathophysiology

During the phase of ketoacidosis and osmotic diuresis the phosphate moves out of cells and is then excreted in the urine. These patients therefore become depleted of phosphate. When they are given insulin the plasma $PO_4$ is taken back into the cells, resulting in a moderate to severe hypophosphataemia.

## ALKALOSIS

Hysterical 41 year old woman examined in the Emergency Department.

| Plasma | Ca | 2.45 | mmol/l | (2.15–2.55) | Blood pH | 7.52 | (7.35–7.45) |
|---|---|---|---|---|---|---|---|
| | $PO_4$ | 0.55 | mmol/l | (0.60–1.25) | $P{CO_2}$ | 3.8 kPa | (4.7–6.0) |
| | Alb. | 41 | g/l | (37–52) | | | |
| | AP | 100 | U/l | (30–120) | | | |
| | $HCO_3$ | 23 | mmol/l | (23–33) | | | |

Prolonged overbreathing causes excessive loss of $CO_2$ and a consequent respiratory alkalosis. It has been suggested that alkalosis stimulates glycolysis, in which case the increased need for $PO_4$ is satisfied by its moving into the cells.

LIVER DISEASE

A 48 year old man with severe alcoholic liver disease.

| Plasma | Ca | 2.01 | mmol/l | (2.05–2.55) |
|---|---|---|---|---|
| | $PO_4$ | 0.50 | mmol/l | (0.60–1.25) |
| | Alb. | 27 | g/l | (30–50) |
| | AP | 425 | U/l | (30–120) |
| | $HCO_3$ | 18 | mmol/l | (23–33) |
| | ALT | 35 | U/l | ($<$35) |
| | Bili. | 240 | µmol/l | ($<$20) |

Severe liver disease has two consequences that will affect phosphate metabolism: (1) respiratory alkalosis, (2) failure of the liver to store adequate glycogen causes increased glucose metabolism by the muscle tissue. Both these situations increase cell utilization of phosphate. Alcoholism is a common cause of hypophosphataemia and may be due to the effects of: (1) starvation, (2) vomiting, (3) malabsorption, (4) hypomagnesaemia, (5) carbohydrate ingestion during treatment (IV dextrose etc.).

DIURETIC THERAPY

A 66 year old woman on thiazide therapy for congestive cardiac failure.

| Plasma | Ca | 2.27 | mmol/l | (2.15–2.55) |
|---|---|---|---|---|
| | $PO_4$ | 0.48 | mmol/l | (0.60–1.25) |
| | Alb. | 43 | g/l | (37–52) |
| | AP | 120 | U/l | (30–120) |
| | K | 2.9 | mmol/l | (3.2–4.8) |
| | $HCO_3$ | 35 | mmol/l | (23–33) |

The resorption of both sodium and phosphate are closely associated in the renal proximal tubule and can be decreased by carbonic anhydrase inhibitors. Thiazides have some such inhibitory activity. The metabolic alkalosis seen in the above case may also contribute to the hypophosphataemia.

### HYPOMAGNESAEMIA

Magnesium depletion is associated with an increase in the renal excretion of phosphate, but the mechanism is unknown.

### RENAL PHOSPHATE LEAK

A 7 year old girl with hypophosphataemic rickets.

| | | | |
|---|---|---|---|
| Plasma Ca | 2.45 | mmol/l | (2.05–2.55) |
| PO$_4$ | 0.65 | mmol/l | (1.30–2.80) |
| Alb. | 43 | g/l | (30–50) |
| AP | 155 | U/l | (50–250) |

This condition arises as a result of an increased renal loss of phosphate that is thought to be due to either: (1) a deficient renal phosphate transport mechanism or (2) a deficient intestinal uptake of $Ca^{2+} \rightarrow \uparrow PTH \rightarrow \uparrow$ renal PO$_4$ excretion.

## HYPERPHOSPHATAEMIA

Hyperphosphataemia may arise from the following situations:

*Factitious* — prolonged contact of plasma with red cells ($\gtrsim$8h).

*Increased intake*   (1) Diet, (2) intravenous therapy to treat hypercalcaemia, (3) excess vitamin D.

*Increased release from cells*   (1) Diabetes mellitus, (2) starvation.

*Increased release from bone*   (1) Malignancy, (2) renal failure ($\uparrow$PTH).

*Decreased excretion*   (1) Renal failure, (2) hypoparathyroidism, (3) excess of growth hormone.

### Case examples/pathophysiology

#### HYPERCALCAEMIA (treatment with IV phosphate)

A 48 year old woman with breast cancer. The associated hypercalcaemia was treated with phosphate infusions.

|  | Pre-treatment | Post-treatment | | |
|---|---|---|---|---|
| Plasma Ca | 3.49 | 2.49 | mmol/l | (2.05–2.55) |
| $PO_4$ | 0.79 | 2.68 | mmol/l | (0.60–1.25) |
| Alb. | 32 | 33 | g/l | (30–50) |
| AP | 112 | 155 | U/l | (30–120) |

Although phosphate is successful at temporarily lowering the plasma [Ca], it does so at the risk of metastatic calcification and also renal damage due to precipitation of calcium phosphate (nephrocalcinosis).

### EXCESS VITAMIN D

For case see Hypercalcaemia.

Pathophysiology

Vit D ↑ → (1) ↑ gut absorption of $PO_4$, (2) ↑ gut uptake of $Ca^{2+}$ → ↓ PTH → ↑ renal retention of $PO_4$

### DIABETES MELLITUS

A 27 year old male diabetic admitted to hospital in a semi-coma.

| Plasma Ca | 2.39 | mmol/l | (2.15–2.55) |
|---|---|---|---|
| $PO_4$ | 2.09 | mmol/l | (0.60–1.25) |
| Alb. | 40 | g/l | (37–52) |
| AP | 104 | U/l | (30–120) |
| Glu. | 52.0 | mmol/l | (3.0–5.5) |

Since phosphate is a component of the lean body mass, any increase in the catabolism of this compartment (e.g. prolonged insulin deficiency) will result in release of phosphate into the ECF, followed by increased renal excretion. Therefore, the hyperphosphataemia is associated with a total body depletion of phosphate. The degree of hyperphosphataemia will depend on the prevailing GFR. If this is not severely reduced the plasma level of $PO_4$ may be normal. Insulin therapy causes ECF phosphate to return to the cells, with a subsequent hypophosphataemia.

### METASTATIC BONE DISEASE

A 54 year old lady with carcinoma of the breast.

| Plasma | Ca | 3.01 | mmol/l | (2.05–2.55) |
|---|---|---|---|---|
| | $PO_4$ | 1.95 | mmol/l | (0.60–1.25) |
| | Alb. | 25 | g/l | (30–50) |
| | AP | 255 | U/l | (30–120) |
| | Creat. | 0.18 | mmol/l | (0.06–0.13) |
| Creatinine clearance | | 0.8 | ml/s | (1.5–2.0) |

Pathophysiology

Local erosion of bone by secondary deposits of tumour.

Hyperphosphataemia associated with the hypercalcaemia of malignancy may be due to renal failure. However, in renal failure the plasma $PO_4$ rarely rises until the creatinine clearance is below 0.3 ml/s (approx. plasma creatinine of 0.3–0.4 mmol/l). In the above case the increased release of $PO_4$ from bone and the mild renal insufficiency is probably responsible for the raised plasma $PO_4$.


RENAL FAILURE

A 66 year old man with chronic renal failure.

| Plasma | Na | 133 | mmol/l | (132–144) |
|---|---|---|---|---|
| | K | 5.0 | mmol/l | (3.2–4.8) |
| | Cl | 101 | mmol/l | (98–108) |
| | $HCO_3$ | 15 | mmol/l | (23–33) |
| | Creat. | 0.95 | mmol/l | (0.06–0.12) |
| Creatinine clearance | | 0.1 | ml/s | (1.5–2.0) |
| | Ca | 1.69 | mmol/l | (2.05–2.55) |
| | $PO_4$ | 2.80 | mmol/l | (0.60–1.25) |
| | Alb. | 36 | g/l | (30–50) |

Reduced excretion of phosphate arises as a consequence of the severely reduced glomerular filtration rate.


HYPOPARATHYROIDISM

A 44 year old man who has undergone a parathyroidectomy for primary hyperparathyroidism. These results are from a blood sample taken two weeks postsurgery.

| Plasma | Ca | 1.89 | mmol/l | (2.05–2.55) |
|---|---|---|---|---|
| | $PO_4$ | 1.99 | mmol/l | (0.60–1.25) |
| | Alb. | 41 | g/l | (30–50) |
| | AP | 67 | U/l | (30–120) |

↓ PTH → ↑ renal reabsorption of $PO_4$

The removal of a parathyroid adenoma is often followed by a period of hypoparathyroidism, because the remaining normal parathyroid glands take some time to return from a suppressed state to normal function after their prolonged exposure to hypercalcaemia.

GROWTH HORMONE EXCESS

High levels of plasma phosphate are usually associated with acromegaly, gigantism and growth in normal children. It appears that growth hormone increases phosphate resorption by the kidney.

## MAGNESIUM

Magnesium has widespread distribution, being particularly concentrated in bone, muscle and heart. It is a cofactor for over 300 enzymes, particularly for those concerned with ATP and energy production. Magnesium is also required for normal DNA function, cell permeability regulation and neuromuscular excitability. Magnesium has a close association with calcium metabolism and is necessary for both the release and function of PTH and the formation of $25\text{-}OHD_3$. A deficiency of magnesium may result in an increased neuromuscular and cardiac excitability, whilst an excess of the cation can lead to depressed muscle performance or paralysis.

## MAGNESIUM METABOLISM

**Body distribution**

| | |
|---|---|
| Total | 1000 mmol |
| Skeleton | 50% |
| Soft tissues | >49% |
| Extracellular fluid | <1% |

**Gut uptake**

About 30% of the dietary intake of 12–15 mmol/day is absorbed in the small intestine. Little is known about the absorptive mechanism of magnesium, but it appears that gut uptake is energy dependent, and that the amount absorbed by the small bowel is inversely related to the lumen concentration of magnesium up to a saturable limit. Uptake is decreased by high levels of phosphate (insoluble magnesium phosphate complexes) or fatty acids (soap formation) in the diet. Steatorrhoea is therefore associated with an increased faecal loss of magnesium and calcium.

### Renal excretion

The kidney filters some 70% of the total plasma magnesium, representing the free and complexed fractions (see below). Approximately 95–97% of the filtered magnesium is reabsorbed, mainly by the loop of Henle and the proximal tubule.

TUBULAR HANDLING OF MAGNESIUM

| Portion of tube | % reabsorbed |
| --- | --- |
| Proximal convoluted tubule | 20–30 |
| Thick ascending loop and distal tubule | 50–60 (PTH increases absorption) |
| Collecting duct | 1–5 |

Magnesium reabsorption is decreased by: (1) increased ECF volume, (2) osmotic diuresis, (3) diuretic therapy, (4) hypercalcaemia (increased renal excretion of calcium), (5) alcohol.

The amount of magnesium excreted in the urine (3–10 mmol/day) is closely related to the dietary intake, so that during deprivation the urinary output can be very low (<0.5 mmol/d).

### Hormones and magnesium metabolism

#### PTH

Experimental studies suggest that magnesium inhibits the release of PTH, but not so effectively as does calcium. PTH has been shown to increase the renal uptake of magnesium, whilst evidence for PTH directly or indirectly inhibiting magnesium uptake by the intestine is complicated by the fact that the effect is dependent on the level of calcium in the diet. Calcium directly opposes both the intestinal and renal transport of magnesium. PTH might therefore be involved as follows:

$$PTH \rightarrow \uparrow 1,25\text{-}(OH)_2D_3 \rightarrow \uparrow Ca^{2+} \text{ uptake} \rightarrow \downarrow Mg^{2+} \text{ uptake}$$

A lack of magnesium increases skeletal resistance to the action of PTH, so that hypomagnesaemia can lead, experimentally at least, to a secondary hypocalcaemia.

### VITAMIN D, CALCITONIN

No good evidence is available that suggests that vitamin D metabolites or calcitonin play a direct role in the metabolism of magnesium.

### MINERALOCORTICOIDS

Acute administration of aldosterone does not appear to alter the renal handling of magnesium, whilst long-term exposure to excess mineralocorticoid has been shown to result in an increased excretion. The reduction in the reabsorption of magnesium is not due to a direct effect of the excess aldosterone, but to the consequent increase in the extracellular volume having an inhibitory effect on tubular uptake.

### THYROID HORMONES

Increased urinary losses of magnesium have been noted in hyperthyroid patients, whilst hypermagnesaemia and low urinary levels are to be found in hypothyroid cases.

## LABORATORY INVESTIGATION OF MAGNESIUM METABOLISM

### PLASMA MAGNESIUM DISTRIBUTION

| | |
|---|---|
| Free ($Mg^{2+}$) | 55% |
| Protein bound | 30% |
| Complexes | 15% |

The $Mg^{2+}$ species is the physiologically active fraction that affects such properties as muscle excitability. The remaining plasma magnesium is either bound to albumin, or is complexed to anions such as phosphate and citrate. Laboratories measure total plasma magnesium, usually by an atomic absorption method.

Primary defects of magnesium metabolism are extremely rare, occurring possibly as a familial disorder. Alterations in magnesium metabolism are nearly always secondary to a variety of disease states, and so it is important to investigate plasma magnesium levels in patients with conditions known to disturb magnesium homeostasis, rather than to screen patients for a rare dysfunction. Plasma and bone magnesium are in equilibrium and, therefore, the total body stores of magnesium are not necessarily reflected by the plasma concentration. Magnesium tends to move in and out of bone with calcium and in and out of cells with potassium.

Hypermagnesaemia is due to either renal or iatrogenic causes and, if caused by Mg containing antacids, may be associated with a metabolic alkalosis. Hypomagnesaemia is often found to be associated with: (1) hypokalaemia, (2) hypocalcaemia, (3) hypophosphataemia.

Initial laboratory investigations of magnesium metabolism should usually include measurement of plasma Mg, Ca, K, $PO_4$, albumin, $HCO_3$.

## HYPERMAGNESAEMIA

Hypermagnesaemia may be caused by:

*Increased intake* (1) Oral, (2) intravenous—Mg supplements added to IV fluids used for hyperalimentation.

*Decreased excretion* Acute or chronic renal failure.

**Case examples/pathophysiology**

EXCESS ALKALI INGESTION

Results from a 72 year old man admitted for investigation of 5 months intermittent abdominal pain, which he had been relieving with the heavy use of proprietary antacids.

| Plasma | Na | 150 | mmol/l | (132–144) |
|---|---|---|---|---|
| | K | 3.3 | mmol/l | (3.2–4.8) |
| | Cl | 96 | mmol/l | (98–108) |
| | $HCO_3$ | 46 | mmol/l | (23–33) |
| | Urea | 8.0 | mmol/l | (3.0–8.0) |
| | Mg | 2.08 | mmol/l | (0.75–1.00) |

Both magnesium carbonate and magnesium hydroxide are to be found in many antacid preparations.

RENAL FAILURE

A 69 year old man with advanced chronic renal failure.

| Plasma | Na | 131 | mmol/l | (132–144) |
|---|---|---|---|---|
| | K | 6.3 | mmol/l | (3.2–4.8) |
| | Cl | 94 | mmol/l | (98–108) |
| | $HCO_3$ | 15 | mmol/l | (23–33) |
| | Creat. | 0.71 | mmol/l | (0.06–0.12) |
| | Mg | 1.44 | mmol/l | (0.75–1.00) |

In the oliguric phase of acute renal failure plasma magnesium is raised, but falls to normal during the diuretic phase. Plasma levels are normal during the early stages of chronic renal failure, but rise as the renal disease progresses to the stage when the GFR has fallen to around 0.25–0.30 ml/s (equivalent to a plasma creatinine of 0.3–0.4 mmol/l).

## HYPOMAGNESAEMIA

Hypomagnesaemia may be due to the following.

*Decreased intake* (1) Starvation, (2) malabsorption syndrome, (3) prolonged gastric suction, (4) hyperalimentation.

*Extrarenal loss* (1) Prolonged diarrhoea, (2) laxative abuse, (3) loss from fistula, (4) excessive lactation.

*Increased renal loss* (1) Alcoholism, (2) diuretic therapy, (3) osmotic diuresis, (4) hypercalcaemia, (5) primary aldosteronism.

**Case examples/pathophysiology**

ALCOHOLISM

A 51 year old man admitted with acute pancreatitis.

| Plasma | Mg | 0.40 | mmol/l | (0.75–1.00) |
|---|---|---|---|---|
| | Ca | 1.92 | mmol/l | (2.15–2.55) |
| | Alb. | 39 | g/l | (37–52) |
| | Amy. | 8000 | U/l | (<300) |

Pathophysiology

The hypomagnesaemia is contributed to by the following:
(1) Alcohol has a magnesuric effect by increasing the GFR. The percentage of tubular reabsorption of magnesium remains unchanged and therefore the raised filtered load results in an increased urinary excretion.
(2) Reduced intake caused by poor diet initially and then by vomiting during the pancreatitis attack.

(3) Pancreatitis. The acute release of free fatty acids caused by increased lipase activity leads to the removal of magnesium in soap form (cf. hypocalcaemia in pancreatitis).

Total body stores of magnesium can be markedly reduced in patients with alcoholic disease, so that supplements may have to be given during treatment.

DIURETICS

Long-term treatment with a variety of diuretic agents will often lead to reduced stores of magnesium. It is presumed that diuretics such as ethacrynic acid and frusemide act by inhibiting magnesium uptake in the loop of Henle. Thiazides have been variously reported as having little or no magnesuric effect.

OSMOTIC DIURESIS

A 36 year old lady admitted with diabetic ketoacidosis.

Plasma Mg    0.55 mmol/l    (0.75–1.00)
       Glu.   60.0   mmol/l    (3.0–5.5)

Increased renal excretion of magnesium is associated with osmotic diuresis (glycosuria). Intestinal absorption of magnesium is thought to be reduced in diabetes, but it is not clear whether this is caused by a lack of insulin.

HYPERCALCAEMIA

A 47 year old woman with carcinoma of the breast.

Plasma Mg    0.42 mmol/l    (0.75–1.00)
       Ca     2.72 mmol/l    (2.15–2.55)
Urine Ca     15.0 mmol/day  (2.5–7.5)

Hypercalcaemia increases the renal excretion of magnesium, possibly by the increased tubular load of calcium interfering with magnesium reabsorption.

PRIMARY ALDOSTERONISM

The increased renal excretion of magnesium that occurs in this condition is due to the expanded ECF volume (salt and water retention) inhibiting magnesium reabsorption.

NEONATAL HYPOMAGNESAEMIA

This condition is rare and is usually associated with hypocalcaemia in premature babies. If the symptoms of hypocalcaemia do not respond to calcium infusions the plasma magnesium level should be checked, since hypomagnesaemia produces similar symptoms.

# 10 Glucose

**Carbohydrate digestion and absorption**

Complex carbohydrates (starches, etc.) are broken down in the intestinal lumen to mono- and disaccharides by salivary and pancreatic amylases. The disaccharides are cleaved to form monosaccharides (hexoses) by enzymes (maltase, lactase, sucrase) situated in the brush border of the intestinal mucosae. The hexoses (glucose, galactose, fructose) are then transported by an energy-dependent process into the portal venous system, and then to the liver. In the liver, galactose and fructose are converted to glucose-1-phosphate and fructose-1-phosphate respectively, utilized if energy is immediately required, otherwise both are converted via glucose-1-phosphate to glycogen for hepatic storage.

**Renal handling of glucose**

The glucose which appears in the glomerular filtrate is actively reabsorbed in the proximal tubule. At plasma glucose levels above approximately 10 mmol/l this transport process approaches saturation, and then glucose appears in the urine. This plasma level is known as the 'renal threshold' for glucose. Occasionally, this renal tubule transport mechanism is defective and glucose appears in the urine at a much lower plasma level (renal glycosuria).

**Glucose metabolism**

Glucose, oxidized through the glycolytic pathway and the TCA cycle, is the most important source of energy for most tissues. Many tissues can also utilize fatty acids (β-oxidation), whilst nervous tissue is unique in that it can only metabolize glucose and ketone bodies. Maintenance of the plasma glucose level is thus essential for nervous tissue function.

In the postprandial state the excess glucose, remaining after the cell requirements for glycolysis are met, is converted and stored by the liver as glycogen and triglyceride. During fasting the plasma glucose level is maintained by glycogen breakdown (glycogenolysis) and gluconeogenesis (glucose derived from glycerol, amino acids and lactate) (Fig. 10.1).

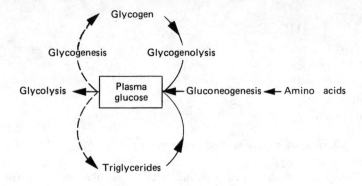

**Fig. 10.1** Metabolic factors affecting plasma glucose levels; — — pathways stimulated by insulin; — pathways inhibited by insulin.

GLYCOLYSIS

This is the oxidation of glucose, with the formation of ATP, through glucose-6-phosphate to pyruvate. The pyruvate is converted to acetyl-CoA and enters the TCA cycle. In the absence of molecular oxygen the TCA cycle cannot proceed and pyruvate accumulates, resulting in a rise in lactate (lactic acidosis). This may occur during generalized anoxia, or in severe muscular exercise where the rate of glucose metabolism outstrips the oxygen supply.

Insulin is necessary for the uptake of glucose by muscle and adipose tissue, but not for brain and liver. It is also responsible for the induction of several of the enzymes responsible for the glycolysis pathway.

GLYCOGENESIS

Storage of glucose as glycogen occurs in the liver and muscle tissue. It is stimulated by insulin and decreases in the absence of this hormone.

GLYCOGENOLYSIS

Degradation of glycogen to glucose-6-phosphate is stimulated by glucagon and adrenalin. The liver contains the enzyme glucose-6-phosphatase, which allows glucose to be formed and added to the plasma. Muscle does not contain this enzyme and therefore cannot supply glucose to the plasma.

## LIPID SYNTHESIS

Triglycerides can be formed in the liver from glucose.

glucose → (1) glycerol-3-phosphate, (2) acetyl-CoA → fatty acids
glycerol-3-phosphate + fatty acids → triglyceride

These reactions are insulin dependent. The triglycerides are stored in adipose tissue.

## GLUCONEOGENESIS

This is the formation of glucose from amino acids, fats and lactate.

*Lipids* Triglycerides → fatty acids + glycerol. Glycerol → glucose.

*Amino acids* Protein, mainly from muscle, can be degraded to L-amino acids, which may then be converted in the liver to glucose. This process is decreased by insulin and increased by glucocorticoids (cortisol).

*Lactate* Lactate produced in the muscle and elsewhere can be converted by the liver back to glucose via pyruvate and phosphoenol pyruvate.

### Hormones and plasma glucose homeostasis

## INSULIN

Insulin is a protein hormone of 51 amino acid residues, arranged as two polypeptide chains. The β-cells of the islets of Langerhans in the pancreas produce proinsulin (84 amino acids). Prior to secretion proinsulin is cleaved to form insulin and an inactive peptide of 33 amino acids, termed *C-peptide*. Both insulin and C-peptide are then secreted in equal amounts into the portal bloodstream. These two molecules have different plasma half-lives (insulin $t_{1/2} < 10$ min, C-peptide $t_{1/2} \sim 15$–20 min).

Control of secretion

The most important stimulus to insulin secretion is hyperglycaemia, but other substances such as small intestinal peptides (motilin), and certain amino acids (leucine, arginine), will also stimulate secretion. Hypoglycaemia suppresses secretion.

Actions

The major action of insulin is the lowering of the plasma glucose level by:
(1) Facilitation of cellular uptake of glucose in muscle and adipose tissue.
(2) Stimulation of glycolysis.
(3) Stimulation of glycogenesis.
(4) Inhibition of gluconeogenesis.
Other actions include:
(1) Increased lipogenesis.
(2) Stimulation of protein synthesis from amino acids.
(3) Inhibition of ketogenesis.
(4) Increased cell uptake of potassium and phosphate.

## GLUCAGON

A polypeptide hormone (29 amino acid residues) synthesized by the α2-cells of the pancreas; significant amounts also derive from other tissues, particularly the intestine. Glucagon opposes the action of insulin by stimulating glycogenolysis, gluconeogenesis and lipolysis, thereby raising blood glucose and free fatty acid levels. Glucagon release is suppressed by high concentrations of plasma glucose but the mechanism of its release is thought to be more complex, involving both low glucose levels and other factors such as catecholamines. It is now thought that both the fasting hyperglycaemia and ketosis seen in uncontrolled diabetes may be caused by the release of inappropriate amounts of glucagon in the presence of hypoinsulinaemia.

## ADRENALIN

In addition to its action on the smooth muscle and cardiovascular system (see p. 496), adrenalin also promotes hyperglycaemia by the following mechanisms: (1) promotion of glycogenolysis, (2) de-

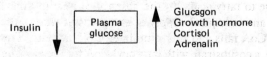

Fig. 10.2 The effect of hormones on the plasma glucose concentration.

pression of insulin secretion by the β-cells of the pancreas. Adrenalin also increases lipolysis, thereby raising plasma fatty acid concentrations.

GLUCOCORTICOIDS

Glucocorticoids (cortisol) tend to promote hyperglycaemia by: (1) increased gluconeogenesis from amino acids, (2) depression of both glycolysis and cellular glucose uptake. Cortisol secretion is increased in the stress situations that are produced by hypoglycaemia.

GROWTH HORMONE (GH)

Growth hormone secretion is stimulated by hypoglycaemia and suppressed by hyperglycaemia. Growth hormone has an anti-insulin effect, in that it suppresses the insulin-mediated cell uptake of glucose. The overall effect of GH is to raise plasma glucose and free fatty acids. Excessive secretion of growth hormone may result in hyperglycaemia (e.g. acromegaly), but this effect is probably offset in most cases by increased insulin secretion.

**Ketone metabolism**

In the absence of insulin, fat deposits are mobilized to provide the body's energy requirements from the β-oxidation of fatty acids. This oxidation process occurs within the liver mitochondria, where fatty acyl CoA is converted to energy via acetyl CoA and the TCA cycle. If the supply of fatty acyl CoA is massive, as in severe diabetes mellitus (↓ insulin → ↑ lipolysis → ↑ fatty acyl CoA), the capacity of the TCA cycle is exceeded and the excess acetyl CoA is converted to ketone bodies, resulting in ketosis.

The control of ketogenesis is poorly understood, but it appears that the process is limited by the rate of transfer of fatty acyl CoA across the mitochondrial membrane. This process is mediated by an enzyme complex, carnitine acyl transferase (CAT), the activity of which is regulated by the tissue levels of malonyl CoA and carnitine. Malonyl CoA is an early intermediate in the conversion of glucose to fatty acids. In the absence of insulin (diabetes mellitus), or an inadequate supply of glucose (starvation), the level of malonyl CoA falls and this stimulates CAT activity. The level of carnitine, a cosubstrate with fatty acyl CoA for the enzyme CAT, is increased in the liver cytosol in diabetes mellitus. This appears to be due to the increased glucagon levels which are a feature of diabetes mellitus (Fig. 10.3). Another mechanism for the control

Chapter 10

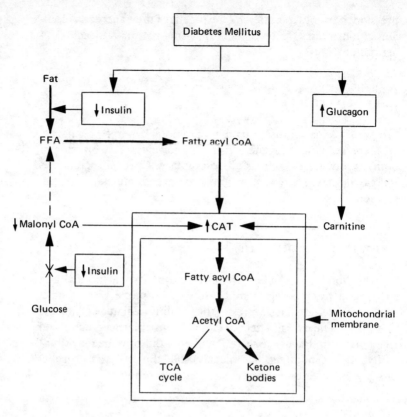

**Fig. 10.3** Ketogenesis in diabetes mellitus (see text for details). CAT, carnitine acyltransferase.

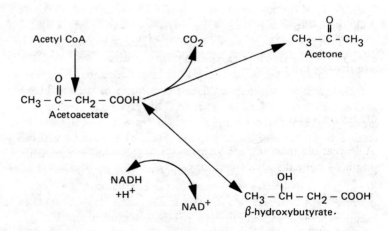

**Fig. 10.4** Ketone body formation.

210

of ketogenesis appears to be the level of liver glycogen. If liver glycogen is high, ketogenesis does not proceed.

The ketone bodies formed in the liver are released into the circulation and cleared (oxidized) by various peripheral tissues (brain, skeletal muscle). However, in diabetes mellitus the rate of production exceeds the rate of clearance and plasma ketone levels rise. There is also evidence that in diabetes mellitus the rate of peripheral oxidation of ketones is inhibited.

Ketone bodies exist in the plasma as acetoacetate (AcAc) (the first ketone body produced from acetyl CoA), β-hydroxybutyrate (β-OHB), produced enzymatically from acetoacetate, and acetone produced from acetoacetate by spontaneous non-enzymatic removal of $CO_2$ (Fig. 10.4).

The conversion of acetoacetate to β-hydroxybutyrate is reversible and is dependent on the state of tissue oxygenation. In tissue hypoxia the levels of $NADH_2$ are increased and the reaction is driven towards β-hydroxybutyrate. In diabetic ketoacidosis the ratio of β-hydroxybutyrate to acetoacetate varies from 3 to 20:1. The ratio of acetone to acetoacetate is approximately 2-5:1. Acetoacetate and β-hydroxybutyrate are cleared fairly rapidly by the kidneys when present in high concentrations. Acetone is slowly excreted via the kidney and lungs.

Ketones may be associated with: (1) diabetes mellitus, (2) starvation, (3) intoxications (a) ethanol, (b) methanol, (c) salicylate.

## LABORATORY INVESTIGATION

*Note* All glucose values quoted were determined by a glucose oxidase method.

#### URINARY GLUCOSE

If a sufficiently sensitive test is used glucose can be detected in the urine of almost all normal people. The commonly used qualitative tests are relatively insensitive, so that positive tests found when using these methods indicate significant glycosuria. Such glycosuria occurs in the fasting state in the following two conditions: (1) hyperglycaemia ($\gtrsim 10$ mmol/l), (2) renal glycosuria—rare cause of glycosuria.

Commercial test sticks based on the glucose oxidase reaction may give misleading results in the following situations:
(1) False positives—oxidative cleansing agents (hypochlorite) used on bed pans.
(2) False negatives—elevated levels of vitamin C in the urine.

Qualitative tests based on the reducing power of glucose (e.g. 'Clinitest', Ames Co.) will give false positive results when other reducing substances are present in the urine, e.g. galactose (galactosaemia), fructose (fructose intolerance), homogentisic acid (alkaptonuria).

FASTING PLASMA GLUCOSE

The glucose concentration in plasma and erythrocyte water is the same, but the amount of water in red cells ($\sim 72\%$) differs from that of plasma ($\sim 94\%$), and therefore plasma glucose levels are higher than those for whole blood. It follows that the blood glucose level is also dependent on the haemotocrit and thus it is preferable to use plasma for glucose estimations. If whole blood is left to stand at room temperature, the red cells will continue to metabolize the glucose, causing its concentration to decrease by 20-50%/hour; therefore glucose levels should be estimated only on plasma taken from blood samples that have had an appropriate enzyme inhibitor added (fluoride, iodoacetate). Estimations of glucose made on serum taken from blood that has been allowed time to naturally clot are not recommended.

The normal fasting plasma glucose level is affected by: (1) length of fast, (2) sex.

Lower limit of normal

For the purpose of diagnosing fasting hypoglycaemia the following figures are a useful guide:

|  | Overnight fast (mmol/l) | 72 hour fast (mmol/l) |
| --- | --- | --- |
| Males | <3.5 | <3.0 |
| Females | <3.5 | <2.5 |

After a glucose load (e.g. oral glucose tolerance test, OGTT) the plasma glucose often drops to a value lower than the initial fasting level. This fall occurs within 3-4 hours of the glucose load and is due to an 'overswing' effect of the insulin secreted in response to the load. Under these circumstances the diagnosis of hypoglycaemia should not be entertained unless the plasma glucose level falls to below 3.0 mmol/l. However, 30-40% of normal subjects will have a plasma glucose <3.0 mmol/l at some period during the 5 hours after an OGTT and remain asymptomatic.

Upper limit of normal

After an overnight fast the plasma glucose level in normal subjects is rarely found to be above 5.5-6.0 mmol/l. A fasting level greater than 7.8 mmol/l is considered abnormal, whether or not it is associated with the clinical signs and symptoms of diabetes mellitus. Such a glucose level is diagnostic of this disease and does not require confirmation by a glucose tolerance test.

**The diagnosis of abnormal glucose tolerance**

An international workgroup sponsored by the National Diabetes Data Group (NDDG) of the National Institutes of Health, Bethesda, USA, in 1979 published recommendations for the classification and diagnosis of diabetes mellitus. These recommendations have been substantially accepted by the WHO Expert Committee on Diabetes, various national diabetes groups, and the International Federation of Clinical Chemists (IFCC) Scientific Committee. In the NDDG report the biochemical criteria are stated in traditional units, i.e. mg/dl, but in the following section the plasma glucose values have been converted to SI units.

Diabetes mellitus is diagnosed if:
(1) A subject has classical symptoms of diabetes (polyuria, polydypsia, ketonuria, rapid weight loss) and unequivocal hyperglycaemia.
(2) A subject without obvious symptoms has a fasting venous plasma glucose ⩾7.8 mmol/l *on more than one occasion*.

*Note* The WHO recommendations suggest that a fasting venous plasma glucose of <6.0 mmol/l or a random value of <8.0 mmol/l excludes the diagnosis.

An oral glucose tolerance test (OGTT) is unnecessary for diagnostic purposes in either of the above situations, and may even be dangerous for the patient. For subjects, with or without diabetic symptoms, who cannot be unequivocally classified by the above criteria, an OGTT is helpful for diagnosis.

ORAL GLUCOSE TOLERANCE TEST (OGTT)

Patient preparation

(1) 3 days of normal diet (carbohydrate >150 g/d).
(2) Overnight fast (10–16 hours)—water allowed.
(3) Seated quietly for 30 min prior to, and during, test.
(4) No smoking or exercise during the test.
(5) Perform test in the morning.
(6) Known not to be taking drugs that interfere with glucose estimation.

*Glucose load* Non-pregnant adult—75 g. Child—1.75 g/kg to a maximum of 75 g. To be taken in water (25g/100ml) in less than 5 min.

*Samples* Venous or capillary blood at 0, 0.5, 1, 1.5, 2 hours for either plasma or whole blood estimation of glucose.

It is no longer recommended that urine samples for qualitative detection of glucose be taken as part of a diagnostic OGTT. This procedure may be more useful when determining the renal threshold of glucose in newly diagnosed diabetics, for the purpose of self-monitoring of glycosuria as a guide to plasma concentration.

Interpretation

Values given below are for venous plasma glucose estimated by a glucose oxidase method. The following criteria apply to non-pregnant adults.

*Normal*
Fasting (basal) plasma glucose    <6.4 mmol/l
2 hour plasma sample              <7.8 mmol/l
0.5, 1, 1.5 hour samples          <11.1 mmol/l

*Diabetic*
2 hour sample + one intermediate sample ⩾11.1 mmol/l

*Impaired glucose tolerance*  Three criteria must be satisfied:
Fasting plasma glucose             <7.8 mmol/l
Intermediate OGTT sample           ⩾11.1 mmol/l
2 hour OGTT sample between         7.8 and 11.1 mmol/l

For the criteria required for the performance and interpretation of an OGTT in other groups, such as pregnant subjects and children, the references given on p. 521 should be consulted.

Many factors, apart from sex, weight, age or family history of diabetes, can affect the fasting plasma glucose level of an individual or their response to an OGTT, e.g. (1) stress, (2) degree of recent physical activity, (3) amount of recent carbohydrate intake, (4) degree of fasting, (5) injury, (6) time of day, (7) drugs, (8) endocrine disease.

Studies have shown that the response of individual subjects to repeated OGTTs under standard conditions over a year varied markedly. It is therefore obligatory to demonstrate a diabetic response to an OGTT on more than one occasion before that individual is diagnosed, with all its social and economic implications, as having diabetes mellitus.

INTRAVENOUS GLUCOSE TOLERANCE TEST

It has been claimed that this test has an advantage over the oral test because it avoids any variability that may be associated with gastro-

intestinal absorption. However, this factor is probably outweighed by the following considerations:
(1) The oral procedure uses the physiological route.
(2) There is more information available about the oral procedure and a wider consensus on what constitutes normal and abnormal responses.
(3) The oral test is more acceptable to patients.

KETONE BODIES

Estimation of plasma or urinary ketones is an important procedure in the diagnostic work up of diabetic coma. They may be estimated in the plasma or urine by qualitative methods (nitroprusside reaction), or quantitatively by enzymatic methods. The latter methods are technically difficult and time consuming and do not offer any great advantage over the qualitative ones with respect to diagnosis and therapy. However it should be noted that the nitroprusside reaction (e.g. Ketostix, Ames Co.) estimates acetoacetate and acetone but not β-hydroxybutyrate. In diabetic ketoacidosis the ratio of hydroxybutyrate to acetoacetate varies from 3:1 to 20:1. The qualitative tests may give a false impression of the degree of ketosis in the following situations:
(1) In tissue hypoxia (shock, pneumonia) the ratio of β-hydroxybutyrate to acetoacetate is increased, and so the nitroprusside test will not indicate the degree of ketosis.
(2) When insulin and fluid therapy is instituted in a patient with tissue hypoxia and a high β-hydroxybutyrate : acetoacetate ratio, the β-hydroxybutyrate is converted to acetoacetate and the nitroprusside reaction will become more positive, falsely suggesting a worsening of the ketosis.
(3) Although acetone has only 1/20 of the activity of the acetoacetate with respect to nitroprusside, acetone is slowly excreted and may remain in sufficient quantities to give a positive nitroprusside reaction after the level of acetoacetate has been reduced.

LACTIC ACID

Lactic acidosis associated with diabetes mellitus most often occurs following biguanide (Phenformin) therapy, and has a high mortality. Although plasma lactate levels are necessary for the definitive diagnosis of this disorder, its estimation is usually unnecessary as the diagnosis can be made on the strength of the plasma ketone and electrolyte results (see below). The plasma level of lactate is normally less than 2 mmol/l and lactic acidosis occurs when the level reaches 5–7 mmol/l. It is important to note that in ketoacidosis and hyperglycaemic coma increased levels of plasma lactate occur in 30–50% of cases.

## PLASMA ELECTROLYTES

In hyperglycaemic diabetic coma it is essential for correct diagnosis and treatment that the following plasma analytes be measured at regular intervals: (1) sodium and potassium, (2) bicarbonate, (3) urea and creatinine (Acetoacetate gives positive interference in the Jaffé method for creatinine. Therefore in ketoacidosis the plasma creatinine concentration will be falsely high.), (4) osmolality.

Diagnosis of hyperglycaemic diabetic coma

|  | Ketoacidosis | Hyperosmolar coma | Lactic acidosis |
| --- | --- | --- | --- |
| Plasma [$HCO_3^-$] | ↓ | N | ↓ |
| Plasma anion gap | ↑ | N | ↑ |
| Ketosis | +ve | −ve | −ve |

*Note* Mixed varieties are common, and the above relationships do not apply in all cases.

*Plasma creatinine* Acetoacetate interferes with the Jaffé reaction for plasma creatinine and gives falsely high results.

## PLASMA INSULIN

Plasma insulin levels in diabetics are variable, ranging from the very low in the insulin dependent diabetic, to high levels in the obese non-insulin dependent type. There is no evidence that a knowledge of plasma insulin levels is of any additional value in the evaluation of diabetes mellitus. However, insulin estimations are useful for the diagnosis of insulinoma. In suspected cases insulin and glucose levels should be measured after the patient has fasted (overnight to 72 h). High insulin levels in the face of hypoglycaemia are not diagnostic (a similar situation is found in overdose with oral hypoglycaemics), but when considered with the clinical picture, it is the most useful diagnostic index in this disorder.

## PLASMA C-PEPTIDE

Assays are now available for this peptide, but its usefulness as a diagnostic test is yet to be fully assessed. It has been found to be helpful in the evaluation of hypoglycaemia associated with hyperinsulinaemia, where surreptitious administration of insulin is suspected, i.e.

|  | Plasma insulin | Plasma C-peptide |
|---|---|---|
| Insulinoma | ↑ | ↑ |
| Insulin administration | ↑ | ↓ |

CSF GLUCOSE

High CSF glucose levels are found in hyperglycaemia, and low levels are associated with hypoglycaemia. Low levels are also found when the CSF contains large numbers of cells and bacteria. It is doubtful whether CSF glucose levels provide any useful information in the above three situations.

HAEMOGLOBIN $A_{IC}$ (GLYCOSYLATED HAEMOGLOBIN)

Haemoglobin $A_{IC}$ (Hb$A_{IC}$) is formed by the non-enzymatic attachment of glucose to the N-terminal amine group of the β-chain of haemoglobin A. It is formed slowly and continuously throughout the average 120 day life span of the red cell and provides an index of the 'average' plasma glucose concentration over the preceding 2–3 months. Normally it represents less than 6% of the total haemoglobin, but in diabetes its concentration may be increased two to three fold.

It has been used in: (1) detection of diabetes, (2) assessment of diabetic control.

Diagnosis of diabetes

A raised plasma level of Hb$A_{IC}$ is highly specific for diabetes, but is a less sensitive parameter than the glucose tolerance test.

Assessment of control

The major clinical application of Hb$A_{IC}$ at the present time is in the assessment of diabetic control, especially in the insulin-labile diabetic, where it gives an index of whether the mean plasma glucose concentration over the preceding 2–3 months has been abnormally elevated for significant periods of time or not.

HYPERGLYCAEMIA

Fasting hyperglycaemia may occur in the following conditions:

*Pancreatic disorders* (1) Pancreatectomy, (2) haemochromatosis, (3) chronic pancreatitis, (4) carcinoma of pancreas.

*Endocrine disorders* (1) Cushing's syndrome (↑ cortisol), (2) phaeochromocytoma (↑ adrenalin), (3) acromegaly (↑ growth hormone), (4) thyrotoxicosis.

*Stress* (↑ cortisol, ↑ adrenalin) e.g. infection, cerebrovascular accident, myocardial infarction.

*Iatrogenic* e.g. IV infusions.

*Drug therapy/overdose* (a) Salicylates, (2) oral contraceptives, (3) steroids, (4) thiazides.

## Diabetes mellitus (DM)

### DEFINITION

A state of chronic hyperglycaemia which may result from genetic or environmental factors, often acting jointly (WHO definition). Hyperglycaemia may be due to a lack of insulin or an excess of one or more factors that oppose the action of insulin.

### CLASSIFICATION

*Primary* (1) Insulin dependent (IDDM)—severe, liable to develop ketoacidosis. Can occur at any age. (2) Insulin independent (NIDDM)—not prone to ketoacidosis. This class may be further subdivided according to whether obesity is present, and the type of treatment required.

*Secondary* (1) Pancreatic disorders, (2) endocrine disorders.

*Gestational* Glucose intolerance during pregnancy.

Potential abnormality of glucose tolerance

Subject who has a substantially higher risk than the general population to develop diabetes mellitus, i.e. (1) identical twin of diabetic, (2) both parents diabetic or one parent diabetic and other has a family history of diabetes, (3) has given birth to a baby of greater than 4.5 kg birth weight.

Previous abnormality of glucose tolerance

Subject with normal glucose tolerance which becomes abnormal during stress (e.g. pregnancy, infection) or during steroid therapy.

## Diabetes mellitus and coma

CAUSES

*Hypoglycaemia*   Insulin, oral hypoglycaemic overdose.

*Hyperglycaemia*   Ketoacidosis, hyperosmolar, lactic acidosis.

*Other causes*   Cerebrovascular accident, drug overdose, etc.

## Case examples/pathophysiology

IMPAIRED GTT

Obese female, aged 40 years, glycosuria found during a routine medical check.

| Time (min) | Plasma glucose (mmol/l) | Urine glucose |
|---|---|---|
| 0   | 5.4  | −ve |
| 30  | 7.7  |     |
| 60  | 10.8 | +ve |
| 90  | 12.7 |     |
| 120 | 9.9  | +ve |

This result is, by definition (2 h sample <11.1 mmol/l), not diabetic and should therefore be classified as impaired glucose tolerance. This type of GTT may be found in any of the conditions associated with glucose intolerance (above list), in obesity and in the early stages of diabetes mellitus.

DIABETIC GTT

Male, aged 45 years, complaining of weight loss and polyuria.

| Time (min) | Plasma glucose (mmol/l) | Urine glucose |
|---|---|---|
| 0 | 6.2 | −ve |
| 30 | 7.0 | |
| 60 | 10.7 | +ve |
| 90 | 12.7 | |
| 120 | 14.0 | +ve |

### FLAT GTT RESPONSE

Male aged 15 years with a family history of diabetes mellitus.

| Time (min) | Plasma glucose (mmol/l) | Urine glucose |
|---|---|---|
| 0 | 5.0 | −ve |
| 30 | 7.1 | |
| 60 | 6.0 | +ve |
| 90 | 5.0 | |
| 120 | 4.6 | +ve |

Seen in normal subjects, especially young people (efficient absorption and rapid clearance), and very occasionally in the malabsorption syndrome.

### LAG STORAGE GTT RESPONSE

Male, aged 50 years, partial gastrectomy for a duodenal ulcer.

| Time (min) | Plasma glucose (mmol/l) | Urine glucose |
|---|---|---|
| 0 | 4.8 | −ve |
| 30 | 11.6 | |
| 60 | 7.2 | +ve |
| 90 | 5.5 | |
| 120 | 3.8 | −ve |

This type of response is due to rapid absorption of glucose. Insulin release may be delayed, and it often 'overshoots', resulting in a temporary mild to moderate hypoglycaemia. It may be seen in: (1) normal subjects (reactive hypoglycaemia), (2) after gastrectomy, (3) thyrotoxicosis, (4) severe liver disease. It may also be an early manifestation of diabetes mellitus.

### DIABETIC KETOACIDOSIS

A 20 year old female, a known insulin-dependent diabetic, brought to Accident and Emergency in semicoma.

|  |  | Pre-treatment | 8 hours post-treatment |  |  |
|---|---|---|---|---|---|
| Plasma | Na | 127 | 134 | mmol/l | (132–144) |
|  | K | 7.0 | 3.1 | mmol/l | (3.2–4.8) |
|  | Cl | 93 | 102 | mmol/l | (98–108) |
|  | HCO$_3$ | 7 | 25 | mmol/l | (23–33) |
|  | Urea | 10 | 6.0 | mmol/l | (3.0–8.0) |
|  | Creat. | 0.18 | 0.07 | mmol/l | (0.06–0.12) |
|  | PO$_4$ | 1.85 | 0.30 | mmol/l | (0.6–1.3) |
|  | Glu. | 41 | 7.2 | mmol/l | (3.0–5.5) |
|  | Anion gap | 34 | 10 | mEq/l | (7–17) |
|  | AcAc | 2.6 | 0.28 | mmol/l | (<0.2) |
|  | β-OHB | 12.0 | 1.2 | mmol/l | (<0.25) |

Pathophysiology

See Fig. 10.5.

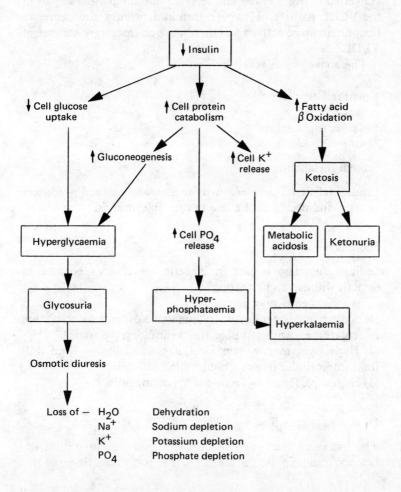

**Fig. 10.5** Pathophysiology of diabetic ketoacidosis.

Carbohydrate metabolism

↓ insulin → ↓ glycolysis
↑ glycogenolysis
↓ cellular uptake of glucose by muscle and adipose tissue
↑ gluconeogenesis → hyperglycaemia → glycosuria
→ osmotic diuresis

Fat metabolism

At the adipose tissue level, insulin deficiency causes increased fat mobilization (lipolysis) and decreased esterification of fatty acids, resulting in increased plasma levels of fatty acids. In the liver most of these fatty acids enter the β-oxidation cycle (stimulated by decreased insulin and perhaps increased glucagon), resulting in excessive production of ketone bodies. Some of the fatty acids are converted to triglycerides and secreted into the plasma as part of the VLDL particles. However, insulin deficiency also depresses lipoprotein lipase activity and therefore decreases the clearance of VLDL.

The above events result in:

↑ plasma fatty acids
ketosis
↑ plasma triglycerides

Protein metabolism

Insulin deficiency accelerates protein catabolism and gluconeogenesis, and decreases the rate of protein synthesis.

Sodium metabolism

Sodium depletion occurs in diabetic ketoacidosis because of osmotic diuresis and vomiting.
Hyponatraemia may be due to:
(1) Sodium and water depletion (due to the osmotic diuresis), and its replacement with salt poor fluids (drinking tap water).
(2) Hyperglycaemia → ↑ extracellular osmolality → fluid shift from intracellular to extracellular fluid (dilution).
(3) Hyperlipidaemia → factitious hyponatraemia.

Potassium metabolism

The increased lean body tissue catabolism due to insulin deficiency results in the release of cellular potassium. Some of this potassium is excreted in the urine (osmotic diuresis), resulting in total body potassium deficiency. Thus the hyperkalaemia is due to increased cellular release and also to the acidosis caused by the ketosis. With the institution of insulin therapy the plasma potassium rapidly re-enters the cells, and this may result in hypokalaemia if supplements are not given.

### Bicarbonate-acid base

The ketone bodies β-hydroxybutyrate and acetoacetate are acids, the presence of which result in the consumption of bicarbonate and the creation of an acidosis. The rate of acid production in severe cases exceeds the ability of the kidney to secrete hydrogen ion and results in a severe high anion gap metabolic acidosis. With the institution of insulin therapy ketogenesis is suppressed, and circulating ketones are oxidized in the peripheral tissues. Ketone bodies are taken up by the cells as $H^+.ketone^-$, and this results in generation of $HCO_3^-$ (i.e. $H^+$ are removed from the extracellular fluid).

### Urea and creatinine

The increased urea and creatinine levels in the above case result from the dehydration caused by the osmotic diuresis. It is important to remember that acetoacetate positively interferes with the measurement of creatinine, giving falsely high values.

### Phosphate

Increased tissue catabolism results in the release of phosphate from cells. Much of this phosphate is excreted in the urine, resulting in phosphate depletion. With the institution of insulin therapy, phosphate is forced back into the cells and hypophosphataemia may result.

## DIABETIC KETOACIDOSIS AND HYPERCHLORAEMIC METABOLIC ACIDOSIS

Female, aged 20 years, who presented with diabetic ketoacidosis. She was treated with insulin and saline infusions. The insulin was given at a rate of 4 U/hour.

| Time | | 10.30 | 19.30 | | |
|---|---|---|---|---|---|
| Plasma | Na | 127 | 139 | mmol/l | (132–144) |
| | K | 4.9 | 3.2 | mmol/l | (3.2–4.8) |
| | Cl | 102 | 123 | mmol/l | (98–108) |
| | $HCO_3$ | 3 | 4 | mmol/l | (23–33) |
| | Urea | 11.7 | 7.0 | mmol/l | (3.0–8.0) |
| | Creat. | 0.29 | 0.17 | mmol/l | (0.06–0.12) |
| | Anion gap | 27 | 15 | mEq/l | (7–17) |
| | Glu. | 26.0 | 9.5 | mmol/l | (3.0–5.5) |
| | AcAc | 3.0 | 1.0 | mmol/l | (<0.2) |
| | β-OHB | 10.0 | 2.1 | mmol/l | (<0.25) |

In this case, following insulin therapy, it can be seen that: (1) plasma [$HCO_3^-$] has not increased, (2) plasma [$Na^+$] and [$Cl^-$] have increased, (3) plasma anion gap has decreased and returned within the reference range, i.e. the original high anion gap acidosis has been converted to a hyperchloraemic (normal anion gap) acidosis (cf. previous case).

When insulin is administered to a patient with diabetic ketoacidosis, the ketone bodies can then be metabolized. Each mole of ketone body that is metabolized generates 1 mole of $HCO_3^-$, resulting in a rise in plasma $HCO_3^-$ concentration (approximately 1–2 mmol/h), and a decrease in the anion gap (see previous case). However, a proportion of patients respond in a different manner (above case); their anion gap decreases, but their plasma $HCO_3^-$ level shows no significant increase, so that a hyperchloraemic metabolic acidosis ensues.

The mechanism for this phenomenon is unclear. It may reflect insulin levels which are sufficient to suppress ketogenesis, but inadequate to stimulate tissue uptake and metabolism of ketone bodies. In this situation the ketone bodies, and accompanying sodium ions, could be cleared by the kidney and replaced by NaCl (saline infusion). Another possible cause is the development of a renal tubular acidosis with associated bicarbonate wasting.

## HYPEROSMOLAR COMA

Female, 80 years of age, presenting in coma. Not a known diabetic.

|  |  | Pre-treatment | 5 hours post-treatment |  |  |
|---|---|---|---|---|---|
| Plasma | Na | 158 | 170 | mmol/l | (132–144) |
|  | K | 4.6 | 4.3 | mmol/l | (3.2–4.8) |
|  | Cl | 118 | 130 | mmol/l | (98–108) |
|  | $HCO_3$ | 33 | 32 | mmol/l | (23–33) |
|  | Urea | 16 | 12 | mmol/l | (3.0–8.0) |
|  | Osmo. | 393 | 371 | mmol/kg | (275–295) |
|  | Glu. | 43 | 8.9 | mmol/l | (3.0–5.5) |
|  | Ketones | −ve | −ve |  | (−ve) |

The pathogenesis of this condition is unclear. It has been suggested that, compared with ketoacidosis, there is a lesser degree of insulinopenia and that this is sufficient to suppress ketosis. However, this has not been substantiated by plasma insulin measurements. Other work has suggested, on the basis of lower circulating levels of fatty acids, that hyperosmolality *per se* inhibits lipolysis and perhaps also ketogenesis.

This patient was treated with insulin infusion and intravenous 0.9% saline. The rise in plasma sodium concentration is due to a shift of water from the extracellular fluid into the intracellular compartment as the ECF osmolality falls consequent to a fall in the glucose level. It is important to note that in treating this condition the clinician tries to lower the extracellular osmolality slowly, as sudden falls will cause large water movements from the ECF (low osmolality) into the ICF (high osmolality), which can result in cerebral oedema.

Hyperosmolar coma tends to occur in patients of middle or advanced age who have mild diabetes which has not been recognized. 30–50% of these cases have an accompanying metabolic acidosis due to the combined effects of lactate accumulation and renal insufficiency.

## LACTIC ACID

### Lactate metabolism

The energy released during the metabolism of glucose to carbon dioxide and water is stored as ATP. In the presence of molecular oxygen one molecule of glucose eventually yields approximately 38 molecules of ATP (Fig. 10.6).

During glycolysis the production of the intermediate pyruvate from glucose requires $NAD^+$ and produces a net increase of two molecules of ATP per molecule of glucose metabolized. The enzyme phosphofructose kinase (PFK) is a rate-limiting enzyme in this process, its activity being influenced by the tissue levels of

ATP (↑ ATP → ↓ activity). In the presence of molecular oxygen the mitochondrial enzyme systems convert pyruvate to $CO_2$ and water (TCA cycle). During this process a further 36 molecules of ATP are produced and $NADH_2$ is oxidized to $NAD^+$. If $O_2$ is not available the mitochondrial system cannot proceed, and therefore ATP can only be produced by the glucose to pyruvate steps of the pathway (anaerobic glycolysis), which, as noted above, produce only two ATP for each glucose molecule converted to pyruvate. The consequent low level of ATP increases the glucose to pyruvate conversion (glycolysis) by stimulating PFK activity, and therefore raises ATP. However, during this process $NAD^+$ will be consumed ($NAD^+ \rightarrow NADH_2$) and the reaction will be suppressed unless adequate levels of this coenzyme are made available. This deficiency is overcome by the action of the cytosol enzyme lactate dehydrogenase (LD).

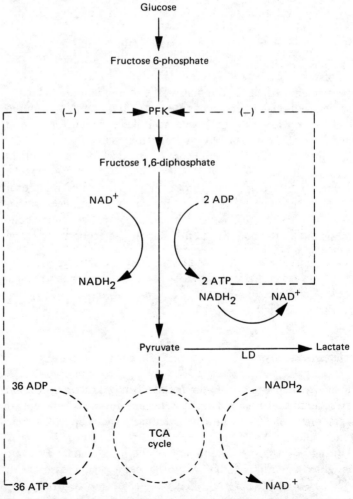

**Fig. 10.6** Glucose metabolism in hypoxia. PFK, phosphofructose kinase; LD, lactate dehydrogenase; TCA, tricarboxylic acid cycle.

Thus in the absence of molecular oxygen (anaerobic state), ATP is produced by the inefficient process of anaerobic glycolysis, which results in the accumulation of lactate. When oxygen becomes available again the accumulated lactate is converted to pyruvate ($NAD^+ \rightarrow NADH_2$), and then the mitochondrial system oxidizes pyruvate and $NADH_2$ to produce ATP and $NAD^+$.

Under normal circumstances the extrahepatic tissues produce more lactate (and pyruvate) than can be metabolized locally, and this excess is added to the circulation (approximately 1000–2000 mmol/day). This excess lactate is removed by the liver, and to a lesser extent by other tissues (e.g. kidney), and converted to $CO_2$ and water and glucose (gluconeogenesis). The addition of lactic acid ($H^+L^-$) to the extracellular fluid results in the consumption of $HCO_3^-$ ($H^+ + HCO_3^- \rightarrow H_2CO_3 \rightarrow H_2O + CO_2$), and if it occurs in excess there will appear a high anion gap metabolic acidosis ($HCO_3^-$ replaced by $L^-$). When the plasma lactate is taken up by the liver it crosses the cell wall as $H^+L^-$ and $HCO_3^-$ is regenerated because the $H^+$ is removed from the ECF. As 1000–2000 mmol of lactate are produced daily under resting conditions, the liver therefore plays an important role in acid-base regulation.

## LACTIC ACIDOSIS

The normal plasma lactate level is $<2$ mmol/l. An increase above this level is termed 'hyperlactataemia' and, if there is an associated metabolic acidosis ($\downarrow HCO_3^-$) 'lactic acidosis' is said to occur. There is no general agreement on what level a plasma lactate must reach before it should be termed a lactic acidosis, but a reasonable compromise would be a plasma level of 4 mmol/l if it is associated with a significant drop in the plasma [$HCO_3^-$]. This type of acidosis is always associated with a high anion gap, the unmeasured anion being lactate ($L^-$).

### Causes

Lactic acidosis can be conveniently divided into those due to overt tissue hypoxia, sometimes called Type A, and those which occur in the absence of hypoxia (Type B). In the Type B group hypoxia may be present but is not clinically detectable.

### OVERT HYPOXIA

(1) Severe hypoxia (asphyxia), (2) shock (a) haemorrhagic, (b) cardiogenic, (c) septic, (3) congestive cardiac failure, (4) carbon monoxide poisoning.

HYPOXIA ABSENT

(1) Drugs (a) biguanides, (b) ethanol/methanol, (c) salicylates, (d) ethylene glycol, (2) Diabetes mellitus, (3) Neoplastic disease, (4) Liver failure, (5) Pulmonary embolism, (6) Sepsis.

**Case examples/pathophysiology**

HYPOXIA

Patient, aged 60 years, with bleeding varices, severe blood loss and a prolonged period of hypotension.

| Plasma | Na | 139 | mmol/l | (132–144) | pH | 7.01 | | (7.35–7.42) |
|---|---|---|---|---|---|---|---|---|
| | K | 4.8 | mmol/l | (3.2–4.8) | $H^+$ | 100 | nmol/l | (35–45) |
| | Cl | 98 | mmol/l | (93–108) | $Po_2$ | 3.3 | kPa | (4.6–6.0) |
| | $HCO_3$ | 6 | mmol/l | (23–33) | $Pco_2$ | 26 | kPa | (10.6–13.0) |
| | Urea | 24 | mmol/l | (3.0–8.0) | $AHCO_3$ | 6 | mmol/l | (24–32) |
| | Creat. | 0.36 | mmol/l | (0.06–0.12) | | | | |
| | Anion gap | 40 | mEq/l | (7–17) | | | | |
| | Lactate | 28 | mmol/l | (<2) | | | | |

Tissue hypoxia due to prolonged hypotension (↓ tissue perfusion) results in anaerobic glycolysis and excessive lactic acid production. The addition of this acid load to the extracellular fluid results in the consumption of bicarbonate and its replacement by lactate. This produces an increased anion gap, since lactate is not measured in the electrolyte profile. The above case therefore shows an increased anion gap metabolic acidosis.

DIABETES MELLITUS

Lactic acidosis may occur in diabetes mellitus in three situations: (1) associated with ketoacidosis or hyperosmolar coma, (2) associated with biguanide therapy, (3) as a spontaneous syndrome.

Ketoacidosis/hyperosmolar coma

An increase in plasma lactate occurs in 30–50% of cases of ketoacidosis and hyperosmolar coma. The level is usually less than 4–5 mmol/l and probably reflects tissue hypoxia.

Biguanide therapy

Phenformin, an oral hypoglycaemic agent (which has recently been withdrawn from the market) lowers the blood sugar by inhibiting the GI absorption of glucose, gluconeogenesis and oxidative metabolism. It also appears to stimulate lactate production and impair its utilization by the liver. Severe lactic acidosis has been associated with its use in diabetic patients.

Spontaneous lactic acidosis

Lactic acidosis has also been observed in diabetic patients who have no evidence of any precipitating cause, e.g. hypoxia.

NEOPLASTIC DISEASE

Lactic acidosis has been described in patients with acute leukaemias and lymphomas. This may be due to excessive production of lactate by the malignant cells, as malignant tissue has a relatively high rate of glycolysis.

LIVER FAILURE

Most patients with advanced liver failure have an increased plasma level of lactate. This may be due to decreased liver uptake of lactate, although other organs (kidney) are also capable of oxidizing the acid. Liver failure may also be associated with respiratory alkalosis, and alkalosis *per se* will increase tissue lactate production (phosphofructokinase is stimulated by a low $[H^+]$).

SEPSIS

The lactic acidosis associated with septicaemia may be due to respiratory alkalosis or, as in the majority of cases, shock.

PULMONARY EMBOLISM

Like sepsis, lactic acidosis in pulmonary embolism may be due to respiratory alkalosis or shock.

DRUGS/TOXINS

Ethanol

The occurrence of moderate hyperlactataemia in ethanol overdose is common and severe lactic acidosis may sometimes occur. The cause is unclear, but may be related to increased levels of $NADH_2$

since alcohol dehydrogenase requires $NAD^+$ to proceed, and the metabolism of ethanol causes accumulation of $NADH_2$.

### Methanol

As well as leading to the production of formaldehyde and formic acid, methanol also inhibits mitochondrial oxidation. This predisposes to lactate accumulation.

### Ethylene glycol

Increased production of lactic, as well as glycolic and oxalic, acid has been reported in ethylene glycol intoxication.

### Salicylate

Salicylate intoxication causes uncoupling of oxidative phosphorylation in the mitochondria. This may result in increased plasma levels of many organic acids, including ketone bodies, citric acid and lactic acid. Salicylate also has a direct stimulating effect on the respiratory centre, causing respiratory alkalosis. Therefore these patients usually have a mixed respiratory alkalosis and metabolic (high anion gap) acidosis.

# HYPOGLYCAEMIA

**Classification**

*Drug induced*   (1) Insulin—iatrogenic, self-administered, (2) oral hypoglycaemics—sulphonylureas, especially glibenclamide, (3) salicylates—mainly children.

*Fasting*   (1) Insulinoma, (2) endocrine disorders (a) hypopituitarism, (b) hypoadrenalism, (3) liver disease (severe), (4) extrapancreatic tumours (a) retroperitoneal fibroma, (b) adrenal carcinoma, (c) hepatoma.

*Reactive*   (1) Glucose (a) functional, (b) postgastrectomy, (c) early diabetes, (2) leucine, (3) fructose, (4) galactose, (5) alcohol.

*Neonatal*

*Infancy and childhood*

Hypoglycaemia in adults is nearly always the result of overdosage with insulin or sulphonylureas. Other causes are relatively uncommon.

**Case examples/pathophysiology**

SURREPTITIOUS ADMINISTRATION OF GLIBENCLAMIDE

A 26 year old female complained of dizziness, palpitations and sweating if she missed a meal or exercised strenuously. She had a past history of glycosuria during pregnancy (?gestational diabetes), for which she was prescribed glibenclamide. Fasting plasma glucose levels of 1.7, 1.5 and 1.3 mmol/l had been recorded. Results after a 12 hour fast (hypoglycaemic symptoms present) were:

| | | | |
|---|---|---|---|
| Plasma Glu. | 1.6 | mmol/l | (3.0–5.5) |
| Insulin | 20 | mU/l | (2.0–6.0) |
| C-Peptide | 2.3 | pmol/l | (0.4–0.8) |

A diagnosis of insulinoma was made, but a laparotomy and partial pancreatectomy revealed no abnormality. After these procedures proved fruitless further tests revealed glibenclamide to be present in the urine.

Glibenclamide is a sulphonylurea which produces hypoglycaemia by: (1) increased pancreatic insulin release, (2) decreased liver glucose release, (3) increased sensitivity of peripheral tissues to the action of insulin.

The increased insulin release will be accompanied by an increased secretion of C-peptide. The above picture, clinical and biochemical, is similar to that of an insulinoma, and can only be distinguished from the latter condition by demonstrating the presence (plasma, urine) of the offending drug.

INSULIN OVERDOSE

This common cause of hypoglycaemia can usually be diagnosed from the clinical story. Biochemically the features are hypoglycaemia associated with hyperinsulinaemia and a low plasma C-peptide level.

SALICYLATE OVERDOSE

Hypoglycaemia due to salicylate overdose has been described in young children. The mechanism is unclear.

## INSULINOMA

A 35 year old female was admitted to a psychiatric ward because of bizarre behaviour. She had a 2 week history of unusual behaviour in the mornings, with total amnesia for the event. Biochemistry results after an overnight fast were:

Plasma Glu.      1.5mmol/l     (3.0–5.5)
       Insulin    45 mU/l       (2.0–6.0)
       C-Peptide  >2 pmol/l     (0.4–0.8)

A screen for the presence of sulphonylureas in the blood was negative. A diagnosis of insulinoma was made, and this tumour was found and removed. The best diagnostic approach to insulinoma is to perform repeated determinations of the plasma levels of glucose and insulin after an overnight fast. Should this fail to give any information, prolong the fast to 72 hours in order to provoke hypoglycaemia. Only about 10% of cases will require a fast longer than 12 hours to demonstrate hypoglycaemia. Provocative tests (leucine, tolbutamide) have been used in the past, but have given false negative tests in up to 25% of cases.

## ENDOCRINE DISORDERS

### Hypopituitarism

Deficiency of growth hormone and cortisol (anti-insulin actions) may lead to severe spontaneous hypoglycaemia in untreated cases.

### Hypoadrenalism

Deficiency of cortisol may cause hypoglycaemia (decreased gluconeogenesis) in the fasting state.

## LIVER DISEASE

In severe liver disease inadequate glycogen storage and depressed gluconeogenesis may lead to hypoglycaemia. However, this condition is unusual except in severe alcoholic liver disease, where alcohol ingestion further suppresses gluconeogenesis (see below).

## EXTRAPANCREATIC NEOPLASMS

Hypoglycaemia may, in rare instances, be associated with retroperitoneal fibromas, adrenal carcinomas and hepatomas. The cause is unclear, but may be due to: (1) tumour secretion of insulin-like material, (2) excessive utilization of glucose by the tumour, (3) ectopic production of insulin (rare).

GLUCOSE INDUCED HYPOGLYCAEMIA

Hypoglycaemia after glucose ingestion may occur in the postgastrectomy state, in early diabetes mellitus and in a condition referred to as functional hypoglycaemia.

Postgastrectomy

Rapid absorption of glucose leads to a large insulin secretion and a consequent excessive fall in the plasma glucose level in 2–3 hours, i.e. lag storage curve.

Early diabetes mellitus

In early diabetes postprandial secretion of insulin may be delayed, but once started continues for a longer period of time than necessary. Thus towards the end of glucose utilization insulin may be present in excess and hypoglycaemia may occur.

Functional hypoglycaemia

These patients have normal fasting levels of glucose but become mildly hypoglycaemic, with symptoms, 2–4 hours after a meal; they do not develop fasting hypoglycaemia. A prolonged OGTT (5 h) often demonstrates the hypoglycaemia, but it must be remembered that 30–40% of normal subjects have a plasma glucose level less than 3.0 mmol/l at some period during a 5 hour OGTT.

*Case example*   A 40 year old female complained of dizzy spells 2–3 hours after meals. Symptoms were relieved by food. Results of a prolonged GTT were:

| Time | Plasma Glu. (mmol/l) |
|---|---|
| 0 | 4.5 |
| 1 | 9.5 |
| 2 | 4.8 |
| 3 | 2.5 |
| 4 | 3.6 |
| 5 | 4.6 |

Although this picture fits functional hypoglycaemia, early diabetes mellitus cannot be ruled out. Again it is important to remember that up to 40% of normal subjects will have a glucose level less than 3.0 mmol/l at some period during a prolonged OGTT and not become symptomatic.

## LEUCINE

A number of amino acids are capable of stimulating the release of insulin. Leucine is the only one that has been associated with severe hypoglycaemia. This condition of hypoglycaemia due to leucine sensitivity has been described in infants and manifests itself after a high protein meal. Some patients with insulinoma will also demonstrate hypoglycaemia after being challenged with leucine.

## GALACTOSE AND FRUCTOSE

Infants with galactosaemia often become severely hypoglycaemic after galactose ingestion, whilst in the rare inborn error of fructosaemia, hypoglycaemia is associated with the ingestion of fructose.

## ALCOHOL

Hypoglycaemia due to excessive alcohol ingestion is relatively common and has an incidence in some studies that is higher than that for insulinoma. It is due to two mechanisms:

*Depleted glycogen reserves*   Due to undernourishment (common in alcoholism).

*Decreased gluconeogenesis*   The metabolism of alcohol (ethanol → acetaldehyde → acetate) converts $NAD^+$ to $NADH_2$. The accumulated $NADH_2$ is oxidized back to NAD by the reduction of pyruvate to lactate. However, the formation of glucose from lactate and alanine (gluconeogenic precursors) requires their conversion to pyruvate. Thus the accumulation of $NAD^+$ interferes with substrate utilization along the gluconeogenic pathway.

## NEONATAL HYPOGLYCAEMIA

In the neonatal period the plasma glucose level falls after birth, reaching its lowest level 4–5 hours postpartum; it then returns to normal over the following 4–5 days. In some infants this fall may be excessive and produce hypoglycaemic symptoms (tremor, con-

vulsions). This condition usually occurs in two forms:
(1) Premature infants and 'small for dates babies'. Hypoglycaemia is probably due to intrauterine malnutrition and glucose substrate deficiency.
(2) Babies of diabetic mothers. Hypoglycaemia due to β-cell hyperplasia and hyperinsulinaemia.

HYPOGLYCAEMIA IN INFANCY AND CHILDHOOD

Hypoglycaemia after the neonatal period may be due to the following causes:
(1) Idiopathic—cause unknown. Diagnosis is arrived at by a process of exclusion.
(2) Leucine sensitivity.
(3) Ketotic. The hypoglycaemia is associated with ketosis and may be demonstrated by giving a high fat (ketotic) diet.
(4) Inborn errors of metabolism—galactosaemia, hereditary fructose intolerance, glycogen storage disease.
(5) Other causes, e.g. insulinoma.

# 11 Proteins

At least 80 different proteins, excluding enzymes and polypeptide hormones, have so far been either isolated from, or identified in, normal plasma. Some of these proteins have yet to have a role ascribed to them, whilst the remainder have been shown to perform a wide range of functions.

(1) Nutrition—source of amino acids for protein synthesis, gluconeogenesis.
(2) Buffers—contribution to plasma and cellular buffer systems.
(3) Plasma water—their size restricts most of the proteins to the intravascular compartment; the resultant oncotic pressure draws water into the compartment.
(4) Transport, e.g. steroid, thyroid hormones, vitamins A, D, lipids—free fatty acids, triglycerides, cholesterol and esters, calcium, trace metals, drugs, wastes—bilirubin.
(5) Enzyme (protease) inhibitors—$\alpha_1$ antitrypsin, $\alpha_2$ macroglobulin.
(6) Immune system—immunoglobulins, complement components.
(7) Clotting factors—prothrombin, fibrinogen.

Despite such an extensive involvement in the physiology and metabolic processes of the body, relatively few of the individual plasma proteins have proved useful as markers of pathological processes, and so few are assayed routinely for the purpose of diagnosing or monitoring disease. As well as in plasma, proteins are often measured in other body fluids, such as urine, cerebrospinal fluid, pleural and ascitic fluids, but those assayed have usually been the same proteins as those measured in plasma.

## Acute phase proteins

The term 'acute phase proteins' describes a group of apparently unrelated proteins whose concentrations alter in a characteristic way following acute trauma to the body. Some of the common injuries that lead to these changes are: myocardial infarct, surgery, burns, major infection, inflammation. The plasma levels of some of the acute phase proteins increase after the injury, so probably the tissue damage releases factors that stimulate their synthesis by the liver.

| Acute phase protein | Plasma change | Peak at (days) | Role |
|---|---|---|---|
| Fibrinogen | Rises | 3–5 | Clot formation |
| Haptoglobin | Rises | 3–5 | Haemoglobin salvage |
| Caeruloplasmin | Rises | 5–10 | Haemoglobin synthesis? |
| $\alpha_1$-antitrypsin | Rises | 3–5 | Inhibits trypsin |
| C-reactive protein | Rises | ? | Promotes phagocytosis |
| Transferrin | Falls | 3–10 | Haemoglobin synthesis |
| Glycoproteins | Rises | 3–8 | Repair mechanism? |
| Albumin | Falls | 3–8 | Repair mechanism? |
| Prealbumin | Falls | 3–10 | Repair mechanism? |

Most of the proteins show a rise of up to twice their normal plasma level, except C-reactive protein, which can rise 20–30 fold from barely detectable levels. Albumin may fall by up to 20%. The overall effect of these changes probably reflects one part of the complex response to injury, that is: (1) minimizing blood and fluid loss, (2) salvage of important components, (3) minimizing further damage by lysosomal enzymes released from damaged cells, (4) removal of damaged cells, (5) repair of damage.

Measurement of these various proteins has no diagnostic value in the context of trauma, but care should be taken in interpretation if any of these are measured for other reasons in the 10–12 days following a traumatic event.

## LABORATORY INVESTIGATION

The following analyses are widely used for the routine biochemical investigation of diseases that directly or indirectly affect either the metabolism or fluid concentration of proteins.

*Plasma*   (1) Total protein, (2) albumin, (3) cryoglobulins, (4) liver function tests.

*Serum*   (1) $\alpha_1$-antitrypsin, (2) caeruloplasmin, (3) electrophoresis.

*Urine*   (1) Total protein, (2) electrophoresis.

*CSF*   Total protein.

*Abnormal fluids*   (1) Total protein, albumin.

**Measurement of proteins**

All proteins have a basic structure that is constructed from amino acid residues linked through peptide bonds. A minimum of two adjacent such bonds (a tripeptide) will react with alkaline copper ions ($Cu^{2+}$) to form a coloured complex. The amount of colour formed is proportional to the number of peptide bonds present, and therefore by using suitable standards this biuret reaction allows complex mixtures of proteins, such as plasma, to be simply assayed as a total sum of protein.

The unique properties of specific proteins have been exploited in assay systems, e.g. the oxidase activity of caeruloplasmin, the binding of haemoglobin by haptoglobins, or the binding of dihydrotestosterone by sex hormone binding globulin. With the improved methods for isolating pure proteins now available, the powerful techniques of immunoassay allow many proteins of low concentration to be easily measured in complex mixtures such as serum.

TOTAL PROTEIN

Plasma total protein is usually measured by the biuret method ($\sim$ 60–85 g/l), and is conveniently considered to comprise two major parts, albumin ($\sim$ 30–50 g/l) and globulins ($\sim$ 30–40 g/l). In some disease processes it is not unusual to find hypoalbuminaemia and hypergammaglobulinaemia coexisting, often resulting in a normal plasma total protein value. It is therefore advisable to always measure the concentration of both plasma total protein and albumin when a protein abnormality is suspected.

In fluids such as CSF or urine the normal concentration of total protein is extremely low compared to that in plasma and, therefore, chemical estimation usually takes place after the total protein has been precipitated and drained of fluid.

ALBUMIN

The most commonly used methods are based on the specific binding by albumin of certain dyes, causing a change in colour. Before such specific methods were available, albumin was often indirectly estimated from a total protein value and serum electrophoresis, the latter providing the percentage contribution of albumin to the total protein. A measurement of plasma albumin also, in conjunction with a value for total protein, provides an indirect estimate of total globulin concentration.

SERUM ELECTROPHORESIS

Their amino acid composition means that proteins carry both positive and negative charges, the prevalence of each for a particular protein being dependent on the physico-chemical environment. At an alkaline pH most proteins possess an overall negative charge, but the excess of negative over positive depends on the amino acid residue structure of each protein.

If a mixture of different proteins in a buffer of pH 8.6 is placed on an inert support (cellulose acetate strip), and a direct electric current applied for a fixed time, most of the proteins migrate towards the positive electrode. The distance travelled by each protein species largely depends on the amount of negative charge carried. After such electrophoresis, the proteins can be fixed in position by precipitation, stained for visualization, and the relative amounts estimated by measuring the density of each of the stained areas of protein.

For routine electrophoresis serum is used because the fibrinogen present in plasma yields a stained area that interferes with interpretation. For serum electrophoresis 1-2 μl of serum is applied in a thin line to the cellulose acetate strip. After electrophoresis and staining, normal serum is seen to separate into five distinct bands; the relative density of the bands can be scanned and recorded to give an electrophoretogram (Fig. 11.1). This characteristic pattern forms the basis of a simple classification of serum proteins, based on the position to which they migrate under standard electrophoretic conditions (Table 11.1).

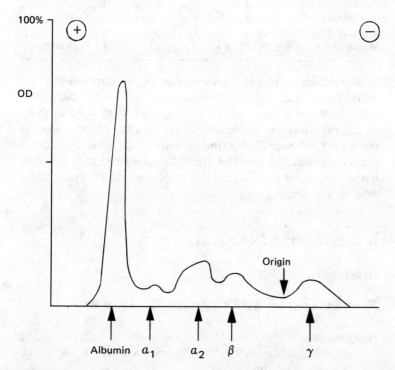

**Fig. 11.1** Electrophoretogram of serum proteins.

**Table 11.1** Major proteins contributing to the bands seen following serum electrophoresis.

| Electrophoretic position | Major protein present | % in normal serum |
|---|---|---|
| Albumin | Albumin | 44–62 |
| $\alpha_1$-globulins | $\alpha_1$-antitrypsin | 3–9 |
| $\alpha_2$-globulins | Caeruloplasmin<br>Haptoglobins<br>$\alpha_2$-macroglobulin | 8–16 |
| $\beta$-globulins | Transferrin<br>Plasminogen<br>Fibrinogen (in plasma) | 11–21 |
| $\gamma$-globulins | IgA<br>IgM<br>IgG | 11–21 |

**Table 11.2** Abnormal electrophoretic patterns seen in various disease states.

|  | Albumin | Globulins $\alpha_1$ | $\alpha_2$ | $\beta$ | $\gamma$ | M band |
|---|---|---|---|---|---|---|
| Acute phase reaction | ↓ | ↑ | ↑ |  |  |  |
| Chronic infection | ↓ |  | ↑ |  | ↑ |  |
| Cirrhosis | ↓ |  |  |  | ↑ |  |
| Nephrosis | ↓ | ↓ | ↑ |  | ↓ |  |
| $\alpha_1$-antitrypsin deficiency |  | ↓ |  |  |  |  |
| Myeloma | N or ↓ |  |  |  | ↓ | Present |

Many disease processes cause distinctive electrophoretic patterns (Table 11.2), but with the development of other diagnostic procedures the serum electrophoretogram is little used in these situations. However, the technique remains the major screening test for the presence of abnormal proteins (M (myeloma) bands) in the serum (paraproteins) and urine (Bence-Jones protein). Urine has to be concentrated 50–100 fold prior to electrophoresis for the detection of Bence-Jones protein.

## HYPERPROTEINAEMIA

Hyperproteinaemia is caused by:

*Haemoconcentration* (1) Dehydration, (2) venous stasis.

*Hypergammaglobulinaemia* (1) Polyclonal (a) chronic liver disease, (b) chronic infections, (2) monoclonal (a) myeloma, (b) macroglobulinaemia.

# HYPOPROTEINAEMIA

Hypoproteinaemia is caused by:

*Haemodilution* (1) Inappropriate IV therapy, (2) syndrome of inappropriate ADH, (3) sample taken from above IV drip needle.

*Hypoalbuminaemia* (1) Decreased synthesis (a) severe hepatic disease, (b) malabsorption, (c) malnutrition, (2) increased loss (a) renal-nephrotic syndrome, (b) skin—burns, (c) gut—protein-losing enteropathy.

*Hypogammaglobulinaemia* Most diagnostic laboratories measure and report both plasma total protein and albumin, but where this is not done it should be noted that it is not uncommon for a normal plasma total protein value to mask a coexisting abnormality of both immunoglobulins (raised) and albumin (reduced).

## Case examples/pathophysiology

### HAEMOCONCENTRATION

A 37 year old woman with aplastic anaemia, found unconscious at home.

|  | On admission | After saline infusion |  |
|---|---|---|---|
| Plasma TP | 89 | 73 g/l | (60–85) |
| Alb. | 54 | 39 g/l | (30–50) |

Haemoconcentration is probably the commonest cause of an elevated total protein and albumin. However, the reference range of both parameters is fairly wide so, depending on the individual's initial levels, marked dehydration can occur without the total protein value reaching beyond the upper reference limit.

### POLYCLONAL HYPERGAMMAGLOBULINAEMIA

A 23 year old man already diagnosed as having rheumatoid arthritis.

Plasma  TP   91 g/l   (60-85)
        Alb. 38 g/l   (30-50)

Note that subtraction allows calculation of the total globulin fraction (53 g/l), which is markedly raised (RR 28-38). Electrophoresis of a sample of this serum revealed a generalized increase in the γ-globulins, suggesting chronic stimulation of the immune system.

MONOCLONAL HYPERGAMMAGLOBULINAEMIA

Man, aged 41 years, with multiple myeloma.

Plasma  TP   101 g/l   (60-85)
        Alb. 39 g/l    (30-50)

TP-Alb. indicates a markedly elevated total globulin value of 62 g/l. Electrophoresis of this sample revealed the presence of an abnormal protein band in the β-γ region. This band represents the presence of large quantities of an abnormal protein synthesized by a clone of malignant plasma cells. Although these proteins have a normal immunoglobulin structure it is not known whether such paraproteins are directed against antigens.

HAEMODILUTION

Blood sample taken for a routine check on a patient recovering from surgery.

Plasma  TP   39    g/l      (60-85)
        Alb. 11    g/l      (30-50)
        Na   150   mmol/l   (132-144)
        K    1.1   mmol/l   (3.2-4.8)
        Cl   >130  mmol/l   (98-108)

The proteins, albumin and other results obtained are clearly not compatible with life. The venepuncture was made just proximally to the site of the indwelling needle of a saline drip.

## ALBUMIN

Some 14-15 g of albumin is synthesized each day by the normal adult liver, but in situations of marked albumin loss from the body the rate of production can be doubled. Albumin is distributed mainly in the extracellular fluid, where almost 40% is located within the intravascular compartment. At a concentration of 30–50 g/l albumin is the most abundant of the plasma proteins, and with a plasma half-life of 19–20 days it is also very stable. The major functions of plasma albumin are:

(1) Control of water distribution between the intra- and extravascular space (oncotic pressure).
(2) Low affinity, high capacity transport protein, e.g. for calcium, magnesium; trace metals—Cu, Zn; hormones—$T_4$, $T_3$, cortisol; waste product—bilirubin; lipids—free fatty acids; amino acids—75% of tryptophan; drugs.
(3) Nutrition—supply of amino acids.
(4) Haemodynamics—good flow properties.

The plasma concentration of albumin is often lower in an individual who has been lying down for a period, compared to when ambulant. In the upright position blood tends to pool in the legs so that plasma fluid leaks into the tissues, thus slightly concentrating the plasma. To account for this diagnostic laboratories should provide a separate plasma albumin reference range for inpatients and outpatients, and also for analytes such as calcium, that are affected by albumin levels. In our hospital plasma albumin for inpatients was 30–50 g/l and for outpatients 37–52 g/l.

It has often been stated that a plasma albumin concentration of 20 g/l or less is invariably associated with oedema. However in practice, the presence of oedema is often observed in patients with albumin levels of 25–30 g/l, whilst in others oedema has not been noted when the albumin concentration has been below 20 g/l. Clearly the development of oedema depends on the response of other homeostatic mechanisms.

## HYPERALBUMINAEMIA

The only known cause is that of haemoconcentration. This may arise through either improper collection (venous stasis), or any process that causes absolute or relative water loss from the vascular compartment.

*Example* Patient with diabetic ketoacidosis and dehydration.

| Plasma | Na | 136 | mmol/l (132–144) | Plasma | Ca | 2.65 | mmol/l (2.15–2.55) |
|---|---|---|---|---|---|---|---|
| | K | 5.8 | mmol/l (3.2–4.8) | | $PO_4$ | 1.76 | mmol/l (0.60–1.25) |
| | Cl | 105 | mmol/l (98–108) | | TP | 96 | g/l (60–85) |
| | $HCO_3$ | 8 | mmol/l (23–33) | | Alb. | 55 | g/l (30–50) |
| | Creat. | 0.31 | mmol/l (0.06–0.12) | | | | |
| | Glu. | 23 | mmol/l (3.0–5.5) | | | | |

Chapter 11    Pathophysiology

(1) osmotic diuresis → water and salt loss → dehydration → haemoconcentration → ↑ plasma [Alb.]

(2) ↑ plasma [Alb.] → ↑ total [Ca]—corrected value is 2.35 mmol/l (see p. 173)

## HYPOALBUMINAEMIA

### Causes

*Haemodilution*  (1) Dilutional state (a) iatrogenic water overload, (b) SIADH, (2) sample taken from IV drip arm.

*Decreased synthesis*  (1) Deficient amino acid supply (a) malnutrition, (b) malabsorption, (2) liver disease (a) acute, (b) chronic, (3) analbuminaemia.

*Loss from body*  (1) Skin (a) burns, (b) exudative skin lesions, (2) kidney—nephrotic syndrome, (3) gut—protein-losing enteropathy.

*Miscellaneous*  e.g. Pregnancy, non-specific illness, malignancy.

### Case examples/pathophysiology

#### HAEMODILUTION

Syndrome of inappropriate secretion of ADH (SIADH) due to carbamazepine therapy.

| Plasma | | | | | | |
|---|---|---|---|---|---|---|
| Na | 113 | mmol/l | (132–144) | Urine Na | 34 | mmol/l |
| K | 3.4 | mmol/l | (3.2–4.8) | Osmo. | 570 | mmol/kg |
| Cl | 81 | mmol/l | (98–108) | | | |
| HCO$_3$ | 24 | mmol/l | (23–33) | | | |
| Urea | 3.0 | mmol/l | (3.0–8.0) | | | |
| Osmo. | 245 | mmol/kg | (275–295) | | | |
| Ca | 1.92 | mmol/l | (2.15–2.55) | | | |
| TP | 53 | g/l | (60–85) | | | |
| Alb. | 25 | g/l | (30–50) | | | |

Pathophysiology

haemodilution → ↓ plasma [Alb.] → ↓ plasma [Ca] (corrected [Ca] = 2.22 mmol/l)

### DEFICIENT AMINO ACID SUPPLY

Malabsorption syndrome due to chronic pancreatitis.

| Plasma | Ca | 1.90 | mmol/l | (2.15–2.55) |
|---|---|---|---|---|
| | TP | 48 | g/l | (60–85) |
| | Alb. | 27 | g/l | (30–50) |
| Three day faecal fat | | 66 | g | (<21 g) |

Pathophysiology

(1) Prolonged decrease in absorption of amino acids → ↓ albumin synthesis.
(2) Hypocalcaemia due to (a) hypoalbuminaemia, (b) ? vitamin D deficiency.

### LIVER DISEASE

Alcoholic liver disease.

| Plasma | TP | 54 | g/l | (60–85) |
|---|---|---|---|---|
| | Alb. | 19 | g/l | (30–50) |
| | AP | 555 | U/l | (30–120) |
| | ALT | 61 | U/l | (<35) |
| | Bili. | 128 | µmol/l | (<20) |
| | Ca | 1.78 | mmol/l | (2.15–2.55) |
| | PO$_4$ | 0.40 | mmol/l | (0.6–1.25) |

Pathophysiology

(1) Hypoalbuminaemia—decreased synthesis.
(2) Hypocalcaemia due to hypoalbuminaemia (corrected value = 2.20 mmol/l).

### ANALBUMINAEMIA

Very rare congenital disorder due to decreased albumin synthesis.

*Chapter 11*    NEPHROTIC SYNDROME

Male, 20 years, with peripheral oedema and proteinuria.

| Plasma | TP | 51 | g/l | (60–85) | Urine TP 12 g/day (<0.15) |
|---|---|---|---|---|---|
| | Alb. | 25 | g/l | (30–50) | |
| | Glob. | 26 | g/l | (30–40) (by subtraction) | |
| Plasma | Chol. | 17.52 | mmol/l | (3.5–7.5) | |
| | Trig. | 2.81 | mmol/l | (0.3–1.7) | |

Pathophysiology

(1) Hypoalbuminaemia—renal loss of albumin (there is also loss of some of the smaller molecular weight globulins).
(2) Hyperlipidaemia—see Chapter 14.

### PROTEIN-LOSING ENTEROPATHY

Male, aged 57 years, with a carcinoma of the stomach.

| Plasma | TP | 39 g/l | (60–85) | Urine TP | 0.01 g/day | (<0.1) |
|---|---|---|---|---|---|---|
| | Alb. | 22 g/l | (30–50) | Faecal $\alpha_1$-AT | 2.5 mg/g | |
| | Glob. | 17 g/l | (30–40) (by subtraction) | | | |
| | AP | 90 U/l | (30–120) | | | |
| | ALT | 19 U/l | (<35) | | | |
| | Bili. | 10 µmol/l | (<20) | | | |

Pathophysiology

Excessive loss of protein (albumin and globulins) from the gut resulting in hypoalbuminaemia may occur in many conditions, including the following: (1) carcinoma of gastrointestinal tract, (2) ulcerative colitis, regional ileitis, (3) intestinal inflammations, (4) intestinal lymphangiectasia.

In this patient there was no evidence of malabsorption or malnutrition. The liver function tests (LFT) were normal, and there was no loss of protein in the urine. A useful test for this condition is to measure the faecal $\alpha_1$-antitrypsin level, which under normal circumstances is low (<0.3 mg/g faeces). However, if there is loss of ECF protein into the gut lumen $\alpha_1$-antitrypsin may be used as a marker because it can pass out into the faeces without being degraded by intestinal trypsin, so that high levels will be found in the faeces.

## Pregnancy

Hypoalbuminaemia is not uncommon in the latter half of pregnancy. This may be due to: (1) haemodilution, (2) inadequate protein intake, (3) increased amino acid utilization by fetus.

## Non-specific illness

In any illness, especially acute infections, it is not unusual to see a rapid drop in the plasma albumin concentration (acute phase reaction). The mechanism for this fall is uncertain. One suggestion is that there is a compartmental shift of albumin due to increased permeability of the capillary walls.

## Malignancy

Hypoalbuminaemia is a common finding in malignant diseases. This could be due to various reasons, e.g. (1) decreased intake of essential amino acids, (2) protein-losing enteropathy, (3) decreased synthesis (a) liver disease (secondaries), (b) increased uptake of amino acids by the tumour.

# GLOBULINS

The electrophoretic pattern (EPP) of serum globulins performed under standard conditions shows four fractions, $\alpha_1$, $\alpha_2$, $\beta$, $\gamma$ (see Fig. 11.1).

### $\alpha_1$-GLOBULINS

Concentration—2–5 g/l
Main component—$\alpha_1$ antitrypsin
Increased—tissue damage (acute phase reaction)
Decreased—$\alpha_1$ antitrypsin deficiency

### $\alpha_1$-antitrypsin (plasma 2–4 g/l)

Protease inhibitor, molecular weight $\sim 50\,000$ Daltons, found throughout the extracellular fluid. It is increased in the acute phase reaction (see p. 237). Decreased levels indicate a deficiency (inborn error), which may be associated with liver cirrhosis (neonates) or emphysema of the lung (young adults), suggesting that a major role of $\alpha_1$-antitrypsin is the protection of healthy cells from attack by circulating protease enzymes. Low levels are also associated with protein-losing states (see p. 244).

### $\alpha_2$-GLOBULINS

Concentration 4–10 g/l
Main components—(1) caeruloplasmin, (2) $\alpha_2$ macroglobulins, (3) haptoglobins.
Increased—(1) acute phase reaction, (2) chronic infection, (3) nephrotic syndrome.

### Caeruloplasmin (plasma 0.2–0.6 g/l)

Complexes most of the plasma copper. Also possesses oxidase activity, the physiological significance of which is uncertain, but may be concerned with oxidizing $Fe^{2+}$ prior to incorporation into haemoglobin. Plasma levels may be raised in acute phase situations and are low in Wilson's disease.

### $\alpha_2$-macroglobulin (plasma 1.5–4.0 g/l)

Large molecular weight protein ($\sim 900\,000$ Daltons) which may be increased in protein-losing states (e.g. nephrotic syndrome), caused possibly by an increased synthesis due to feedback stimulation by reduced levels of other proteins. $\alpha_2$-Macroglobulin functions as a proteinase inhibitor.

### Haptoglobins (plasma 4–8 g/l)

A group of proteins which bind any haemoglobin that has been released into the plasma by intravascular haemolysis or red cell damage. After such binding the haptoglobin–haemoglobin complex is taken up by reticuloendothelial tissue and catabolized, thus ensuring salvage of iron and amino acids. Low levels of plasma haptoglobins are therefore associated with haemolytic conditions. High levels are found with major tissue damage (acute phase reaction).

## β-GLOBULINS

Concentration 6–12 g/l
Main components are β-lipoproteins and transferrin.

## γ-GLOBULINS

Concentration 8–16 g/l
These comprise the major component of the serum globulins, so that increased and decreased levels of the γ fraction are usually clearly reflected by similar changes in the total globulin level. All the γ-globulins are immunoglobulins, but not all the immunoglobulins migrate to the γ region following serum electrophoresis.

### Immunoglobulins

Proteins which have antibody activity. There are 5 classes (G, A, M, D, E), based on the type of heavy chain present.

### Structure

The basic immunoglobulin unit can be considered as a Y-shaped molecule consisting of four polypeptide chains (two heavy (H) chains, and two light (L) chains), linked by disulphide bonds. In a single immunoglobulin molecule the H chains are the same (γ, α, μ, δ, ε occur in IgG, IgA, IgM, IgD and IgE respectively). The L chains are of two types, κ and λ, and only one type occurs in a single molecule (Fig. 11.2).

### *IgG*
Molecular weight—~160 000 Daltons
Concentration (adult)—8–13 g/l
Adult level reached—by 3–5 years of age
Function—synthesis stimulated by soluble antigens; protects tissue spaces

### *IgA*
Molecular weight—~160 000 Daltons (secretory IgA = dimer IgA + secretory piece + J (joining chain) mol.wt. ~400 000 Daltons)
Concentration (adult)—2–3 g/l
Adult level reached—by 15 years of age
Function—synthesized in lamina propria of respiratory and intestinal tracts; protects body surfaces against infection

Chapter 11

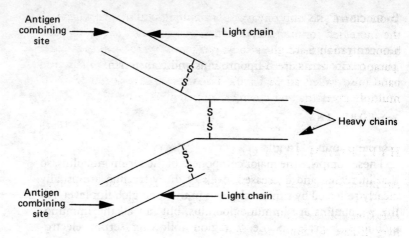

**Fig. 11.2** Schematic structure of the basic immunoglobulin unit.

*IgM*
Molecular weight—$\sim$ 900 000 Daltons
Concentration (adult)—$\sim$ 1 g/l
Adult level reached—$\sim$ 9 months of age
Function—protects bloodstream against particulate antigens, e.g. bacteria

*IgE*
Molecular weight—$\sim$ 200 000 Daltons
Concentration (adult)—0.0005 g/l
Adult level reached—15 years of age
Function—concerned with immediate hypersensitivity

*IgD*
Molecular weight—$\sim$ 190 000 Daltons
Concentration (adult)—0.05 g/l
Adult level reached—15 years of age
Function—unknown for the free plasma molecule. Associated with immature B cell membranes

Paraproteins

Each immunoglobulin producing cell (plasma cell) manufactures only one type of immunoglobulin. Cells producing identical immunoglobulin (directed against the same antigen, i.e. determinant) are collectively termed a 'clone'. If there is proliferation of several clones, as occurs in a chronic infection, the response is termed 'polyclonal'. On the electrophoretogram this will be seen as a diffuse increase in the $\gamma$-globulin fraction. On the other hand, if only one cell type or clone proliferates the response is termed 'monoclonal', i.e. only one type of immunoglobulin is produced. If the increased immunoglobulin is the result of a clone of cells becoming malignant, the protein produced by the clone is called a 'paraprotein' and is seen on the EPP as a dense narrow discrete band (also called an M band). This course of events occurs in multiple myeloma and macroglobulinaemia.

## HYPERGAMMAGLOBULINAEMIA

An increased plasma globulin level can be due to the following.

*Polyclonal increase* (diffuse)  (1) Chronic infections, (2) chronic liver disease, (3) autoimmune disease, (4) sarcoidosis.

*Monoclonal increase* (paraprotein, M band) (1) Malignant (a) myelomatosis, (b) macroglobulinaemia, (c) heavy chain disease, (d) lymphoreticular malignancy, (2) benign (a) chronic infections, (b) autoimmune disease, (c) cirrhosis of liver, (3) essential.

**Case examples/pathophysiology**

### CHRONIC INFECTIONS

Pulmonary tuberculosis in a patient aged 65 years.

| | | | |
|---|---|---|---|
| Plasma TP | 83 | g/l | (60–85) |
| Alb. | 33 | g/l | (30–50) |
| AP | 96 | g/l | (30–120) |
| ALT | 30 | U/l | (<35) |
| Bili. | 15 | μmol/l | (<20) |
| Glob. | 50 | g/l | (28–38) (by subtraction) |

EPP: diffuse increase in γ globulins

Most chronic infections will result in an increased γ-globulin fraction, but it occurs especially with the following infections: (1) chronic bronchitis, (2) bronchiectasis, (3) lung abcess, (4) osteomyelitis, (5) subacute bacterial endocarditis, (6) malaria.

### CHRONIC LIVER DISEASE

Chronic active hepatitis in a female aged 42 years.

Plasma TP   93 g/l      (60–85)
     Alb.   36 g/l      (30–50)
     AP    450 U/l     (30–120)
     ALT   248 U/l     (<35)
     Bili.  12 µmol/l   (<20)
     Glob.  57 g/l     (28–38) (by subtraction)
EPP: diffuse increase in γ-globulin fraction

For interpretation of the liver function tests see Chapter 13. The normal plasma albumin level in this case suggests that liver function is not yet severely impaired. Most cases of cirrhosis also have an increase in the γ globulin fraction, and this is usually associated with a hypoalbuminaemia (see p. 311).

### AUTOIMMUNE DISEASE

Rheumatoid arthritis in a female, aged 45 years.

Plasma TP   90 g/l      (60–85)
     Alb.   38 g/l     (30–50)
     AP     83 U/l    (30–120)
     ALT    20 U/l    (<35)
     Bili.   3 µmol/l  (<20)
     Glob.  52 g/l    (28–38) (by subtraction)
EPP: diffuse increase in γ-globulin fraction

Other autoimmune diseases associated with hypergammaglobulinaemia are systemic lupus erythematosis (SLE) and scleroderma.

### SARCOIDOSIS

This is a chronic granulomatous disease of unknown aetiology. The plasma globulin levels are increased in the majority of cases and the albumin level is often decreased. The EPP may show the 'sarcoid step pattern', i.e. stepwise increase of $\alpha_2$, β and γ fractions.

### MALIGNANT PARAPROTEINAEMIA

Multiple myeloma in a male, aged 49 years.

| | | | | | |
|---|---|---|---|---|---|
| Plasma | TP | 118 | g/l | (60–85) |
| | Alb. | 39 | g/l | (30–50) |
| | AP | 51 | U/l | (30–120) |
| | ALT | 10 | U/l | (<35) |
| | Bili. | 6 | μmol/l | (<20) |
| | Glob. | 79 | g/l | (28–38) (by substraction) |
| | Ca | 2.76 | mmol/l | (2.15–2.55) |
| | Urate | 0.52 | mmol/l | (0.2–0.45) |

EPP: paraprotein present (65 g/l), normal γ-globulins decreased
Urinary Bence-Jones proteins +ve

Malignant proliferation of plasma cells results in disordered immunoglobulin synthesis, and increased production of one immunoglobulin type (monoclonal). In 60% of cases the paraprotein is IgG, 16% are IgA and 15% are IgM. IgD and IgE types are very rare. Very often the production of the normal immunoglobulin is reduced so that the γ region of the EPP is fainter than normal, against which the sharp paraprotein band is clearly seen.

Urinary Bence-Jones proteins (immunoglobulin light chain) are present in the majority of cases and may be the only abnormality, other than decreased plasma γ-globulins, in up to 10% of cases. The plasma urate is raised (increased nucleic acid turnover) in 50–60% of cases, whilst hypercalcaemia, as a result of increased bone resorption due to tumour deposits, occurs in 25–50% of cases.


## MACROGLOBULINAEMIA

In this condition the paraprotein is of the IgM type. It is produced by malignant cells which resemble lymphocytes rather than plasma cells. Bence-Jones protein may be detected if urine is concentrated ∼100 fold.


## HEAVY CHAIN DISEASE

This is a rare disorder in which the paraprotein is due to excessive production of part of a heavy chain.


## BENIGN PARAPROTEINAEMIA

A number of conditions that are usually associated with a polyclonal increase in globulins may occasionally present with a paraproteinaemia. These are termed benign paraproteinaemias and may occur in the following conditions: (1) autoimmune diseases, (2) chronic liver disease, (3) chronic infections.

These benign conditions usually have the following features.
(1) Paraprotein is less than 20 g/l and does not increase over a long period.
(2) Plasma albumin level is normal.
(3) Bence-Jones protein is absent.

ESSENTIAL PARAPROTEINAEMIA

Paraproteins are often found in patients who are otherwise normal. These paraproteins are more common in the older age groups ($>60$ years), and in some series have accounted for up to 20% of all paraproteins.

## HYPOGAMMAGLOBULINAEMIA

Hypogammaglobulinaemia is due to the following.

*Decreased synthesis* (1) transient, (2) primary, (3) secondary.

*Protein loss* Nephrotic    lymphoma

**Case examples/pathophysiology**

TRANSIENT HYPOGAMMAGLOBULINAEMIA

The circulating IgG of the newborn is derived from placental transfer from the maternal circulation (passive immunity). During the first 3–4 months after birth the level of $\gamma$-globulin decreases (half-life $\sim$ 28 days), and then rises as endogenous production increases. During this transition babies are sometimes more prone to infection. Occasionally endogenous production may be delayed, producing a physiological hypogammaglobulinaemia.

PRIMARY HYPOGAMMAGLOBULINAEMIA

There are a number of rare syndromes that are due to an inability of the immune system to produce immunoglobulins. These defects may be so severe that there is a complete absence of $\gamma$-globulins (agammaglobulinaemia). These children suffer from severe recurrent infections.

## SECONDARY HYPOGAMMAGLOBULINAEMIA

Male, aged 50 years, with a widespread lymphoma.

Plasma  TP    55 g/l   (60–85)
        Alb.   34 g/l   (30–50)
        Glob.  21 g/l   (28–38) (by subtraction)

Hypogammaglobulinaemia is commonly associated with malignancy, especially those of the haemopoietic and lymphoreticular systems. It may be due to replacement of the immunoglobulin producing cells by the tumour, or related to therapy causing damage to healthy plasma cells (chemotherapy, irradiation). A similar picture may be seen in uraemia and diabetes mellitus.

### PROTEIN LOSS

Nephrotic syndrome in a boy aged 14 years.

Plasma  TP    47 g/l   (60–85)
        Alb.   26 g/l   (30–50)
        Glob.  21 g/l   (28–38) (by subtraction)
Urine   TP    12 g/l   ($<0.15$)

In protein losing states the loss of the small molecular weight IgG molecules often results in hypogammaglobulinaemia. As well as in the nephrotic syndrome this situation may be seen in protein-losing enteropathy and severe burns.

## URINARY PROTEINS

The glomerular membrane allows the passage of only small amounts of low molecular weight proteins ($<50\,000$–$60\,000$ Daltons, depending on shape). In normal subjects the glomerular filtrate contains approximately 10–30 mg of protein (mainly albumin) per 100 ml of filtrate, i.e. the glomeruli can filter up to 50 g of protein a day. The major portion of this protein is reabsorbed in the proximal renal tubule and catabolized. Only 100–150 mg/day escapes in the urine (the upper reference limit is generally considered to be 150 mg/day). In some disease states small molecular proteins other than albumin may appear in the urine. Such proteins include Bence-Jones protein, haemoglobin, myoglobin and certain nucleoproteins. If the glomerular permeability is increased (e.g. nephrotic syndrome) massive amounts of protein may appear in the urine (up to 30 g/day). Proteinuria can also be due to the formation of protein within the urinary tract, e.g. urinary infections.

Chapter 11    # PROTEINURIA (>0.15 g/day)

*Nephrotic syndrome*

*Benign proteinuria*   (1) Fever, (2) severe exercise, (3) anaemia, (4) orthostatic proteinuria.

*Escape of small molecular weight proteins in disease states*   (1) haemoglobinuria, (2) myoglobinuria, (3) Bence-Jones protein.

*Protein produced within the urinary tract*   e.g. infections.

## Case examples/pathophysiology

### ORTHOSTATIC PROTEINURIA

Boy, aged 11 years. Proteinuria picked up at a school health clinic. Lumbar lordosis was prominent. Subject otherwise healthy, taking part in a lot of sporting activities at school.

|  | 24 hour collection 9.00 pm–9.00 pm | Spot urine 8.00 am (on rising) |
|---|---|---|
| Urinary TP | 4 g/24 h (<0.15 g/24 h) | 0.10 g/l (<0.15) |

This condition, which has been variously described as essential albuminuria, postural proteinuria and orthostatic proteinuria, is benign and is usually associated with lumbar lordosis. The proteinuria is cyclic in nature, occurs more commonly in boys than girls, and can be present in several members of a family. The pathophysiology is unclear but may be caused by 'kinking' of the renal vein as a consequence of the upright posture. It is an important condition in that it may be confused with the nephrotic syndrome. The diagnosis is made by demonstrating a normal urinary protein level in a sample produced immediately after the subject arises from bed in the morning.

### NEPHROTIC SYNDROME

Male aged 14 years, pale, oedematous (legs and ascites) and generally unwell.

| Date: | 6/9 | 13/9 | 10/10 | 9/12 | 24/12 | |
|---|---|---|---|---|---|---|
| Urine TP (g/24 h) | 11.6 | 9.6 | 7.8 | 6.7 | 1.6 | (<0.15) |
| Plasma Alb. (g/l) | 24 | 25 | 27 | 28 | 32 | (30–50) |

Steroids (prednisolone) were commenced on the 7/9

The causes of the nephrotic syndrome are multitudinous, ranging from the idiopathic type (usually treatable) to the secondary types (diabetes mellitus, amyloid, renal, toxins, etc.) in adults. These secondary types do not have a favourable prognosis and respond poorly to treatment with steroids.

The basic cause in all cases is increased filtration of protein by the glomeruli due to increased glomerular permeability. In the above case the patient had a 'minimal' glomerular lesion, i.e. glomeruli appeared normal with a haemotoxylin and eosin stain, but electron microscopy showed minimal changes. The primary type is most often found in children and usually responds to steroid therapy, but unfortunately the adult (secondary) types are usually progressive and resistant to treatment. 'Minimal change' types occur occasionally in adults and respond to treatment, but these cases are rare.

*Note* One of the dilemmas of renal medicine is deciding when to investigate proteinuria by a renal biopsy. Nephrologists generally suggest that a biopsy should not be performed unless the proteinuria exceeds 0.5 g/24 h.

## PROTEINS OF OTHER FLUIDS

*Serosal fluids*

*Cerebrospinal fluid (CSF)*

**Serosal fluids**

Tapping of body cavity (peritoneal, pleural, pericordium) fluids is usually done with two objects in mind: (1) relief of symptoms and (2) diagnosis of the cause of the fluid accumulation.

In the diagnostic area the clinician is usually concerned with resolving whether the fluid is a transudate (oedema fluid) or an exudate (infections, malignancy, etc.). The laboratory may be helpful in this area by quantitating the total amount of protein in the fluid. The TP of a transudate is usually less than 40 g/l, whereas that of an exudate is usually greater than 40 g/l. In many cases this analytical procedure may not be helpful, and better diagnostic discrimination is usually obtained by culturing for bacteria (or viruses) or examining the fluid for malignant cells by microscopy.

## CEREBROSPINAL FLUID

Cerebrospinal fluid (CSF) is a secretion that is produced by the choroid plexus, a tissue that is situated in the lateral and fourth ventricles of the brain. From these ventricles the fluid flows in two directions:

(1) Outwards and across the outer surface of the cerebrum through to the subarachnoid space. The CSF is absorbed in this area by structures called the arachnoid villi.

(2) Downwards through the spinal canal. Absorption of the CSF in this area is probably via the local capillaries and lymphatics.

The mechanism of CSF production is not clearly understood, but it seems probable that the secretion represents a fairly modified ultrafiltrate of plasma. A comparison of the compositions of CSF and plasma reveals that:

(1) CSF protein concentration is about one hundredth that of plasma.

(2) CSF calcium levels are lower and closely approximate the plasma ionized concentration.

(3) CSF chloride concentration is higher than in plasma, probably due to a Donnan equilibrium effect.

(4) Glucose levels are lower in CSF, usually by about 1 mmol/l.

(5) $CO_2$ diffuses rapidly, and $HCO_3^-$ slowly, from the plasma into CSF, so that in certain situations the pH of the CSF and plasma can markedly differ.

### Laboratory investigation

SAMPLING

After the needle has been inserted between the third and fourth vertebrae and into the spinal canal the fluid pressure is usually measured, followed by the taking of three separate samples:

(1) 1-3 ml—discarded.

(2) 1-3 ml—for biochemical tests (0.5-1.0 ml in a plain tube for total protein estimation, and 0.5 ml in a fluoride-oxalate tube for glucose analysis).

(3) 1-3 ml—for microbiological analysis.

## APPEARANCE

The appearance of each sample should be carefully noted, comparing against a tube of distilled water.

*Clear and colourless*   Normal.

*Bright red*   Blood present. Providing the operator is certain that he has tapped the spinal canal, the uniform appearance of fresh blood in all samples strongly supports a diagnosis of subarachnoid haemorrhage of recent occurrence. Sometimes, in the process of pushing the needle through to the canal, blood from damaged vessels enters the needle. In such situations blood may be found in the first sample, which usually quickly disappears from subsequent samples.

*Xanthochromic (yellow)*   Suggests a haemorrhage has occurred in the recent past (days), the colour being due to the breakdown of haemoglobin to bilirubins or (2) jaundice.

*Turbid*   Can indicate the presence of white cells and is suggestive of bacterial or viral infection.

Total protein

Reference range:

| Age | g/l |
| --- | --- |
| <1 m | <1.94 |
| 1–3 m | <0.96 |
| 3–12 m | <0.47 |
| 1–10 y | <0.36 |
| >10 y | <0.75 |

Increased levels may be found in: (1) infection (pus, white cells, etc.), (2) blood contamination, (3) chronic inflammatory disorders of the CNS (a) tuberculosis, (b) syphilis, (c) multiple sclerosis, (d) chronic alcoholism, (e) Guillain-Barré (4) Froin's syndrome—in this situation there is a blockage of the spinal canal (neoplasm, infection, prolapsed vertebral disc) and the fluid is usually xanthochromic, has a very high protein content and often clots spontaneously.

Immunoglobulins

This measurement has become a popular procedure, but its diagnostic value is yet to be proven. An increased IgG is often associated with multiple sclerosis.

Glucose

Normal range: 1.7–3.9 mmol/l. Decreased levels occur in: (1) infection (local metabolism by ↑ white cells), (2) hypoglycaemia.

Increased levels occur in hyperglycaemia; unfortunately some cases of diabetes mellitus are brought to the clinician's attention because of the fortuitous finding of a high CSF glucose concentration. CSF glucose concentrations are normally lower than the plasma level, often by up to 1 mmol/l, therefore it is essential to measure the patient's plasma glucose level at the same time that a CSF specimen is taken, so as to allow the CSF glucose value to be interpreted.

CSF chloride

Usually about 20 mmol/l higher than in plasma, and changes in parallel with the plasma level, e.g. the low CSF levels sometimes found in cerebral tuberculosis is caused by the fall in plasma chloride following vomiting.

SELECTION OF TESTS

The most important diagnostic procedures on CSF are: (1) appearance—compare with distilled water, (2) white cell counts—polymorphs, lymphocytes, pus cells, (3) bacteriological microscopy and culture, (4) detection of malignant cells.

In general the measurement of CSF protein and glucose has proved to be of little additional diagnostic value when investigating possible infectious pathologies of the CNS. However, the estimation of total protein can be useful in cases of inflammatory processes in the CNS, whilst the analysis of immunoglobulins have shown that abnormal immunoglobulins are detectable in demyelination situations such as multiple sclerosis.

**Case examples/pathophysiology**

PNEUMOCOCCAL MENINGITIS

A 21 month old girl, generally unwell with a cough and mild neck stiffness.

| | | | | |
|---|---|---|---|---|
| CSF | Appearance | Turbid | | (clear) |
| | Polymorphs | 48 | per µl | (<2) |
| | Lymphocytes | 60 | per µl | (<2) |
| | TP | 0.88 | g/l | (<0.36) |
| | Glu. | 2.5 | mmol/l | (1.7–3.9) |
| Plasma | Glu. | not taken | mmol/l | (3.0–5.5) |

Microbiology

Gram-positive diplococci seen on microscopy. The white cell counts and the microscopy gives an adequate diagnosis for the appropriate treatment to be commenced. Note that without a plasma glucose value it is difficult to comment on whether the CSF glucose is lower than might be expected. Streptococcus pneumoniae was later cultured from this CSF specimen.

### GUILLAIN-BARRÉ SYNDROME

A 23 year old man with increasing peripheral neuropathy.

| Date | | 13/11 | 27/11 | |
|---|---|---|---|---|
| Appearance | | Clear | Clear | |
| CSF | Red cells | 2 | 0 | (0) |
| | Lymphocytes | 0 | 2 | (<2) |
| | Polymorphs | 0 | 0 | (<2) |
| | TP | 0.27 | 1.4 g/l | (<0.75) |
| | Glu. | 3.3 | 3.2 mmol/l | (1.7–3.9) |

The elevated CSF total protein present in the second tap fluid in the absence of a raised white cell count strongly supported a clinical diagnosis of Guillain-Barré syndrome.

### VIRAL MENINGITIS

A 16 year old boy complaining of headache, sore throat, nausea and slight fever (38.8°C). No neck stiffness.

| CSF | Appearance | Clear | |
|---|---|---|---|
| | Red cells | 16 | (0) |
| | Lymphocytes | 48 | (<2) |
| | Polymorphs | 1 | (<2) |
| | TP | 0.66 g/l | (<0.75) |
| | Glu. | 3.5 mmol/l | (1.7–3.9) |

No organisms seen or cultured. No viruses isolated. An elevated lymphocyte count is strongly suggestive of viral meningitis and allowed the conservative treatment of bed rest and aspirin.

# 12 Plasma enzymes

Enzymes are proteins that act as biological catalysts, altering the rate, but not the final equilibrium, of metabolic reactions. They are most easily measured in terms of their biochemical activity rather than their mass. Although a great diversity of enzymes are found in the tissues and fluids of the body, the assays of relatively few have found a regular place in the routine laboratory.

When cells die or are damaged some of their intracellular enzymes are eventually washed into the plasma, where their activities are readily measured; the normal levels of most of the routinely assayed plasma enzyme activities are derived from the natural leakage and turnover of the body tissues. When an organ or tissue is damaged by a disease process there may be increased cellular destruction, and a consequent increased release of cellular enzymes into the bloodstream.

Thus in principle, the measurement of appropriate plasma enzyme activities could yield organ-specific diagnostic information. However in practice, most commonly used enzymes have a widespread distribution throughout the body tissues and are therefore not organ-specific. The measurement of plasma enzyme activities are thus of most value in confirming a diagnosis (e.g. myocardial infarct) and for monitoring the course of a disease process (e.g. infectious hepatitis).

Isoenzymes are different molecular forms of the same enzyme, usually deriving from different tissues, and may often be separately measured or detected by exploiting differing biological or physicochemical properties. However, the assay methods are often complex and, as isoenzyme estimations do not routinely improve diagnostic information, their use has so far been reserved for determining the source of a raised plasma enzyme activity when the clinical findings and other biochemical tests do not suggest the tissue of origin.

Enzymes are present in low mass concentrations in normal plasma ($<2$ g/l). Increased levels of a plasma enzyme activity may be due to:

(1) Increased cellular release (a) cell damage, e.g. raised creatine kinase in myocardial infarct, (b) ↑ cell population, e.g. raised alkaline phosphatase during growth.

(2) Enzyme induction, e.g. γ-glutamyltransferase in alcoholism.

(3) Decreased excretion or degradation, e.g. raised amylase in renal failure.

Decreased levels of plasma enzyme activity may be caused by:
(1) Reduced cell population, e.g. severe liver failure.
(2) Genetic deficiencies, e.g. hypophosphatasaemia.
(3) Reduced activity, e.g. organophosphate poisoning of pseudocholinesterase.

## AMYLASE—ACUTE PANCREATITIS

*Source*  Pancreas, salivary glands, lung, fallopian tubes, ovary, prostate.

*Isoenzymes*  (1) p—pancreatic, (2) s—salivary.

*Plasma half-life*  ~12 hours.

*Excretion/degradation*  (1) kidney—25% (clearance = 1-4 ml/min), (2) extrarenal—75%.

### Laboratory investigation

Plasma amylase estimations are almost exclusively used to confirm or exclude a diagnosis of acute pancreatitis. Acute pancreatitis is a condition of multiple aetiology, characterized by severe acute abdominal pain which often radiates through to the back. The pathology varies from pancreatic oedema to severe necrosis and haemorrhage of the organ. Such damage allows the release of pancreatic enzymes (amylase, lipase, proteases) which then act on, and cause further damage to, the local healthy tissue (e.g. fat necrosis). The enzymes are absorbed into the bloodstream, where very high levels of activity may often be demonstrated. There is no specific biochemical test for the diagnosis, but an attack of acute pancreatitis can be associated with a number of abnormal parameters. Some or all of the following tests are helpful in the evaluation of hyperamylasaemia.

| | |
|---|---|
| Plasma amylase | Liver function tests |
| Amylase isoenzymes | Plasma electrolytes and urea |
| Urinary amylase | Plasma lipids |
| Plasma lipase | Plasma calcium |
| Amylase/creatinine ratio | Plasma glucose |

PLASMA AMYLASE (Amy.)

Hyperamylasaemia is not specific for acute pancreatitis, but extremely high levels of plasma amylase activity (>5000 U/l) make the diagnosis likely. During an acute attack maximal plasma levels are reached in 24–48 hours, with a return to normal in 3–6 days. Since the amylase is cleared rapidly from the plasma the time of blood sampling in relation to the symptoms should be taken into account when interpreting the results.

Intra-abdominal emergencies of a non-pancreatic nature (e.g. perforated peptic ulcer) may also be associated with high plasma amylase activity, but rarely above 5000 U/l. Thus moderate rises in plasma amylase activity associated with an acute abdomen presents a diagnostic dilemma, which can only be resolved by careful clinical assessment and observation of the patient.

It should be noted that:
(1) The level of plasma amylase activity does not correlate with the severity of the pancreatitis.
(2) Plasma amylase levels in severe haemorrhagic pancreatitis may be normal, or only slightly elevated.
(3) A high plasma level ($>$ 5000 U/l) supports, but does not confirm, the diagnosis of acute pancreatitis.
(4) Plasma amylase levels are often not helpful in distinguishing acute pancreatitis from diseases masquerading as pancreatitis.

### AMYLASE ISOENZYMES

Although techniques are becoming available for the separate estimation of amylase isoenzymes (pancreatic, salivary), the measurement of isoenzymes is unlikely to resolve the major diagnostic problem of differentiating acute pancreatitis from non-pancreatic abdominal emergencies (perforated peptic ulcer, gut obstruction, etc.). This is because, in the latter situation, the pancreatic amylase that is normally present in the gut lumen leaks through the intestinal wall into the peritoneal cavity and from there it is absorbed into the bloodstream.

### URINARY AMYLASE

With normal renal function amylase is rapidly cleared from the plasma and, therefore, the measurement of urinary amylase does not provide additional diagnostic information where the plasma amylase is found to be raised. Since the rate of urinary excretion of amylase may remain high for some hours after the plasma amylase level has returned to normal, the estimation of urinary amylase can be of value in cases where a strong clinical suspicion of pancreatitis is unsupported by a plasma amylase estimation, particularly if the abdominal pain occurred several days prior to the investigation.

A low or normal urinary level of amylase, in association with a raised plasma activity, can arise in renal failure and in the condition termed 'macroamylasaemia'.

AMYLASE-CREATININE CLEARANCE RATIO

This ratio takes into account the glomerular filtration rate and is calculated from the following formula:

$$\text{clearance ratio} = \frac{Ua \times Pc}{Pa \times Uc} \times 100\%$$

where $Ua$ = urinary amylase (U/l); $Pa$ = plasma amylase (U/l); $Pc$ = plasma creatinine (mmol/l); $Uc$ = urinary creatinine (mmol/l).

Normally this value is less than $\sim 7\%$. Increased levels are said to occur in acute pancreatitis. The increased clearance of amylase in acute pancreatitis is thought to be due to either the rapid clearance of the p-isoenzyme, or to renal tubular dysfunction secondary to acute pancreatitis. Recent studies have shown that the ratio is elevated in a variety of abdominal conditions and therefore this test is an unreliable indicator of acute pancreatitis.

Increased ratios also occur in: severe burns, diabetic ketoacidosis, after surgical exploration of the common bile duct.

PLASMA LIPASE

Pancreatic lipase is released during attacks of pancreatitis, but the assay of this enzyme generally does not provide additional diagnostic information. Plasma lipase estimations can be of value in situations when the source of a hyperamylasaemia may be in doubt (e.g. mumps).

LIVER FUNCTION TESTS

Biliary tract disease (especially cholecystitis, cholelithiasis, and stones in the common bile duct) is associated with 30–75% of cases of acute pancreatitis. These conditions may also be associated with abnormal liver function tests. The other major condition associated with acute pancreatitis is alcoholism and therefore in some cases there will be abnormalities of the liver function tests where alcoholic liver disease is present.

PLASMA ELECTROLYTES AND UREA

Moderate hyperamylasaemia (up to 1000 U/l) occurs in renal failure, whilst acute pancreatitis itself can cause acute renal failure. Plasma electrolyte abnormalities often occur in acute pancreatitis due to vomiting and the consequent dehydration.

## PLASMA LIPIDS

Two forms of hyperlipidaemia may be associated with acute pancreatitis.
(1) Familial hyperlipidaemia. Some of these conditions are associated with a high incidence of acute pancreatitis (see p. 268).
(2) A temporary hyperlipidaemia may occur during an episode of acute pancreatitis. This condition can arise in alcohol associated pancreatitis and may therefore reflect chronic alcoholism. Alternatively, there may be a genetic disorder of lipid metabolism which has been unmasked by the stress of the acute attack of pancreatitis.

## PLASMA CALCIUM

*Hypocalcaemia*   Over half of the cases of acute pancreatitis have an associated hypocalcaemia (see p. 189 for possible mechanisms). The fall in plasma calcium occurs between the first and fifth days after the onset of symptoms, and may persist for up to 10 days. The plasma calcium may be reduced to levels of the order of 1.6–1.9 mmol/l. Clinical manifestations of hypocalcaemia (e.g. tetany) are rare.

*Hypercalcaemia*   Reports from the literature suggest that the incidence of acute pancreatitis in primary hyperparathyroidism is higher than that in the general population. Hypercalcaemia *per se* has also been invoked by some authorities as a possible cause of pancreatitis.

## PLASMA GLUCOSE

Plasma glucose levels in acute pancreatitis are variable but are often increased, probably due to stress or a temporary decrease in insulin secretion. Some 30–60% of cases of severe ketoacidosis will have an associated hyperamylasaemia, the plasma amylase levels often approaching those seen in acute pancreatitis. The cause is unclear, and there is no other evidence that these patients have a concurrent pancreatitis.

# HYPERAMYLASAEMIA

Hyperamylasaemia may occur in: (1) pancreatic disease, (2) non-pancreatic abdominal emergencies, (3) miscellaneous conditions.

## Pancreatic disease

Diseases of the pancreas which may result in hyperamylasaemia are: (1) acute pancreatitis, (2) carcinoma of the pancreas. In chronic pancreatitis the plasma amylase levels are generally normal, or decreased due to the loss of functional exocrine tissue.

## Case examples/pathophysiology

### ACUTE PANCREATITIS

This may be associated with: (1) biliary tract disease, (2) alcohol ingestion, (3) hyperlipidaemia, (4) hyperparathyroidism, (5) drug therapy (a) thiazides, (b) frusemide, (c) azathioprin, (d) glucocorticoids, (6) hereditary pancreatitis, (7) pancreatic trauma.

Alcohol ingestion

A 38 year old male with a history of alcoholism, who presented at Accident and Emergency with severe abdominal pain that radiated through to the back. Onset of pain occurred 3 hours after a heavy bout of drinking.

|  | 18 hours postsymptoms | 3 days postsymptoms | | |
|---|---|---|---|---|
| Plasma Amy. | 10 800 | 280 | U/l | (<300) |
| Ca | 1.95 | 2.37 | mmol/l | (2.15–2.55) |
| Alb. | 38 | 40 | g/l | (30–50) |

Alcohol is responsible for a large proportion of all cases of acute pancreatitis. Alcohol associated pancreatitis is more common in males, with the first attack usually occurring before the age of 40 years. This case also illustrates how rapidly the plasma amylase levels can return to normal. Note the moderate hypocalcaemia (see p. 189).

Chapter 12    Biliary disease

A 44 year old female with a past history of biliary disease, who presented with a history of severe acute central abdominal pain which was preceded by several bouts of biliary colic. There was moderate jaundice.

| Plasma | Amy. | 6850 | U/l | (<300) |
|---|---|---|---|---|
| | Alb. | 44 | g/l | (30–50) |
| | AP | 285 | U/l | (30–120) |
| | ALT | 305 | U/l | (<35) |
| | Bili. | 124 | µmol/l | (<20) |

This patient had acute cholecystitis and a stone impacted in the common bile duct. The liver function tests reflect hepatocellular damage (↑ ALT) and cholestasis (↑ AP), caused by the bile duct obstruction and probable ascending cholangitis. Biliary tract disease is present in 30–75% of all cases of acute pancreatitis. This type of pancreatitis is usually mild, and tends not to recur once the biliary disease has resolved.

Hyperlipidaemia

A 22 year old woman with acute abdominal pain. She had a past history of similar attacks since the age of 16 years. The blood specimen taken on admission was noted by the laboratory to be severely lipaemic.

| Plasma | Amy. | 5425 | U/l | (<300) |
|---|---|---|---|---|
| | Chol. | 29.5 | mmol/l | (<6.5) |
| | Trig. | 103.7 | mmol/l | (0.3–1.7) |

As noted above, there are two forms of hyperlipidaemia associated with acute pancreatitis, the familial form, and the transitory type. In the familial type (above case) the onset of pancreatitis occurs in childhood or early adult life and takes the form of recurrent episodes. The hyperlipidaemia will not disappear on clinical improvement, as it does in the transitory type. The cause of the pancreatitis in familial hyperlipidaemia is unclear.

CARCINOMA OF THE PANCREAS

A male, aged 52 years, with severe abdominal pain radiating through to the back. At laparotomy a carcinoma of the head of the pancreas was found.

| Date | 27/5 | 30/5 | 4/6 | | *Plasma enzymes* |
|---|---|---|---|---|---|
| Plasma Amy. (U/l) | 1500 | 500 | 800 | (<300) | |

Carcinoma of the pancreas is only rarely associated with a raised plasma amylase. If present, the hyperamylasaemia is usually moderate, fluctuates and persists. These high plasma levels may be due to: (1) associated pancreatitis, (2) destruction of acinar tissue by the carcinoma, (3) duct obstruction by the carcinoma, (4) synthesis of amylase by the tumour.

Persistently elevated plasma amylase levels may be due to: (1) persistent pancreatitis, (2) pancreatic cysts, (3) carcinoma of the pancreas, (4) salivary gland disease, (5) tumours, e.g. carcinoma of bronchus, (6) renal failure, (7) macroamylasaemia.

### Non-pancreatic abdominal emergencies

Acute abdominal pain and hyperamylasaemia occurs not only in acute pancreatitis, but also in the following conditions: (1) perforated peptic ulcer, (2) intestinal obstruction, (3) ischaemia and infarction of the small intestine, (4) ruptured ectopic pregnancy, (5) salpingitis.

In these conditions there is leakage of amylase from the diseased organ (intestine, fallopian tubes) into the peritoneal cavity, where the enzyme is absorbed into the bloodstream. The plasma levels may be quite high (up to 5000 U/l) and approximate those found in acute pancreatitis.

*Example* A male, aged 36 years, with a history of duodenal ulceration for 5 years. He had been vomiting blood for 6 hours, and on admission had severe diffuse abdominal pain. The plasma amylase level was 5800 U/l (<300). A laparotomy revealed the presence of a perforated duodenal ulcer. This case illustrates the point that high plasma amylase levels are not always due to acute pancreatitis, and that an acute abdomen associated with hyperamylasaemia requires careful clinical assessment.

### Miscellaneous conditions

Moderate hyperamylasaemia (<3000 U/l) may be associated with:
(1) Salivary gland disease—mumps, obstruction of salivary gland ducts.
(2) Tumours—carcinoma of bronchus, colon, pancreas and papillary cystadenoma of the ovary.
(3) Renal insufficiency.

(4) Diabetic ketoacidosis.
(5) Macroamylasaemia.
(6) Drug therapy—opiates.

SALIVARY GLAND DISEASE

A female, aged 7 years, with mumps, abdominal pain and vomiting.

Plasma Amy.   2700 U/l   (<300)

In mumps the hyperamylasaemia may be due to either parotitis or pancreatic involvement, the latter situation being a potentially serious complication of the disease. The abdominal pain in this patient suggests pancreatic involvement, although proof would require the estimation of plasma lipase levels (normal in uncomplicated parotitis), or isoenzyme determinations.

TUMOURS

Many reports in the literature have indicated that hyperamylasaemia may occur in association with malignant tumours, especially carcinoma of the bronchus. Many cell types normally synthesize amylase, but it is not clear if the hyperamylasaemia is due to 'ectopic' production of the enzyme, or whether the raised amylase levels reflect 'normal' secretion by the tumour. Other tumours that may be associated with high plasma amylase levels include carcinomas of the colon and pancreas, and cystadenoma of the ovary.

RENAL INSUFFICIENCY

A 49 year old male with chronic renal failure.

Plasma Amy.   600   U/l        (<300)
       Creat.   0.91 mmol/l   (0.06–0.12)

A major route of amylase excretion is via the kidney, so it is to be expected that increased plasma levels occur in renal failure. The levels rarely rise above 1000 U/l in renal failure.

DIABETIC KETOACIDOSIS

A female, aged 26 years, with severe diabetic ketoacidosis.

Plasma Amy. 2425 U/l (<300)
Glu. 77.5 mmol/l (3.0–5.5)

Hyperamylasaemia occurs in up to 60% of cases of diabetic ketoacidosis, the plasma level often being above 1000 U/l. The cause is unknown, and there has been no other evidence to incriminate pancreatitis in these cases.

MACROAMYLASAEMIA

In this condition, which has been estimated to be present in up to 1.5% of the population, hyperamylasaemia is associated with a low renal secretion of the enzyme. The plasma enzyme is of high molecular weight and represents either binding of amylase to a plasma protein, or the formation of high molecular weight polymers. The condition is harmless, but it is important to recognize in the differential diagnosis of hyperamylasaemia.

DRUG THERAPY

Plasma amylase activity can rise after the administration of opiate drugs. The levels may reach 1000–2000 U/l and are due to drug induced constriction of the choledochal sphincter and pancreatic ducts, which in turn causes a rise in intraductal pressure and regurgitation of the enzyme into the plasma. This effect may cause confusion in the interpretation of hyperamylasaemia in a patient with abdominal pain who has been administered morphine.

# CARDIAC ENZYMES

Although cardiac muscle contains numerous enzymes, only three, creatine kinase (CK), aspartate aminotransferase (AST) and lactate dehydrogenase (LD), have been widely used for the routine diagnosis and evaluation of suspected cardiac disease.

Following a myocardial infarct the plasma levels of activity of these three enzymes follow a predictable pattern (Fig. 12.1).

Since the majority of myocardial infarcts can be diagnosed from the clinical findings and ECG changes, the plasma enzyme activities are of confirmatory value only. However, enzyme estimations can be of diagnostic value in those cases where the clinical findings are atypical, or the ECG is equivocal.

Using plasma enzyme activities for the evaluation of chest pain presents two major problems:

# Chapter 12

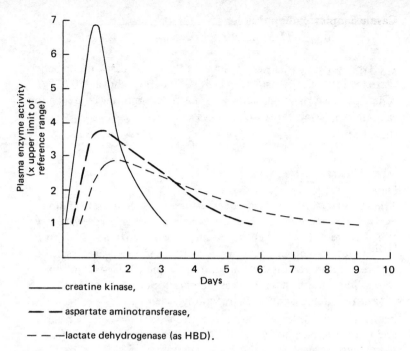

**Fig. 12.1** Typical plasma enzyme pattern following a myocardial infarct.

(1) After myocardial infarction the plasma levels reached by individual enzymes are dependent on the time of blood sampling (see Fig. 12.1).
(2) None of the above three enzymes is specific to cardiac muscle.

These difficulties can be overcome to a certain extent by: (1) selecting the appropriate enzyme on a time basis, in relation to clinical symptoms, (2) being aware of other causes of raised enzyme activity, (3) the use of isoenzymes.

## CREATINE KINASE (CK)

*Source* Cardiac muscle, skeletal muscle, brain.

*Isoenzymes* MM—skeletal muscle (also found in heart muscle), BB—brain (also found in a variety of other tissues), MB—cardiac muscle (also found in striated muscle).

**Plasma creatine kinase**

This may be elevated in: (1) cardiac disease (a) myocardial infarction, (b) myocarditis, (2) skeletal muscle damage—muscular dystrophy, (3) miscellaneous causes.

**Case examples/pathophysiology**

MYOCARDIAL INFARCTION

A 64 year old man admitted whilst complaining of substernal chest pain that radiated down the left arm. The ECG pattern confirmed this classical clinical picture of myocardial infarction (MI).

| Date | 12/01 | 13/01 | 14/01 | 15/01 | 16/01 | |
|---|---|---|---|---|---|---|
| Time | 16.00 | 08.00 | 08.30 | 08.00 | 08.00 | |
| CK | 48 | 1550 | 518 | 148 | 92 | U/l (30-140) |
| HBD | 163 | 826 | 985 | 750 | 615 | U/l (125-250) |

In acute myocardial infarction the level of plasma CK activity begins to rise within 3-6 hours, reaches a peak around 24 hours, and returns to normal by the third day. Plasma AST activity starts to increase some 6-12 hours post MI, whilst HBD (see p. 278) begins to increase by 12-24 hours, and may remain elevated for up to 10 days. The differences in the times at which these enzymes start to appear in plasma following infarct are due to at least the following factors:
(1) Relative concentration of the enzymes in the cardiac tissue.
(2) Intracellular site of the enzyme (mitochondria, cytoplasm, etc.).

The differences between the time periods for the enzymes to reach their peak activities and return to normal mainly depend on the rate of clearance of the individual plasma enzymes.

If a sample of blood is taken at the optimum time following an MI, a raised CK will be found in 98% of cases. Peak values of 1000-2000 U/l are commonly seen, but it is unusual for levels to surpass 5000 U/l. From the foregoing and the above case, it is clear that plasma CK values are of little use in confirming a diagnosis of MI after the third day, or within 3-6 hours, from the onset of symptoms. However, if both CK and HBD are assayed, then useful information is generally obtained, regardless of when the sample was taken during the first few days after symptoms, and the measurement of AST as an 'intermediate time' enzyme cannot be justified.

Many studies have attempted to correlate the peak CK value reached post MI with the degree of cardiac tissue damage, and although there does seem to be some correlation, no simple routine method for such calculation has yet emerged. Therefore, a single plasma sample for CK and HBD (or LD) each day until the levels start to decrease, provides the maximum available diagnostic information. Once the plasma levels start to fall there is no point in continuing daily tests, unless there is a clinical suggestion of infarct extension, since the known plasma half-lifes of the enzymes would allow a reasonable prediction of when normal levels will probably be reached (plasma half-lives: CK = ~1.5 d, $LD_1$ = ~4.5 d). A check on plasma CK and HBD levels just prior to discharge of uncomplicated cases of MI should be quite adequate.

In angina of effort the plasma CK levels are usually normal, but 5-10% of patients show marginal increases, presumably because the enzyme is leaking from anoxic cells.

MYOCARDITIS

Elevated plasma activity levels of CK, as well as AST and LD, have been reported in myocarditis.

SKELETAL MUSCLE DISORDERS

The highest plasma levels of CK activity are seen in cases of primary skeletal muscle disease (muscular dystrophies, myoglobinuria).

*Example*  An 8 year old boy with lower limb weakness and apparent hypertrophy of the calf muscles. A diagnosis of pseudohypertrophic muscular dystrophy was made.

| | | |
|---|---|---|
| Plasma CK | 1770 U/l | (<140) |
| AST | 477 U/l | (<35) |
| LD | 3000 U/l | (<150) |

LD and AST are also present in skeletal muscle, and therefore will also rise to high plasma values in this condition. As the disease progresses and finally 'burns out' there is widespread loss of muscle cells, and so the plasma enzyme levels will fall back towards normal.

Increased CK levels due to disorders of skeletal muscle occur in: (1) muscular dystrophy, (2) myoglobinuria, (3) dermatomyositis, (4) after generalized convulsions, (5) after severe exercise, (6) after trauma and surgical procedures involving muscle, (7) after IM injections.

The levels reached after intramuscular (IM) injections are variable, but may approach 2-3 times the upper limit of the reference range.

MISCELLANEOUS

Increased levels of CK occur in:

*Hypothyroidism* may be due to generalized increase in cell permeability.

*Cerebrovascular accidents* may be due to release of brain enzymes (BB isoenzymes), or to injury to skeletal muscle at the time of the accident.

*Hyperthermia and hypothermia* due to increased cell permeability, e.g. malignant pyrexia.

*Intramuscular injections* CK levels of up to 600 U/l are often observed.

Hypothyroidism

Female, aged 50 years, with typical symptoms and signs of myxoedema. There was no chest pain or evidence of skeletal muscle injury.

Plasma CK    582  U/l      (30–140)
       $T_4$    <10  nmol/l   (60–160)

60–80% of patients with hypothyroidism will show increased plasma CK activity. Values can reach eight to 10 times the upper limit of the reference range. Isoenzymes studies suggest that skeletal muscle is the origin of the enzyme (MM). The exact mechanism of enzyme release is uncertain, but it is thought to be due to increased muscle cell membrane permeability.

## CK isoenzymes

About 15% of the heart CK is in the form of the MB isoenzyme, the remaining 85% being MM. It should be noted that skeletal muscle also contains some MB isoenzyme (<0.4%). The amount of MB isoenzyme increases in the plasma after a myocardial infarct, and therefore in theory provides a fairly specific test. However, it has not gained widespread acceptance as a routine test for the assessment of chest pain, because of the following.
(1) The great majority of myocardial infarcts can be diagnosed on the strength of the clinical picture and the ECG pattern.
(2) Increased plasma levels of MB are invariably accompanied by an increase in the total CK, i.e. if other causes of an increased CK can be ruled out, the raised plasma CK must originate from the heart muscle.

The most useful application of estimating plasma MB activity is in the investigation of those patients suspected of having an infarct, but who already have increased plasma CK levels, e.g. postoperative state.

Chapter 12  ASPARTATE AMINOTRANSFERASE (AST)

*Source* Heart, liver, skeletal muscle, pancreas, erythrocytes.

*Isoenzymes* Heart and liver contain two types of AST, which are located in the (1) mitochondria, (2) cytosol.

**Plasma aspartate aminotransferase**

Plasma aspartate aminotransferase may be raised in: (1) myocardial disease, (2) liver disease, (3) skeletal muscle disease, (4) miscellaneous.

**Case examples/pathophysiology**

MYOCARDIAL DISEASE

Increased plasma levels occur after a myocardial infarct, or during myocarditis.
Myocardial infarct in a 55 year old man. Blood sample was taken 2 days after the onset of pain.

Plasma CK   1760 U/l   (<140)
       AST    180 U/l   (<35)
       LD     600 U/l   (<150)

After a myocardial infarct the plasma activity of AST begins to rise in 6–12 hours, reaches a peak level in 24–36 hours and returns to normal in 5–6 days. 95–97% of patients with a myocardial infarct show raised plasma AST levels with peak levels reaching from two to 10 times the upper reference limit. If CK and HBD (or LD) activities are measured, then AST is unlikely to provide additional diagnostic information in these cases.

LIVER DISEASE

In hepato-biliary disease AST levels parallel the changes seen in plasma ALT (see Chapter 13).

SKELETAL MUSCLE DISEASE

Skeletal muscle is rich in AST, and therefore plasma levels rise when the muscle is damaged.

Increased plasma levels are mainly seen in: (1) trauma, (2) muscular dystrophies, (3) dermatomyositis, (4) myoglobinuria.

*Example*  Myoglobinuria of uncertain aetiology in a 30 year old male.

Plasma CK   3200 U/l  (<140)
       AST    322 U/l  (<35)
       LD     598 U/l  (<150)

MISCELLANEOUS

Increased AST levels may be seen in: (1) haemolysis (*in vivo* and *in vitro*), (2) renal infarction, (3) acute pancreatitis.

## LACTATE DEHYDROGENASE (LD)

*Source*  Widespread, with the highest concentrations being found in heart, skeletal muscle, liver, kidney, erythrocytes.

*Isoenzymes*  Five types: $LD_1$, $LD_2$, $LD_3$, $LD_4$, $LD_5$ (based on electrophoretic mobility). (1) Heart—$LD_1$ and $LD_2$. (2) Liver—$LD_5$. (3) Skeletal muscle—variable.

**Plasma LD**

Plasma LD may be elevated in: (1) myocardial disease, (2) liver disease, (3) skeletal muscle disease, (4) miscellaneous.

**Case examples**

MYOCARDIAL DISEASE

Increased levels are found after infarction and in myocarditis, e.g. myocardial infarct, blood sample taken 4 days after onset of pain.

Plasma CK   103 U/l  (<140)
       AST    96 U/l  (<35)
       LD    965 U/l  (<150)

After myocardial infarction the plasma LD activity begins to rise in 12-24 hours, reaches maximum levels in 24-48 hours and returns to normal within 8-10 days. 90-95% of patients with infarction will show increased plasma levels of $LD_1$, generally of the order of two to 10 times the upper reference limit.

LD isoenzymes in cardiac disease

The heart isoenzymes ($LD_1$, $LD_2$) exhibit three features that enable them to be easily measured in the laboratory.
(1) Distinct electrophoretic mobilities—electrophoretograms show that in normal serum the $LD_2$ peak has a greater area (activity) than $LD_1$, whilst after a myocardial infarct $LD_1 > LD_2$, often termed 'flipped LD' (Fig. 12.2). LD electrophoresis is of value when the source of a raised total LD is unclear from clinical or other diagnostic tests.
(2) Heat stability.
(3) Ability to utilize the substrate α-hydroxybutyrate (hence these isoenzymes are collectively known as hydroxybutyrate dehydrogenase, (HBD)).

It has been suggested that HBD activity is a more specific test for heart muscle damage than total LD estimation. However, the other LD isoenzymes ($LD_{3,4,5}$) can also utilize HBD substrate to a

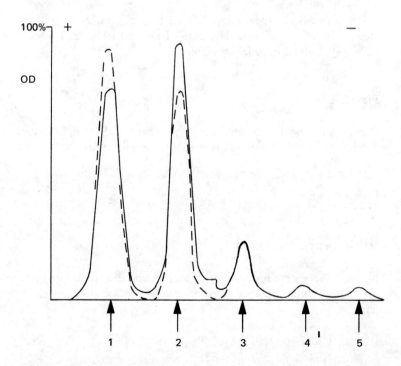

**Fig. 12.2** Electrophoretograms of LDH isoenzymes. ——— normal isoenzyme pattern, ——— post myocardial infarct.

certain extent, so that any condition that markedly raises the plasma levels of these isoenzymes will result in increased plasma HBD activity. Also, as $LD_1$ and $LD_2$ are present in small amounts in other tissues, plasma levels of HBD activity are increased when these tissues are damaged. Plasma HBD activity may be raised in the following disorders: (1) renal infarction, (2) haematological disorders, e.g. blast cell leukaemias, (3) skeletal muscle disease, (4) liver disease.

Therefore HBD cannot be considered as being a specific test for heart muscle damage.

### LIVER DISEASE

Plasma LD is increased in hepatocellular disease, due to the release of the isoenzyme $LD_5$. However, total LD estimation has not found wide acceptance as a test for the diagnosis of hepatobiliary disease.

### SKELETAL MUSCLE DISEASE

LD plasma levels are increased in the same diseases of skeletal muscle that result in rises in the plasma CK and AST (see p. 274). Isoenzymes of LD are not very helpful in determining the skeletal muscle origin of LD because all five isoenzymes have been reported as being increased in various muscular conditions.

### MISCELLANEOUS

Increased levels of plasma LD activity may also be seen in:

*Haematological disorders* (1) Haemolysis—*in vivo* and *in vitro*, (2) leukaemia—very high, (3) pernicious anaemia—very high, (4) myeloproliferative disorders.

*Malignancy* All types.

*Renal infarction*

*Pulmonary embolus*

## ALKALINE PHOSPHATASE (AP)

*Source* Bone (osteoblasts), liver, intestine, kidney, placenta.

*Isoenzymes* With careful gel electrophoresis four bands can be seen which correlate with the isoenzymes from bone, liver, intestine and placenta.

## Laboratory investigation

In the majority of cases of hyperphosphatasaemia (↑ AP) the source of the alkaline phosphatase is obvious from the clinical findings. However, in some patients a high level plasma AP activity may be the only biochemical abnormality detected, with the clinical findings proving unhelpful in determining its origin. In the majority of these cases the question to be resolved is whether the enzyme is of bone or liver origin. In this situation the measurement of one of the following enzymes can be helpful: (1) γ-glutamyltransferase (GGT), (2) 5'-nucleotidase (5'NT), (3) isoenzymes of alkaline phosphatase.

### GGT

This enzyme is present in liver, but not in bone, and therefore increased plasma levels suggest a liver pathology. However, plasma levels of GGT may be increased in the absence of overt liver disease, e.g. GGT is induced by both alcohol and certain drugs (barbiturates, phenytoin).

### 5'NT

This enzyme, although found in a variety of tissues, is usually only increased in the plasma in hepato-biliary disorders. It is not affected by drugs and alcohol. Unfortunately, the assay procedure for this enzyme is technically difficult and relatively insensitive, and so has not gained wide acceptance.

### ISOENZYMES OF ALKALINE PHOSPHATASE

On careful electrophoresis of plasma four isoenzyme bands of AP can be distinguished (liver, placenta, intestine, bone), but unfortunately there is a major difficulty in distinguishing the bone enzyme from the liver enzyme, as they migrate very close together in the support systems used. With these difficulties AP isoenzyme detection has so far had to remain as a semiresearch procedure.

PLASMA CALCIUM

A high plasma AP may be associated with hypocalcaemia in: (1) renal failure, (2) vitamin D deficiency.

A high plasma AP can be associated with hypercalcaemia in: (1) malignancy, (2) primary hyperparathyroidism.

## Plasma alkaline phosphatase

This may be raised due to: (1) physiology (a) children/young adults, (b) pregnancy, (2) liver disease—cholestasis, (3) bone disease—osteoblastic activity, (4) malignancy (a) Regan isoenzyme, (b) bone tumours, (c) liver tumours.

## Case examples/pathophysiology

PHYSIOLOGICAL

Shortly after birth the plasma AP level is approximately two and a half to three times the normal adult level. It remains at this level during bone growth until about the age of 12–15 years, when it begins to fall, reaching adult levels by 18–20 years of age.

### Children

*Example* A male aged 7 years. Biochemical screen prior to operation for acute appendicitis. There was no clinical evidence of bone or liver disease.

|        |       |         | Adult         |
|--------|-------|---------|---------------|
| Plasma Ca | 2.50 | mmol/l | (2.15–2.55)  |
| PO$_4$ | 1.45 | mmol/l  | (0.6–1.25)    |
| TP     | 73    | g/l     | (60–85)       |
| Alb.   | 46    | g/l     | (37–52)       |
| AP     | 205   | U/l     | (30–120)      |
| ALT    | 10    | U/l     | (<35)         |
| Bili.  | 6     | µmol/l  | (<20)         |

This high level of alkaline phosphatase reflects the increased osteoblastic activity that occurs during bone growth.

In the authors laboratory the reference range for alkaline phosphatase in children is:

| Age | U/l |
|---|---|
| Birth | 30–100 |
| 1 month | 70–250 |
| 1–3 year | 70–220 |
| 3–10 year | 70–180 |
| 10–16 year | 100–280 |

For similar reasons the plasma phosphate in children is higher.

| Age | $[PO_4]$ mmol/l |
|---|---|
| Neonate | 1.2–2.8 |
| 7 year | 1.3–1.8 |
| 15 year | 0.6–1.25 |

Pregnancy

During the third trimester of pregnancy the plasma AP level rises to two to two and a half times the normal adult level. The excess AP originates from the placenta. The placental AP isoenzyme is heat stable, and can be separately estimated using this property. However, the use of AP as a placental function test is no longer recommended.

Female, aged 26 years, at 36 weeks gestation.

| Plasma | | | |
|---|---|---|---|
| | TP | 63 g/l | (64–84) |
| | Alb. | 35 g/l | (37–52) |
| | AP | 195 U/l | (30–120) |
| | ALT | 12 U/l | (<35) |
| | Bili. | 9 µmol/l | (<20) |

LIVER DISEASE (see Chapter 13)

AP is located in the bile canaliculi and at the sinusoidal surface of the liver cells. The plasma level is raised in all liver conditions that are associated with cholestasis, the highest levels being seen in obstructive lesions (>3 times upper reference limit). In hepatocellular disease the plasma level may be normal or slightly raised (<3 times upper reference limit) due to concurrent cholestasis.

The mechanism for this increase in cholestasis is two fold: (1) increased enzyme synthesis, (2) regurgitation from obstructed ducts.

### Hepatocellular disease

A female aged 26 years, jaundiced, with a dark urine that gave a positive qualitative test for bilirubin. She had a history of contact with infectious hepatitis.

| Plasma | TP | 79 g/l | (64–84) |
|---|---|---|---|
| | Alb. | 42 g/l | (37–52) |
| | AP | 212 U/l | (30–120) |
| | ALT | 1495 U/l | (<35) |
| | Bili. | 69 μmol/l | (<20) |

This liver function test pattern is classical for infectious hepatitis. The raised AP is due to local obstruction caused by the infected and swollen liver cells. A small number of patients with infectious hepatitis will show a cholestatic picture with very high levels of alkaline phosphatase, presumably due to a greater degree of cellular swelling.

### Cholestatic disease

A 61 year old lady with deep jaundice, biliary colic and a past history of cholelithiasis.

| Plasma | TP | 67 g/l | (60–85) |
|---|---|---|---|
| | Alb. | 36 g/l | (30–50) |
| | AP | 1075 U/l | (30–120) |
| | ALT | 181 U/l | (<35) |
| | Bili. | 340 μmol/l | (<20) |

Very high levels of plasma alkaline phosphatase activity (>600 U/l) are commonly found in: (1) obstructive biliary disease, (2) secondary carcinoma of the liver, (3) Paget's disease.

## BONE DISEASE

AP is produced by the osteoblasts, and therefore increased activity of these cells results in further synthesis of the enzyme. Extremely high levels (up to 2000 U/l) may be found in Paget's disease, whilst the levels associated with bone malignancy (primary and secondary), osteomalacia (vitamin D deficiency) and primary hyperparathyroidism are usually only moderately increased (up to 200 ~ 300 U/l).

### Paget's disease

A 72 year old man with clinical and radiological evidence of Paget's disease.

| Plasma | Ca | 2.31 | mmol/l | (2.15–2.55) |
|---|---|---|---|---|
| | $PO_4$ | 1.07 | mmol/l | (0.6–1.25) |
| | Alb. | 43 | g/l | (37–52) |
| | AP | 1150 | U/l | (30–120) |

Paget's disease is characterized by an increased resorption of bone that is accompanied by increased bone formation. The new bone is laid down in a haphazard fashion, resulting in bone deformities that have a characteristic appearance on X-ray. The only detectable biochemical abnormality is a raised plasma alkaline phosphatase. The plasma [Ca] is usually normal, but hypercalcaemia may occur if the patient is immobilized; hypocalcaemia occasionally occurs if treatment with calcitonin is instituted.

Secondary carcinoma of bone

A female, aged 57 years, with advanced carcinoma of the breast. There was radiological evidence of metastases being present in the spine, ribs, pelvis and the head of the left humerus. At postmortem no evidence of liver involvement was found.

| Plasma | Ca | 2.28 | mmol/l | (2.15–2.55) |
|---|---|---|---|---|
| | Alb. | 43 | g/l | (37–52) |
| | AP | 620 | U/l | (30–120) |
| | ALT | 15 | U/l | (<35) |
| | Bili. | 6 | µmol/l | (<20) |
| | GGT | 24 | U/l | (<30) |

In this patient the plasma GGT activity is normal, which suggests that the increased AP is not of liver origin. The level of plasma AP in this patient is higher than the levels usually found in cases where secondaries are present only in bone (see above).

Primary hyperparathyroidism

A male, 55 years, with recurrent bilateral renal stones and hypercalcaemia.

| Plasma | Ca | 2.85 | mmol/l | (2.15–2.55) |
|---|---|---|---|---|
| | $PO_4$ | 0.62 | mmol/l | (0.6–1.25) |
| | Alb. | 41 | g/l | (37–52) |
| | AP | 188 | U/l | (30–120) |

Only 40–50% of cases of primary hyperparathyroidism will present with a raised plasma alkaline phosphatase, probably because nowadays the disease is often detected at an earlier stage of its course. Increased plasma levels of AP suggest significant bone involvement.

Chronic renal failure—secondary hyperparathyroidism

A female, 44 years, with chronic renal failure secondary to analgesic abuse.

| Plasma | Ca | 1.93 | mmol/l | (2.15-2.55) |
|---|---|---|---|---|
| | PO$_4$ | 2.96 | mmol/l | (0.6-1.25) |
| | Alb. | 34 | g/l | (37-52) |
| | AP | 274 | U/l | (30-120) |
| | Creat. | 0.72 | mmol/l | (0.06-0.12) |
| | PTH (C-term) | 16.0 | µg/l | (2.0-6.0) |

The high alkaline phosphatase in this case is a manifestation of a renal osteodystrophy that can be due to a variety of mechanisms, particularly:
(1) ↑ PTH secondary to a low ionized calcium
(2) ↓ production of 1,25 dihydroxycholecalciferol

## MALIGNANCY

The high plasma levels of AP associated with malignant conditions may be due to:
(1) Increased osteoblastic activity—either primary or secondary.
(2) Malignancy of bone.
(3) Cholestasis—primary and secondary malignancy of liver.
(4) AP produced by tumour. This is known as the Regan isoenzyme.
(5) Like the placental AP isoenzyme the Regan isoenzyme is heat stable; it has been found to be associated with tumours of the bronchus, ovary, pancreas, cervix and colon.

Carcinoma of colon

A 66 year old male with weight loss, blood loss *per rectum*, anaemia and hepatomegaly. At laparotomy a large fungating carcinoma of the sigmoid colon, associated with involvement of the regional lymph nodes and secondaries in the liver, was found.

| Plasma | TP | 73 g/l | (60-85) |
|---|---|---|---|
| | Alb. | 40 g/l | (30-50) |
| | AP | 900 U/l | (30-120) |
| | ALT | 95 U/l | (<35) |
| | Bili. | 16 µmol/l | (<20) |
| | GGT | 982 U/l | (<45) |

This case shows the typical biochemical features of a space-occupying lesion of the liver, i.e.

(1) ↑ AP due to intrahepatic cholestasis
(2) moderate ↑ of ALT due to destruction of liver cells by the expanding tumour
(3) ↑ GGT also reflecting the liver pathology
(4) normal plasma bilirubin. In space-occupying lesions of the liver the plasma bilirubin level is usually normal but will rise if: (a) there is obstruction of the major biliary passages, (b) there is massive destruction of liver tissue

## TRANSAMINASES

The two transaminases that are widely used in diagnostic laboratories are: (1) aspartate aminotransferase (AST), (2) alanine aminotransferase (ALT). For a discussion of these two enzymes see p. 298 (ALT) and p. 276 (AST).

## γ-GLUTAMYLTRANSFERASE (GGT)

*Source*   Kidney, liver, pancreas, prostate.

*Isoenzymes*   Several different fractions may be detected by starch-gel electrophoresis, but the value of this procedure remains to be established.

### Increased plasma GGT

Although GGT is widespread throughout the tissues of the body, raised plasma levels usually occur only in liver disease. GGT is increased in those conditions which also cause increases in liver AP.

#### CAUSES

(1) Hepato-biliary disease (a) cholestasis, (b) hepatocellular, (2) alcohol ingestion (a) alcoholic liver disease, (b) enzyme induction, (3) drugs, enzyme induction (a) barbiturates, (b) phenytoin, (4) miscellaneous (a) myocardial infarct (?due to liver involvement), (b) diabetes mellitus (?due to liver involvement), (c) pancreatitis.

Since plasma GGT levels are unaffected by bone disease, and are elevated only in liver disease, plasma GGT assays are very useful for determining the origin of an isolated raised plasma alkaline phosphatase. However, it is important to remember that the plasma GGT can be raised in the absence of overt liver disease by enzyme induction caused by either alcohol or drugs.

# Case examples/pathophysiology

## CHOLESTASIS

A 62 year old man with prolonged cholestasis due to carcinoma of the head of pancreas. Liver secondaries were also present.

| Plasma | TP | 68 g/l | (64–84) |
|---|---|---|---|
| | Alb. | 41 g/l | (37–52) |
| | AP | 525 U/l | (30–120) |
| | ALT | 180 U/l | (<35) |
| | Bili. | 180 µmol/l | (<20) |
| | GGT | 560 U/l | (<45) |

High plasma levels of GGT, often greater than 1000 U/l, are found in obstructive biliary disease and space-occupying lesions of the liver. In this case the high AP and GGT are due to intrahepatic (secondaries) and extrahepatic (common bile duct) obstruction. The high plasma ALT may be due to back pressure of bile in the bile ducts, and/or expansion of the liver secondary, causing hepatocellular damage.

## HEPATOCELLULAR DISEASE

Acute infectious hepatitis in a male, aged 25 years.

| Plasma | TP | 68 g/l | (64–84) |
|---|---|---|---|
| | Alb. | 37 g/l | (37–52) |
| | AP | 150 U/l | (30–120) |
| | ALT | 1050 U/l | (<35) |
| | Bili. | 200 µmol/l | (<20) |
| | GGT | 65 U/l | (<45) |

In uncomplicated infectious hepatitis the level of GGT tends to parallel the AP level, and therefore the GGT does not provide additional diagnostic information. In this case both are only minimally elevated.

## ALCOHOLISM

A 45 year old patient with chronic alcoholism. These liver function tests were performed one day after admission to hospital for 'drying out'.

| Plasma | TP | 67 g/l | (60–85) |
|---|---|---|---|
| | Alb. | 39 g/l | (30–50) |
| | AP | 94 U/l | (30–120) |
| | ALT | 17 U/l | (<35) |
| | Bili. | 18 μmol/l | (<20) |
| | GGT | 95 U/l | (<45) |

Elevated plasma levels of GGT are found in up to 80% of chronic alcoholics. The levels return to normal 10–14 days after cessation of drinking. In this case the biochemical results (other than GGT) suggest the absence of severe liver disease. Thus, this increase in GGT may represent enzyme induction, although the possibility of early alcoholic liver disease cannot be ruled out.

## ACID PHOSPHATASE (ACP)

Acid phosphatase is a phosphomonoesterase which exhibits its optimal activity at pH 4.9 (range 4.5–5.5).

*Source* Prostate (100–400 times the concentration found in any other tissue), erythrocytes, platelets, Gaucher's cells.

*Isoenzymes* Up to 14 bands have been separated on electrophoresis. The only isoenzyme of clinical importance is prostatic acid phosphatase. This may be estimated by one of the following methods: (1) L-tartrate inhibition, (2) specific substrates—thymolphthalein phosphate, α-naphthylphosphate, (3) radioimmunoassay.

*Note* Acid phosphatase is unstable at pH 7.0, and therefore storage requires the addition of acid to the sample.

### Total plasma acid phosphatase

Total plasma acid phosphatase is estimated if either phenyl phosphate or *p*-nitrophenyl phosphate is used as a substrate in the assay. Increased levels of total plasma acid phosphatase may occur in: (1) prostatic carcinoma, (2) after prostatic manipulation (e.g. *per rectum* examination), (3) haemolysis, (4) thrombocytopenic purpura, (5) Gaucher's disease, (6) metastatic disease of bone, (7) Paget's disease.

### Plasma prostatic acid phosphatase

Elevated plasma levels of this enzyme occur in 10–20% of patients with carcinoma of the prostate who do not have demonstrable evidence of metastases, and in 70-80% of patients with evidence of metastatic disease. Thus a high plasma level of this isoenzyme indicates the presence of prostatic carcinoma, but a normal level does not exclude the diagnosis. Various studies have indicated that digital examination and biopsy of suspicious areas of the prostate are more efficient methods of diagnosing carcinoma of the prostate than the measurement of plasma acid phosphatase levels.

After oestrogen therapy (and castration) the high plasma levels of acid phosphatase associated with carcinoma fall rapidly over a period of 3-4 weeks.

Some studies suggest that digital examination of the prostate or the passage of instruments *per urethra* result in increased plasma levels of acid phosphatase, whilst other studies indicate that acid phosphatase levels are unaffected. It is therefore advisable to obtain a blood sample for acid phosphatase estimation before performing any physical examinations in the region of the prostate.

The recent introduction of a radioimmunoassay for prostatic acid phosphatase held out hope that the increased sensitivity of the method would allow earlier detection of prostatic tumours, particularly before the tumour had moved beyond the prostatic capsule, but results so far have been disappointing.

## Case example

A male aged 74 years complaining of urinary retention and pelvic bone pain. Physical examination revealed a hard enlarged prostate. Radiological examination showed multiple lytic lesions in the pelvic bones and neck of the left femur.

| Plasma | Ca | 2.24 | mmol/l | (2.15-2.55) |
|---|---|---|---|---|
| | $PO_4$ | 0.92 | mmol/l | (0.6-1.25) |
| | TP | 67 | g/l | (64-84) |
| | Alb. | 40 | g/l | (37-52) |
| | AP | 1600 | U/l | (30-120) |
| | ALT | 14 | U/l | (<35) |
| | Bili. | 9 | μmol/l | (<20) |
| | ACP | 77 | U/l | (0.1-0.6) |

This case is unusual in that there is a very high plasma AP level (osteoblastic activity), as most patients with bony secondaries due to prostatic carcinoma have normal, or slightly elevated, plasma alkaline phosphatase levels. The other unusual feature of this case are the osteolytic lesions, the bony lesions of this disease being usually osteosclerotic. After surgical intervention and treatment with an oestrogen preparation, the acid phosphatase level fell to 4 U/l, and the alkaline phosphatase level fell to 200 U/l, within a period of 2 months.

## CHOLINESTERASE

*Source*  Plasma, liver, nervous tissue, red cells.

*Isoenzymes*  Two main types are found in human tissues: (1) pseudocholinesterase—plasma, liver, (2) 'true' cholinesterase—nervous tissue, red cells.

The use of the muscle relaxant suxamethonium has led to the recognition of the existence of variants of pseudocholinesterase in plasma. These variants can be classified on the basis of their inhibition by dibucaine (dibucaine number = % inhibition by dibucaine).

| Type | Dibucaine number |
|---|---|
| Normal homozygote | 75–85 |
| Atypical homozygote | 15–30 |
| Heterozygote | 40–70 |

### Suxamethonium sensitivity (scoline apnoea)

The muscle relaxant suxamethonium is destroyed by cholinesterase. In some subjects this enzyme activity is deficient, and prolonged apnoea follows administration of suxamethonium.

These subjects have a plasma cholinesterase with the following characteristics: (1) low plasma levels, (2) dibucaine number of 15–30.

They are homozygous for the atypical gene. It is important to recognize this condition, and to check all close relatives of the patient in case future surgical operations may require muscle relaxants. The heterozygotes are usually not affected, but occasional patients will have some respiratory depression when given scoline (neuromuscular blocking agent) (see case below).

Low plasma levels of pseudocholinesterase may also occur in (1) organophosphate poisoning, (2) severe liver disease.

### Case example

A female, age 23 years, who exhibited 3 hours of apnoea following scoline administration for an obstetrical procedure.

Plasma  pseudocholinesterase  191 U/l  (800–1500)
           dibucaine number          53       (75–85)

This case is an example of a heterozygote scoline sensitivity (dibucaine number 40–70). Not all heterozygotes will exhibit scoline apnoea and, if they do, it is usually of short duration. This patient's sister also showed a similar abnormality:

Plasma  pseudocholinesterase  446 U/l  (800–1500)
         dibucaine number       56     (75–85)

All other members of her immediate family, including her mother, were found to be normal. Her father was deceased. It was assumed that this patient and her sister inherited the abnormal gene from their father.

ORGANOPHOSPHATE TOXICITY

In this condition plasma and red cell cholinesterase activity is depressed. Enzyme measurements on all subjects at risk should be performed at regular intervals.

LIVER DISEASE

In severe liver disease plasma levels are low, and tend to parallel the level of albumin.

## ANGIOTENSIN CONVERTING ENZYME (ACE)

This enzyme converts angiotensin I to angiotensin II (see p. 136).

*Source* Widely distributed in the body, with the greatest concentration being found in the capillary endothelial cells of the lung. It is also present in measurable amounts in the plasma.

SARCOIDOSIS

Although the main action of ACE is the conversion of angiotensin I to angiotensin II (it also inactivates circulating bradykinin), measurement of this enzyme has found its main clinical application in the diagnosis of sarcoidosis. The reason for an increased plasma activity of ACE in sarcoidosis is unclear, but investigators have found that in this condition there is increased synthesis of the enzyme by the capillary endothelial cells of the lung.

The plasma level of this enzyme may also be increased in: (1) leprosy, (2) asbestosis, (3) silicosis, (4) Gaucher's disease.

In sarcoidosis approximately 8% of patients will have normal levels, thus the levels must be interpreted with care.

*Example*  Female, 35 years, generally unwell, chest X-ray suggestive of sarcoidosis.

Plasma   ACE   56 nmol/ml/min   (16–34)

Patient settled on steroid treatment, is now generally well and the radiological abnormalities have subsided. Nine months after the commencement of treatment the plasma ACE level was 30 nmol/ml/min.

# 13 Liver function tests

The liver plays a central role in many of the metabolic processes that occur in the body. The major functions of the liver are summarized below:

*Carbohydrate metabolism*   (1) Synthesis and storage of glycogen, (2) maintenance of fasting plasma glucose level by glycogenolysis and gluconeogenesis.

*Protein metabolism*   (1) Synthesis and degradation of proteins (except immunoglobulins); important examples are albumin, prothrombin, clotting factors (II, VII, IX, X), transport proteins, (2) amino acid metabolism and urea formation.

*Lipid metabolism*   (1) Synthesis of lipoproteins, phospholipids, cholesterol, (2) fatty acid metabolism, (3) bile salt synthesis.

*Excretion and detoxification*   (1) Bilirubin and bile acid excretion, (2) drug detoxification and excretion, (3) steroid hormone inactivation and excretion.

*Miscellaneous*   (1) Iron storage, (2) vitamin (A, D, E, $B_{12}$) storage and metabolism.

Despite the liver holding such a central position in body metabolism the commonly used diagnostic biochemical tests of the liver have tended to focus on investigating the structural integrity of the organ rather than on its metabolic competence, and so the general term of 'liver function tests' should be viewed as encompassing tests of both liver damage and liver function. The dearth of routine biochemical tests of true liver function reflects the fact that the liver possesses an immense reserve of metabolic capacity, so that the presence of hepatic disease is often not reflected by a proportional, or even detectable, disturbance in any of its functions until the pathological process is fairly advanced. The most widely used test of liver function is the measurement of plasma bilirubin, so some understanding of the metabolism of bilirubin is required before its use as a diagnostic tool is discussed.

Chapter 13     **Bilirubin metabolism** (Figs 13.1, 13.2)

SPLEEN

Red cells have an average life-span of 120 days, at the end of which they are broken down by the reticulo-endothelial system (mainly the spleen). The released haemoglobin is further degraded into its protein (globin) and iron-containing (haem) components. The globin amino acids are reutilized in the synthesis of new protein, whilst the iron is salvaged from the haem; the haem is converted to bilirubin. Some 25% of bilirubin is generated at the sites of red cell production (bone marrow) and from the catabolism of haemoglobin-related molecules such as myoglobin.

SPLEEN TO LIVER

Bilirubin is water-insoluble so it therefore has to be bound to albumin before it is transported from the spleen, marrow and other sites, to the liver.

LIVER

*Uptake*   On reaching the surface of the liver cells the bilirubin separates from its carrier albumin and is passed through the cell membrane to be accepted by specific intracellular proteins (Y and Z proteins) that will transport the bilirubin molecule across the hepatic cell.

*Conjugation*   On reaching the smooth endoplasmic reticulum the bilirubin is rendered water-soluble by being conjugated to glucuronic acid by the enzyme uridyl diphosphate (UDP) glucuronyl transferase.

*Secretion*   Following conjugation the bilirubin diglucuronide is secreted from the liver cell into the bile canaliculi and bile ducts, and thence into the gall bladder to be excreted as bile.

INTESTINE

After entering the upper gut as a constituent of bile the conjugated bilirubin passes largely unchanged to the colon, where resident bacteria remove the glucuronide portion. The bilirubin is then further metabolized to a group of compounds collectively termed urobilinogen. Most of the water-soluble urobilinogen is oxidized by bacteria to urobilin and is excreted in the faeces, whilst the rest is reabsorbed by the gut and is returned by the portal system to the liver. The majority of this urobilinogen is excreted in the bile, but some escapes to the systemic circulation and is excreted by the kidney (enterohepatic circulation, Fig. 13.2).

URINE

Urobilinogen is colourless, but it will oxidize to a brown pigment, urobilin, if the urine is left standing.

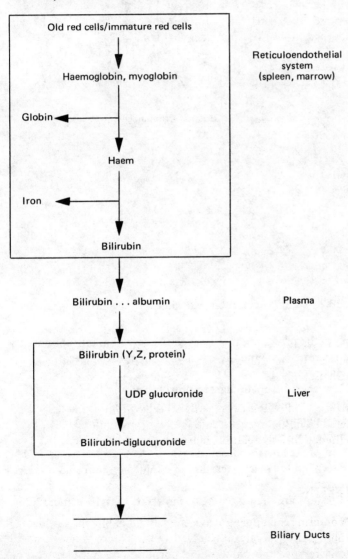

**Fig. 13.1** Outline of bilirubin formation.

Chapter 13

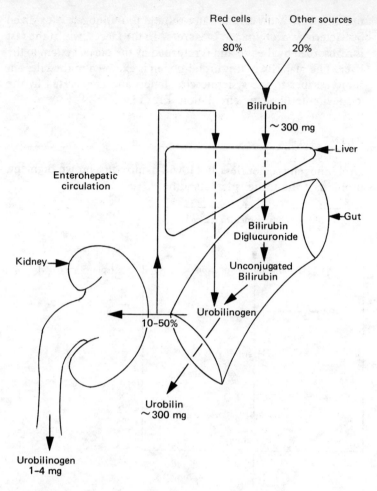

**Fig. 13.2** Outline of bilirubin excretion.

From the preceding outline it can be seen that bilirubin metabolism can be conveniently considered in seven stages:
(1) Haem breakdown.
(2) Transport of unconjugated bilirubin to the liver.
(3) Hepatic uptake of unconjugated bilirubin.
(4) Conjugation.
(5) Secretion of conjugated bilirubin into bile.
(6) Gut breakdown to urobilinogen.
(7) Faecal and urinary excretion of conjugated bilirubin and urobilinogen.
The steps are diagramatically shown in Figs 13.1 and 13.2.

## LABORATORY INVESTIGATION

Routine liver function tests are as follows.

*Plasma* (1) Bilirubin, (2) total protein, (3) transaminases, (4) alkaline phosphatase, (5) γ-glutamyltransferase.

Miscellaneous tests are as follows.

*Tests of liver function* (1) Bromsulphthalein (BSP) excretion, (2) plasma bile acids, (3) coagulation factors.

*Tests to determine aetiology* (1) $\alpha_1$-antitrypsin, (2) α-fetoprotein, (3) hepatitis B antigen, (4) specific antibodies, (5) caeruloplasmin.

PLASMA BILIRUBINS

Bilirubins are present in the plasma in two forms:
(1) Unconjugated bilirubin is water insoluble and must be carried by albumin to the liver. In the normal individual this fraction forms about 85% of total plasma bilirubins. Based on its method of estimation this fraction is sometimes referred to as 'indirect' bilirubin.
(2) Conjugated bilirubin is water soluble and reaches the plasma following its reabsorption from the gut and entry into the enterohepatic circulation; it may also regurgitate directly back into the plasma from the liver cells in small amounts. This fraction is often referred to as 'direct' reacting bilirubin.

Both fractions are normally measured together by laboratories as total bilirubins, whilst paediatric services will also estimate the conjugated bilirubin separately in certain clinical circumstances. Jaundice is clinically evident at a plasma concentration of approximately 40 μmol/l. Hyperbilirubinaemia *per se* indicates the degree of jaundice, but is of little help in determining aetiology.

URINE BILIRUBIN

Unconjugated bilirubin circulates in the blood bound to albumin, and it is therefore not filtered by the glomeruli. Conjugated bilirubin is water soluble and small, and thus it can be filtered by the glomeruli. Hence the presence of bilirubin in the urine indicates increased plasma levels of the conjugated form. A positive urinary bilirubin test may be the earliest indication of infectious hepatitis.

UROBILINOGEN

Urinary urobilinogen levels are raised when increased amounts of bilirubin enter the gut (haemolysis), or if there is impaired re-

## ALBUMIN

Albumin is synthesized in the liver, and has a plasma half-life of 19–20 days. With such a long half-life the concentration of plasma albumin is generally a poor indicator of acute hepatic disease, whilst in chronic liver conditions hypoalbuminaemia may be the only detectable biochemical abnormality.

## GLOBULINS

Total plasma globulin levels are increased in chronic hepatitis and cirrhosis.

## IMMUNOGLOBULINS

Estimation of plasma immunoglobulin levels is not a particularly useful diagnostic procedure in the investigation of liver disease. However, increases of specific immunoglobulins occur in some hepatic diseases, e.g. IgG increased in chronic active hepatitis, IgA increased in early cirrhosis, IgM increased in primary biliary cirrhosis.

## PLASMA TRANSAMINASES

Two transaminase enzymes found in the liver have been widely used for diagnostic purposes. Aspartate aminotransferase (AST) is present in both the mitochondria and cytosol of liver cells, whilst alanine aminotransferase (ALT) is found in the cytosol. Liver cell damage releases these enzymes into the extracellular fluid, and results in increased plasma levels of transaminase activity. ALT is relatively specific to the liver whereas AST is also found in, and can be released from, other tissues (skeletal and cardiac muscle, pancreas); either enzyme has proved to be a useful indicator of hepatocellular damage. Some laboratories measure the plasma activities of both enzymes as a part of their routine liver function tests as it has been suggested that the relative activities give some index of the underlying pathology and severity of the disease processes; e.g. in hepatitis the ALT level is usually greater than that of the AST; if the AST is found to be greater than that of the ALT it suggests that there is widespread hepatitic necrosis (release of mitochondrial enzymes) and a poorer prognosis. Generally the additional testing has not proved of proportional diagnostic value and so most laboratories measure only one or other of the plasma enzyme activities.

In hepatocellular disease (e.g. hepatitis) plasma levels of transaminases reach levels of 10–100 hundred times the upper reference limit. In uncomplicated cholestasis (obstruction) ALT and AST levels may be increased, but usually to less than 10 times the upper reference limit.

## PLASMA ALKALINE PHOSPHATASE (AP)

In the liver alkaline phosphatase (AP) is found at two distinct sites, in the sinusoidal surface of the hepatocyte, and in the microvilli of the bile canaliculi and ducts. AP is also found in a number of other body tissues (see p. 279), thus it is not a specific liver enzyme.

During obstruction of the bile passages (cholestasis) the plasma alkaline phosphatase level rises. This is due mainly to an increased synthesis of AP (mechanism unknown), but the obstruction itself also plays a part by causing regurgitation of the enzyme back into the bloodstream.

In obstructive jaundice the plasma alkaline phosphatase is usually increased to levels greater than three times the upper reference limit. In hepatocellular disease the level may be normal or slightly increased, but any increase is usually less than three times the upper reference limit.

Occasionally patients present with an isolated high level of alkaline phosphatase. In these cases it is important to determine the origin of the enzyme, i.e. from liver or from an extrahepatic source. The most satisfactory way to determine the tissue origin of a high plasma alkaline phosphatase is to identify the isoenzymes present in plasma, but unfortunately current methods can be fickle and are not yet widely available as a routine test. Other tests helpful in identifying the source of a raised plasma AP level is the estimation of the plasma activity of one of the following enzymes: (1) 5′-nucleotidase, (2) leucine aminopeptidase, (3) γ-glutamyltransferase.

These enzymes have a similar liver origin to alkaline phosphatase, and are increased when liver alkaline phosphatase is increased and, although they are not specific to the liver, the plasma activity of these enzymes are rarely raised in extrahepatic disease.

## γ-GLUTAMYLTRANSFERASE (GGT)

This enzyme is present in the bile canaliculi, the epithelial cells lining the bile ducts and, to a certain extent, in the periportal hepatocytes. High plasma levels of GGT activity are found in obstructive liver disease, but it is also raised in most other varieties of liver disease. GGT is rarely raised in extrahepatic disease, and so this enzyme has been found to be a sensitive indicator of liver disease. It is important to note that plasma GGT levels can be raised by alcohol ingestion or drug therapy (e.g. barbiturates). This elevation is due to enzyme induction.

## BSP EXCRETION

The ability to clear the plasma of the dye bromsulphthalein (BSP) after an intravenous injection is a sensitive measure of hepatic function. The major use of this test is in the investigation of patients suspected of liver disease, but whose routine liver function tests are normal. It is particularly useful in the evaluation of patients who have systemic diseases which are known to affect the liver, e.g. ulcerative colitis, malignancy.

BSP should be administered with care as extravascular injection causes tissue necrosis; several deaths have been reported due to anaphylaxis.

## BILE SALTS

The primary bile acids (cholic acid, chenodeoxycholic acid) are synthesized in the liver from cholesterol, conjugated with glycine and taurine, and then excreted into the bile. In the intestine some secondary bile acids (deoxycholic, lithocholic), are formed by the action of bacteria. Of the 15–30 g of bile acids excreted daily over 95% are reabsorbed in the ileum (enterohepatic circulation). In hepato-biliary disease excretion is decreased, and plasma levels rise. Although plasma bile acid levels would appear to be often a useful index of liver disease this estimation has not gained wide acceptance due to technical difficulties with the assay.

## COAGULATION FACTORS

The coagulation factors II, VII, IX and X are synthesized in the liver and require the presence of vitamin K. The plasma levels of these factors may be decreased in the following conditions:
(1) Vitamin K deficiency. Vitamin K is a fat soluble vitamin and therefore requires the presence of bile salts for absorption. Cholestasis reduces the levels of bile acids in the intestine and therefore will lower vitamin K absorption.
(2) Severe hepatocellular disease.
  The 'prothrombin time' is an indirect measure of the plasma level of factors II, VII and X. With adequate amounts of the factors

and vitamin K the clotting time of blood will be normal, whilst in cholestasis and in severe hepatocellular disease the prothrombin time is increased. The prothrombin time response following an injection of vitamin K has been used to discriminate between these two conditions.

## $\alpha_1$-ANTITRYPSIN

$\alpha_1$-antitrypsin deficiency is associated with neonatal hepatitis, neonatal cholestasis and cirrhosis in infancy, childhood and adult life.

## $\alpha$-FETOPROTEIN ($\alpha$FP)

$\alpha$-fetoprotein is synthesized by fetal liver cells, and is the major serum protein in early fetal life. It reaches a peak plasma concentration at about the 12th week of gestation ($\sim$ 30 mg/l) and decreases to very low levels by the age of one year. Very low levels continue to be found in the normal adult ($<50$ µg/l).

Raised levels of plasma $\alpha$FP are found in about 75% of cases of primary hepatocellular carcinoma, suggesting dedifferentiation by the tumour cells, resulting in their synthesizing a fetal product. However, increased levels may also be found in non-neoplastic liver disease, e.g. chronic hepatitis, alcoholic liver disease. This limits its usefulness as a diagnostic test for hepatocellular carcinoma.

## HEPATITIS B ANTIGEN ($HB_sAg$)

Examination of the plasma for $HB_sAg$, which is a marker for infection with hepatitis B virus, is an important part of the investigation of any patient with suspected liver disease.

## SPECIFIC ANTIBODIES

In 90–95% of cases of primary biliary cirrhosis antimitochondrial antibodies are present. Antismooth muscle antibodies may be associated with chronic active hepatitis.

## CAERULOPLASMIN

Wilson's disease (hepatolenticular degeneration), which is due to a defect in copper metabolism, is associated with cirrhosis of the liver. In this disease the plasma caeruloplasmin, a copper containing protein, is low.

## Interpretation of liver function tests

It is helpful to consider each test result in isolation and build up a picture of the possible underlying pathology.

↑ *Total bilirubins (Bili.)* may indicate: (1) ↑ production, e.g. haemolysis, or (2) ↓ excretion (a) hepatocellular damage, e.g. hepatitis, (b) cholestasis, e.g. obstruction.

↑ *ALT* (transaminases) indicates hepatocellular damage, e.g. hepatitis, prolonged obstruction, cirrhosis, infiltrations.

↑ *AP* indicates cholestasis, e.g. obstruction, cirrhosis, infiltrations.

↓ *ALB* (in context of liver disease) indicates severe hepatocellular dysfunction, e.g. cirrhosis, liver failure.

↑ *Plasma globulins* (in context of liver disease) indicates: (1) cirrhosis, or (2) chronic hepatitis.

↑ *GGT* indicates: (1) cholestasis, or (2) enzyme induction, e.g. alcohol ingestion.

## Disorders of the liver

CLASSIFICATION

*Congenital* (1) Gilbert's disease, (2) Crigler-Najjar syndrome, (3) Dubin-Johnson syndrome.

*Hepatitis* (1) acute, (2) chronic.

*Cholestasis* (1) extrahepatic, (2) intrahepatic.

*Cirrhosis*

*Anoxia*

*Infiltrations*

The following case examples are loosely classified under the headings of hyperbilirubinaemia, anoxia, cirrhosis and infiltrations, and alcoholic liver disease. This is a classification of convenience; there is considerable aetiological and biochemical overlap between the groups.

## HYPERBILIRUBINAEMIA

The earliest clinical manifestation of hepatobiliary disease is often jaundice, but jaundice need not necessarily indicate liver pathology (e.g. haemolysis), and liver pathology can be present without jaundice (e.g. space-occupying lesions). However, it is convenient to classify liver disease in terms of jaundice and to this end it is helpful to divide hyperbilirubinaemia into three categories: (1) prehepatic—liver disease not present, (2) hepatic—hepatocellular disease, (3) posthepatic—cholestasis (obstruction).

There is often some overlap between these three and this will be discussed below.

**Prehepatic jaundice**

CAUSES

(1) Increased bilirubin production (haemolysis).
(2) Congenital hyperbilirubinaemia.

**Case examples/pathophysiology**

HAEMOLYSIS

Spherocytosis in a female aged 16 years. Patient was jaundiced, with enlargement of the spleen. Bilirubin was not present in the urine.

| | | | | | |
|---|---|---|---|---|---|
| Plasma total Bili. | 54 | µmol/l | (<20) | Urine Bili. −ve | (−ve) |
| direct Bili. | 10 | µmol/l | (<10) | | |
| Alb. | 40 | g/l | (30–50) | | |
| ALT | 14 | U/l | (<35) | | |
| AP | 106 | U/l | (30–120) | | |

In this disease there is a defect in the red cell membrane, which results in spherocytosis (spherical red cells) and increased haemolysis. The abnormal erythrocytes are removed by the spleen and destroyed (splenomegaly). Thus there is increased production of unconjugated bilirubin, resulting in hyperbilirubinaemia (inability of the liver to cope with the excess bilirubin). No bilirubin appears in the urine because the unconjugated bilirubin is water insoluble and bound to plasma albumin. If pigment (bilirubin) stones form in the biliary system an obstructive biochemical picture will emerge, and conjugated hyperbilirubinaemia (due to regurgitation of bile unable to escape normally) will ensue, resulting in the bilirubin appearing in the urine.

## GILBERT'S DISEASE
## (CONGENITAL HYPERBILIRUBINAEMIA)

Healthy male aged 24 years with intermittent jaundice.

| Plasma | total Bili. | 42 | µmol/l | (<20) | Urine Bili. | −ve | (−ve) |
|---|---|---|---|---|---|---|---|
| | direct Bili. | 10 | µmol/l | (<10) | | | |
| | Alb. | 46 | g/l | (37–52) | | | |
| | ALT | 30 | U/l | (<35) | | | |
| | AP | 75 | U/l | (30–120) | | | |

Gilbert's disease is defined as mild chronic unconjugated hyperbilirubinaemia associated with: (1) normal values for other plasma liver function tests, (2) normal hepatic histology, (3) no overt haemolysis.

The plasma levels of bilirubin are usually less than 50 µmol/l, and fluctuate with time. There is a decreased hepatic clearance of bilirubin as a consequence of defective hepatic uptake, and partial deficiency of UDP glucuronyl transferase. The incidence of the disease in the community is of the order of 1–5%.

Other causes of congenital hyperbilirubinaemia are:

*Crigler-Najjar syndrome*   A dangerous disorder due to UDP glucuronyl transferase deficiency. The hyperbilirubinaemia is of the unconjugated type.

*Dubin-Johnson syndrome*   A mild disorder due to the defective excretion of conjugated bilirubin into the hepatic canaliculi.

**Hepatocellular jaundice**

The purest form of this type of jaundice is that due to liver cell destruction (e.g. hepatitis), but it may also be associated with a bilirubin conjugation defect (Crigler-Najjar syndrome), or with intrahepatic obstruction (cirrhosis, space-occupying lesions, etc.).

CAUSES

*Hepatitis* (1) acute (a) viral—type A, type B, non A non B, infectious mononucleosis, (b) hepatotoxins—alcohol, paracetamol, (2) chronic (a) chronic active, (b) chronic persistent.

*Anoxia*

*Space-occupying lesions*

*Cirrhosis*

ACUTE HEPATITIS

Three main types can be identified from the clinical and biochemical features: (1) typical, (2) anicteric, (3) cholestatic.

**Case examples/pathophysiology**

Typical acute hepatitis

An 11 year old boy who was jaundiced and felt generally unwell. He passed dark urine the day before he became jaundiced.

| Plasma | TP | 78 g/l | (60–85) | Urine Bili. +ve (−ve) |
|---|---|---|---|---|
| | Alb. | 41 g/l | (30–50) | |
| | AP | 144 U/l | (30–120) | |
| | ALT | 3140 U/l | (<35) | |
| | Bili. | 85 μmol/l | (<20) | |

The biochemical features of acute hepatitis (hepatocellular damage) are due to cell swelling and necrosis. Cell necrosis causes the release of intracellular enzymes, which are then absorbed by the local blood vessels, leading to increased plasma enzyme activity. The plasma level of ALT rises early in the disease, prior to the onset of jaundice, reaches a peak (10–100 times upper reference limit) when jaundice appears, and in uncomplicated cases returns to normal in 3–4 weeks. The swollen hepatocytes distort the small intrahepatic bile passages and produce cholestasis, resulting in the retention of bile pigments and increased plasma AP levels (enzyme induction). The raised plasma bilirubin is mainly conjugated because the unaffected hepatocytes are able to function normally, but because of the obstructive element (cholestasis), the liver is unable to excrete all of the pigment. Some of this conjugated bilirubin will be excreted by the kidneys, resulting in bilirubinuria. The hepatic excretion of urobilinogen will also be decreased, and therefore increased levels will appear in the urine. Depending on the severity of the cholestasis the plasma AP levels may be normal or increased, but they rarely rise above three times the normal reference limit.

### Anicteric hepatitis

A 15 year old girl complaining of an influenza-like illness. She attended a boarding school where an outbreak of hepatitis had recently occurred.

| Plasma | TP | 65 | g/l | (60–85) | Urine Bili. | −ve | (−ve) |
|---|---|---|---|---|---|---|---|
| | Alb. | 39 | g/l | (30–50) | | | |
| | AP | 118 | U/l | (30–120) | | | |
| | ALT | 325 | U/l | (<35) | | | |
| | Bili. | 17 | μmol/l | (<20) | | | |

In approximately 50% of cases of acute viral hepatitis (up to 90% in epidemics) jaundice is absent, and the only biochemical abnormality is an increased ALT. This is a mild illness, and the ALT rarely rises to more than 10 times the upper reference limit. In contrast the ALT levels are extremely high in the preicteric stage of typical acute hepatitis.

### Cholestatic acute hepatitis

A 35 year old man, recently returned from SE Asia, presented with 5 days of increasing jaundice.

| Plasma | TP | 66 | g/l | (60–85) | Urine Bili. | +ve | (−ve) |
|---|---|---|---|---|---|---|---|
| | Alb. | 37 | g/l | (30–50) | | | |
| | AP | 500 | U/l | (30–120) | | | |
| | ALT | 2640 | U/l | (<35) | | | |
| | Bili. | 200 | μmol/l | (<20) | | | |

Usually in acute hepatitis the plasma AP does not rise above three times the upper reference limit. Some patients, however, show evidence of marked intrahepatic cholestasis with high AP levels. This cholestasis may be due to precipitation of bile pigments in the bile ductules. Differentiation from the other causes of cholestasis (see below) can usually be made on the grounds of the extremely high ALT levels and clinical history.

# CHRONIC HEPATITIS

Hepatitis lasting more than 6 months is considered to be chronic. There are two varieties: (1) chronic active hepatitis (CAH), (2) chronic persistent hepatitis (CPH).

## Case examples/pathophysiology

### Chronic active hepatitis

A 36 year old woman who had hepatitis 9 months before. The condition had not settled and she was admitted for further investigation.

| | | | |
|---|---|---|---|
| Plasma TP | 113 | g/l | (60–85) |
| Alb. | 30 | g/l | (30–50) |
| AP | 400 | U/l | (30–120) |
| ALT | 350 | U/l | (<35) |
| Bili. | 45 | µmol/l | (<20) |

CAH is a chronic progressive disease which, if untreated, results in cirrhosis and liver decompensation. It may follow acute hepatitis (hepatitis B and hepatitis non A non B), or it may develop insidiously.

There is an active, continuing inflammation of the liver associated with liver cell necrosis and fibrosis. The biochemical features are variable, but the following plasma abnormalities are usually found.
(1) ↑ ALT (2–10 times normal)—liver cell necrosis
(2) ↑ AP—cholestasis due to fibrosis
(3) ↑ Bili. (mild to moderate in 70% of cases)—cholestasis
(4) ↑ Plasma globulins—chronic inflammatory reaction
(5) ↓ Alb.—severe liver dysfunction

### Chronic persistent hepatitis

This is a chronic inflammatory disease of the liver with a prolonged course. Unlike CAH it does not progress to cirrhosis or liver decompensation. It is important to differentiate between these two diseases, because the treatment of the two conditions differs. Differentiation depends on the results of a liver biopsy, but the plasma biochemistry often gives good clues.

| | CAH | CPH |
|---|---|---|
| ALT | 2–10 times normal | 2 times normal |
| Globulins | Increased | Normal |
| Alb. | Hypoalbuminaemia | Normal |

*Chapter 13*    **Posthepatic (cholestatic) jaundice**

Cholestasis is defined as an obstruction to bile flow at any point from the excretory surface of the hepatocyte to the duodenum.

The characteristic plasma abnormality in cholestasis is a raised alkaline phosphatase (AP) level, mainly because of increased synthesis due to enzyme induction, the cause of which is unknown. The decreased excretion due to bile ductule obstruction may also play a part.

Hyperbilirubinaemia, if it is present, is of the conjugated variety (regurgitation of bile back into the blood stream). The plasma level of bilirubin will depend on the extent of the obstruction, e.g. if only a few small ducts are obstructed, as in tumour infiltration, the bilirubin level will be normal (excreted by the remaining patent ducts). In this situation the plasma alkaline phosphatase remains elevated because the hepatocytes are unable to excrete this enzyme.

Classification

(1) Extrahepatic, (2) intrahepatic.

### EXTRAHEPATIC CHOLESTASIS

Causes

(1) Cholelithiasis, (2) tumours, (3) stricture, (4) biliary atresia, (5) cholangitis.

The commonest causes of extrahepatic obstruction are impacted gall stones and tumours (pancreatic, bile duct). Early in the process complete obstruction of the common bile duct produces the typical features of cholestasis, i.e. ↑AP, ↑Bili., normal or slightly increased ALT.

### Case examples/pathophysiology

*Example*   A 45 year old woman with a past history of cholelithiasis, now presenting with right upper abdominal pain. At operation several stones were removed from the common bile duct.

| Plasma | TP | 77 | g/l | (60–85) |
|---|---|---|---|---|
| | Alb. | 41 | g/l | (30–50) |
| | AP | 550 | U/l | (30–120) |
| | ALT | 34 | U/l | (<35) |
| | Bili. | 139 | µmol/l | (<20) |

The raised plasma AP, which is usually greater than three times normal, is due mainly to increased enzyme synthesis, but obstruction and regurgitation may also play a part. If the obstruction is not relieved the back pressure of bile causes hepatocellular necrosis, and the plasma ALT will also rise.

*Example* Carcinoma of the head of the pancreas in a 65 year old man. At operation the tumour did not appear to involve the liver.

| Plasma | TP | 78 g/l | (60–85) |
|---|---|---|---|
| | Alb. | 43 g/l | (30–50) |
| | AP | 1075 U/l | (30–120) |
| | ALT | 381 U/l | (<35) |
| | Bili. | 340 µmol/l | (<20) |

In this case the possibility of liver secondaries cannot be ruled out (see p. 313).

In extrahepatic cholestasis the plasma bilirubin level may be within normal limits in the following situations: (1) mobile stone causing a 'ball valve effect', (2) relief of obstruction, (3) only one hepatic duct involved, (4) cholangitis. That is bilirubin is excreted, but AP remains raised due to continued increased synthesis.

INTRAHEPATIC CHOLESTASIS

Causes

(1) Viral hepatitis (cholestatic variety), (2) drugs—chlorpromazine, steroids, (3) infiltrations—tumours, (4) cirrhosis, (5) cholangitis, (6) biliary atresia.

Intrahepatic cholestasis cannot be distinguished from the extrahepatic variety on the basis of routine liver function tests.

**Differential diagnosis of jaundice**

In the differentiation of acute viral hepatitis from extrahepatic obstruction routine liver function tests are useful only in the early stages of the disease. Initially viral hepatitis is associated with a high ALT, and a normal to slightly increased AP. In obstructive jaundice the AP is high, and the ALT low, e.g.

| | Acute hepatitis | Obstruction |
|---|---|---|
| AP | Normal → <3 times normal | >3 times normal |
| ALT | 10 → 100 times normal | <10 times normal |

In viral hepatitis the ALT falls rapidly, and the AP may rise slightly during the second to third week. With persistent obstruction (cholestasis) the ALT rises due to back pressure of bile, causing hepatocellular destruction. Therefore, if there is a delay in blood sampling the enzyme levels may fall into overlapping zones, and not be useful in differentiating between the two diseases.

*Example* Obstructive jaundice of long standing due to carcinoma of the head of the pancreas. No liver secondaries were found at autopsy.

| Plasma | TP | 72 | g/l | (60–85) |
|---|---|---|---|---|
| | Alb. | 42 | g/l | (30–50) |
| | AP | 345 | U/l | (30–120) |
| | ALT | 363 | U/l | (<35) |
| | Bili. | 200 | µmol/l | (<20) |

## ANOXIA

If the arterial blood supply to the liver is compromised anoxia may result, and liver cell necrosis occur. This may be associated with prolonged hypotension (shock) and lesions of the hepatic artery (embolus, aneurysm, etc.).

### ABDOMINAL ANEURYSM INVOLVING COELIAC AXIS

Male, aged 70 years, with an abdominal aneurysm. These results were obtained from a blood sample taken prior to surgery.

| Plasma | TP | 50 | g/l | (60–85) |
|---|---|---|---|---|
| | Alb. | 30 | g/l | (30–50) |
| | AP | 53 | U/l | (30–120) |
| | ALT | 885 | U/l | (<35) |
| | Bili. | 84 | µmol/l | (<20) |

A similar situation (cell necrosis) occurs in right-sided heart failure which, if prolonged, may lead to cirrhosis.

## CIRRHOSIS

Cirrhosis is a chronic, usually progressive, disease of the liver, characterized by widespread fibrosis and nodules of regenerated parenchyma cells. From a clinical and biochemical view three phases may be identified:

*Quiescent (compensated) phase*  Disease is temporarily halted and associated with minimal disturbances in liver function tests.

*Active phase*  There is progressive liver cell necrosis (↑ ALT) and fibrosis which often results in cholestasis (↑ AP, ↑ Bili.)

*Decompensation phase*  Severe liver dysfunction with gross hypoalbuminaemia and hyperbilirubinaemia (liver failure).

CAUSES

(1) Cryptogenic (idiopathic), (2) alcohol, (3) chronic active hepatitis, (4) viral hepatitis, (5) miscellaneous (a) haemochromatosis, (b) prolonged biliary obstruction, (c) Wilson's disease, (d) $\alpha_1$-antitrypsin deficiency, (e) primary biliary cirrhosis.

The diagnosis of cirrhosis is made on clinical and histological (biopsy) grounds, and not on the results of biochemical tests. Depending on its rate of progress and severity cirrhosis may present a wide spectrum of biochemical pathology, varying from an isolated increase in BSP retention to abnormalities in all of the routine liver function tests. A consideration of individual test results gives some indication of the underlying pathology.

ALBUMIN

In the majority of cases this is decreased, indicating impaired protein synthesis (liver cell dysfunction). Early in the disease the remaining hepatocytes may be able to increase synthesis sufficiently to maintain normal plasma levels.

*Example*  Biliary cirrhosis in a 48 year old female. Proven at biopsy.

| Plasma | TP | 78 | g/l | (60–85) |
|---|---|---|---|---|
| | Alb. | 38 | g/l | (30–50) |
| | AP | 450 | U/l | (30–120) |
| | ALT | 99 | U/l | (<35) |
| | Bili. | 12 | µmol/l | (<20) |

GLOBULINS

Most cases show an increase in the gamma globulin fraction. This may be due to reticuloendothelial cell hyperplasia, or to an autoimmune process.

## ALKALINE PHOSPHATASE

The plasma level of this enzyme is usually increased, especially in the active and late phases, indicating cholestasis due to progressive fibrosis. In the quiescent phase the plasma levels may be normal.

*Example*   Alcoholic cirrhosis in a 50 year old man.

| | | | |
|---|---|---|---|
| Plasma TP | 60 | g/l | (60–85) |
| Alb. | 32 | g/l | (30–50) |
| AP | 709 | U/l | (30–120) |
| ALT | 38 | U/l | (<35) |
| Bili. | 25 | µmol/l | (<20) |

### ALT

In the active phase, during liver cell necrosis, this enzyme is increased. The level reached reflects the extent and severity of the ongoing damage. Normal levels may be found in the quiescent phase, and just prior to terminal liver failure (few remaining undamaged hepatocytes).

*Example*   Active cirrhosis (?alcoholic) in a male aged 70 years.

| | | | |
|---|---|---|---|
| Plasma TP | 59 | g/l | (60–85) |
| Alb. | 30 | g/l | (30–50) |
| AP | 500 | U/l | (30–120) |
| ALT | 700 | U/l | (<35) |
| Bili. | 53 | µmol/l | (<20) |

### BILI

Early in the disease the plasma levels are variable, but are often increased. In the active phase and during hepatic decompensation, very high levels may be reached.

*Example*   Cirrhosis of unknown aetiology in a 56 year old female.

| | | | |
|---|---|---|---|
| Plasma TP | 65 | g/l | (60–85) |
| Alb. | 28 | g/l | (30–50) |
| AP | 185 | U/l | (30–120) |
| ALT | 72 | U/l | (<35) |
| Bili. | 294 | µmol/l | (<20) |

In the early and quiescent phases, increased BSP retention may be the only abnormality.

## INFILTRATIONS

**Causes**

(1) Primary carcinoma, (2) secondary carcinoma, (3) granulomas—sarcoid, tuberculosis, (4) abscess, (5) cysts.

Localized infiltrations and space-occupying lesions of the liver often produce a characteristic biochemical picture of increased plasma AP, associated with BSP retention and a normal plasma Bili.

*Example* Hepatic metastases from carcinoma of the colon.

| | | | |
|---|---|---|---|
| Plasma | TP | 70 g/l | (60–85) |
| | Alb. | 38 g/l | (30–50) |
| | AP | 800 U/l | (30–120) |
| | ALT | 26 U/l | (<35) |
| | Bili. | 13 µmol/l | (<20) |
| BSP retention at 45 minutes: | | 15% | (<5%) |

The increased AP is due to local intrahepatic cholestasis. Both AP and bilirubin are retained, but whereas the bilirubin can be excreted by the non-obstructed areas, the AP cannot.

Widespread, rapidly increasing infiltrations are associated with liver cell necrosis ($\uparrow$ALT), as well as cholestasis.

*Example* Secondary carcinoma of the liver, primary lesion in the stomach.

| | | | |
|---|---|---|---|
| Plasma | TP | 67 g/l | (60–85) |
| | Alb. | 35 g/l | (30–50) |
| | AP | 725 U/l | (30–120) |
| | ALT | 111 U/l | (<35) |
| | Bili. | 110 µmol/l | (<20) |

AP levels are useful indicators of secondary involvement of the liver by malignant tumours, being elevated in greater than 80% of cases. The finding of a high plasma AP (3–20 times the upper reference limit), in the absence of jaundice, should alert the clinician to the possibility of hepatic metastases. Non-malignant space-occupying lesions of the liver (abscess, cyst, etc.) will give a similar picture to malignant infiltration.

*Chapter 13*  ALCOHOLIC LIVER DISEASE

The development of alcoholic liver disease depends on the duration and amount of alcohol consumed. Only about 10% of alcoholics will develop cirrhosis, whilst in the remainder lesser degrees of liver damage are common.

Excessive alcohol consumption may result in any of the four following changes: (1) plasma enzyme changes without overt liver disease, (2) fatty liver with hepatomegaly, (3) alcoholic hepatitis, (4) cirrhosis.

PLASMA ENZYME CHANGES WITHOUT OVERT LIVER DISEASE

Acute alcohol consumption

Plasma enzyme changes associated with acute alcohol consumption are variable.
AST may increase to twice the upper reference limit.
ALT rarely increases.
GGT may increase to twice the upper reference limit.

Chronic alcohol consumption

In 'heavy drinkers' and chronic alcoholics the following changes have been described:
AST increased to up to four times the upper reference limit in ~40% of cases.
ALT increased up to four times the upper reference limit in ~20% of cases.
AP increased in <10% of cases, usually to levels less than two times the upper reference limit.
GGT increased in 80% of cases. Levels are usually less than two times the upper reference limit, but may be increased to 10–20 times this level. This enzyme is useful in monitoring alcohol intake in these patients (see p. 288).

FATTY LIVER

This disorder, due to triglyceride accumulation in the liver, may produce hepatomegaly. It is believed to be reversible, and not to lead on to cirrhosis. Plasma enzyme levels vary from normal to those described above.

## ALCOHOLIC HEPATITIS

This disease is associated with focal hepatocellular necrosis, and may vary in severity from a disorder resembling acute infectious hepatitis to severe hepatic decompensation. The liver function tests may reveal raised transaminase levels (usually <350 U/l), which indicate hepatocellular destruction, and raised alkaline phosphatase levels (usually <350 U/l) indicating cholestasis. These enzyme changes can vary markedly from patient to patient. Plasma bilirubin and albumin levels are also variable, and depend on the severity of the disease process.

*Example* Acute alcoholic hepatitis in a 50 year old male.

| Plasma | TP | 74 | g/l | (60–85) |
|---|---|---|---|---|
| | Alb. | 34 | g/l | (30–50) |
| | AP | 307 | U/l | (30–120) |
| | ALT | 215 | U/l | (<35) |
| | Bili. | 42 | µmol/l | (<20) |
| | GGT | 940 | U/l | (<45) |

In the above case the liver function tests indicate hepatocellular disease (↑ ALT) and cholestasis (↑ AP), but reveal nothing about the aetiology. These results, for example, would also be consistent with infectious hepatitis, cirrhosis, or secondary carcinoma of the liver. The diagnosis of alcoholic hepatitis depends on the clinical story and/or the results of a liver biopsy.

## ALCOHOLIC CIRRHOSIS

Cirrhosis, whatever the cause, is characterized by diffuse hepatic fibrosis and formation of nodules of regenerating hepatic tissue. The symptoms, signs and biochemistry of alcoholic cirrhosis do not differ from those due to other causes.

*Example* Male, aged 49 years, with severe alcoholic liver disease, portal hypertension and ascites.

| Plasma | TP | 57 | g/l | (60–85) |
|---|---|---|---|---|
| | Alb. | 19 | g/l | (30–50) |
| | AP | 750 | U/l | (30–120) |
| | ALT | 113 | U/l | (<35) |
| | Bili. | 150 | µmol/l | (<20) |

In this case the plasma ALB level suggests severe liver dysfunction (↓ albumin synthesis); the plasma AP reflects cholestasis (ductule obstruction by fibrotic tissue); and the plasma ALT suggests hepatocyte destruction. The difference between this picture and simple cholestasis (e.g. impacted gall stone) is the plasma ALB level. In simple cholestasis the liver cell reserve would be adequate, and the plasma albumin level normal.

In liver failure, whatever its cause, the plasma transaminase levels will depend on the amount of residual hepatic tissue, and its rate of destruction. For example, if there is very little residual normal tissue the plasma ALT level may be within the reference range.

# 14 Plasma lipids

Lipids, other than unesterified (free) fatty acids which are transported by albumin, are carried in the plasma as lipoprotein complexes. All lipoproteins contain protein (apoprotein) and varying amounts of triglyceride, cholesterol and phospholipid. The lipoprotein constituents are arranged so that the polar (water soluble) constituents of protein, phospholipid and free cholesterol are orientated on the outside around a core of lipophilic non-polar groups (cholesterol esters, triglyceride). The major role of these lipoproteins is to transport cholesterol and triglyceride.

Plasma lipoproteins are classified into four major groups, according to their density, which can be determined by their behaviour on ultracentrifugation.

| Lipoprotein | Major Lipid | Role |
| --- | --- | --- |
| Chylomicron | Triglyceride (Trig.) | Transport of Trig. from gut |
| Very low density (VLDL) | Triglyceride | Transport of Trig. from liver to tissues |
| Low density (LDL) | Cholesteryl ester | End product of VLDL catabolism Transport of cholesterol to tissues |
| High density (HDL) | Cholesteryl ester | Transport of cholesterol from tissues |

Each of the above classes represents a heterogenous group of particles. About 60% of plasma cholesterol is normally found in the LDL fraction, about 25% in the HDL and the remainder in the VLDL, though it should be remembered that there is wide diversity in individual lipoprotein patterns. In the fasting state chylomicrons are absent from plasma, in which case 50–80% of plasma triglyceride is found in the VLDL.

## Triglyceride metabolism

Dietary triglyceride is an important source of energy, about 40% of the daily caloric intake being derived from fat. The dietary triglycerides (exogenous Trig.) are emulsified in the gut, and converted by the action of pancreatic lipase to free fatty acids (FFA) and

2-monoglyceride. The FFA, monoglycerides and other lipids are solubilized by bile acids into micelles, which are then absorbed by the intestinal mucosa. Within the intestinal cells, triglycerides and cholesteryl esters are reformed and together with cholesterol, phospholipids and locally synthesized apoprotein are formed into chylomicrons, and released into the lymphatic system. Within the capillaries of the peripheral circulation the chylomicrons are catabolized by lipoprotein lipase, an enzyme associated with the endothelial cells lining the vessel walls, to FFA, glycerol and cholesterol-rich chylomicron remnants. The FFA are taken up by the local cells or bound to plasma albumin for transport elsewhere. The chylomicron remnants are taken up and degraded by the liver.

Triglycerides are also synthesized in the liver (endogenous Trig.). The FFAs required to make this triglyceride are either synthesized locally, or are derived from adipose tissue triglyceride (see below). In the liver triglycerides, phospholipid, cholesterol and apoprotein are combined to form VLDL particles.

The VLDL particles are catabolized in the peripheral tissues by lipoprotein lipase, the removal of triglyceride taking place in stages. This results in the formation of intermediate density lipoproteins (IDL), and finally the cholesterol-rich LDL which is taken up and catabolized by both hepatic and extrahepatic cells.

FREE FATTY ACIDS

The plasma free fatty acids (FFA) in the fasting state are mostly derived from the action of hormone-responsive intracellular lipase on triglyceride in adipose tissue. The release of fatty acids from adipose tissue is inhibited by insulin and stimulated by catecholamines. After a meal, FFA are formed by the action of lipoprotein lipase on chylomicrons and VLDL. The free fatty acids are mostly oxidized by tissues as a source of energy, but some are taken up by the liver, where they may be incorporated into triglycerides and secreted as VLDL.

## Cholesterol metabolism

The body derives cholesterol from: (1) diet (0.2–2.0 g/day), (2) synthesis, mainly in the liver (endogenous).

On a cholesterol-free diet about 900 mg of cholesterol is synthesized daily by the liver, and this approximates the amount lost from the body: (1) $\sim$ 250 mg as bile acids, (2) $\sim$ 550 mg as free cholesterol in the bile, (3) $\sim$ 100 mg from the skin (cell loss).

DIETARY CHOLESTEROL

Approximately 1–3 g of cholesterol passes into the small bowel each day, contributed to by the diet (400–700 mg) and the bile (750–1250 mg); of this, 30–60% is absorbed. In the gut any dietary cholesterol esters are converted to free cholesterol by pancreatic lipases, solubilized by bile salts into micelles and absorbed. After absorption most of the cholesterol is re-esterified and is incorporated, along with the free cholesterol, into chylomicrons. After the removal of triglycerides by lipoprotein lipase, the cholesterol-rich chylomicron remnants are taken up and degraded by the liver.

ENDOGENOUS CHOLESTEROL

In the liver the cholesterol that is derived from *de novo* synthesis and from lipoprotein catabolism enters one of three pathways: (1) incorporation into VLDL, (2) conversion to bile acids, (3) excretion as free cholesterol in bile.

The VLDL, after secretion by the liver, is acted upon as described earlier by lipoprotein lipase to form LDL particles. These contain apoprotein B (an apolipoprotein), which can bind to specific receptors on the surface of extrahepatic, and probably hepatic, cells. LDL is then taken up by these cells and, after lysosomal hydrolysis of the lipoproteins, the cholesterol is either esterified and stored, or utilized for steroid production or membrane synthesis. Thus VLDL and LDL are responsible for the transport of cholesterol from the liver to peripheral tissues.

Cholesterol in tissue membranes is in the free (unesterified) form. It appears that this can be taken up by HDL to form part of the lipoprotein pool of free cholesterol. Some HDL cholesterol is esterified by lecithin: cholesterol acyltransferase (LCAT), an enzyme associated with HDL. Cholesterol esters can then be transferred to VLDL and chylomicrons via cholesteryl ester transfer protein (a protein which mediates the bidirectional transfer of cholesteryl ester between lipoproteins). This postulated mechanism may explain the observation that HDL is a negative risk factor for ischaemic heart disease (↑ HDL, ↓ risk).

**Atheroma and ischaemic heart disease (IHD)**

Details of the relationship between plasma lipoproteins and the pathogenesis of atherosclerosis are unclear. At the present time the following two statements have gained general acceptance.
(1) High plasma levels of LDL, and possibly VLDL, are associated with an increased risk of premature atherosclerosis and ischaemic

Chapter 14

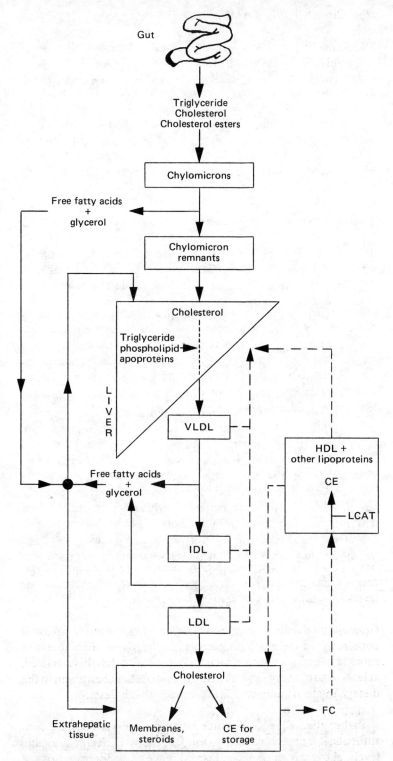

**Fig. 14.1** Outline of the role of lipoproteins in lipid metabolism. FC, free cholesterol; CE, cholesterol ester. — — — postulated pathways; ---- complex steps.

heart disease (IHD). This relationship appears to be a continuous one.
(2) Plasma HDL cholesterol concentration is a negative risk factor, so that high levels appear to protect against IHD and low levels are associated with an increased risk of IHD.

## LABORATORY INVESTIGATION

### PLASMA CHOLESTEROL AND TRIGLYCERIDE

If blood samples are to be taken for the investigation of lipid disorders the subject should:
(1) Take a normal balanced diet for at least seven days prior to the test.
(2) Fast for 12–14 hours prior to blood sampling.
(3) Not be suffering from severe stress (e.g. myocardial infarction, burns, etc.), which is associated with the mobilization of free fatty acids and increased plasma levels of triglycerides. At the same time the plasma cholesterol may change, e.g. with a myocardial infarction the plasma cholesterol concentration usually falls the first day postinfarct, and does not return to the original level until perhaps 3–4 weeks later.

Relevant reference ranges are difficult to define. Many factors influence plasma lipid concentrations, i.e.

*Age*   LDL concentrations are low in cord blood, but rise rapidly in the months following birth. In Western society plasma LDL rises slowly from childhood until the sixth or seventh decade, after which levels may fall slightly.

*Sex*   Plasma HDL concentrations are generally higher in premenopausal women than in men, but after the menopause the levels in women fall and are similar to those in men.

*Geography*   Plasma cholesterol varies from low levels in the rural population of developing countries to relatively high levels in subjects living in a Western society. It is thought that this is related, at least in part, to differences in the prevalence of obesity and in the dietary intake of saturated fat and probably cholesterol.

Rather than have a reference range for the plasma lipid constituents, it is probably better to think in terms of ideal or desirable levels, above which there is clear evidence of an increased risk of premature IHD. For plasma cholesterol a level of 6.5 mmol/l for adults has been suggested in this context. However, studies reveal that in some Western societies about 40% of older age-groups would be classified as being hypercholesterolaemic using this criterion.

PRIMARY OR SECONDARY HYPERLIPIDAEMIA

If a patient is found to be hyperlipidaemic the possibility that the lipid abnormality may be secondary to some other condition should be considered. The need for further laboratory tests therefore depends on clinical findings. Some of the common causes of a secondary hyperlipidaemia are shown below.

| Hypercholesterolaemia | Hypertriglyceridaemia | Combined hyperlipidaemia |
| --- | --- | --- |
| Hypothyroidism | Diabetes mellitus | Diabetes mellitus |
| Nephrotic syndrome | Obesity | Renal failure |
| Obstructive jaundice | Alcoholism | |
| | Pancreatitis | |
| | Renal failure | |

The possibility that a lipid abnormality is familial should also be borne in mind, and so measurement of plasma lipids in family members of a hyperlipidaemic patient may be appropriate.

Based on the fasting plasma cholesterol and triglyceride levels the hyperlipidaemias may be classified as: (1) fasting hypertriglyceridaemia, (2) fasting hypercholesterolaemia, (3) fasting hypertriglyceridaemia and hypercholesterolaemia (combined hyperlipidaemia).

This simple classification is adequate for most clinical purposes, but if the patient fails to respond to therapy as expected then further tests and classification may be necessary.

ELECTROPHORESIS/FREDRICKSON CLASSIFICATION

Plasma lipoproteins are readily separated by electrophoresis on a support medium such as agarose or cellulose acetate. After separation, the various fractions are visualized by staining with a lipophilic dye, and then qualitatively assessed for the presence of bands. Using normal plasma, three bands appear (Fig. 14.2):
(1) $\alpha$ band—high density lipoproteins (HDL).
(2) $\beta$ band—low density lipoproteins (LDL).
(3) pre-$\beta$ band—very low density lipoproteins (VLDL).

Chylomicrons, which are not present in the plasma of fasting normal patients, remain at the application line.

In recent years a classification of the hyperlipidaemias, first put forward by Fredrickson and co-workers, has been widely used for

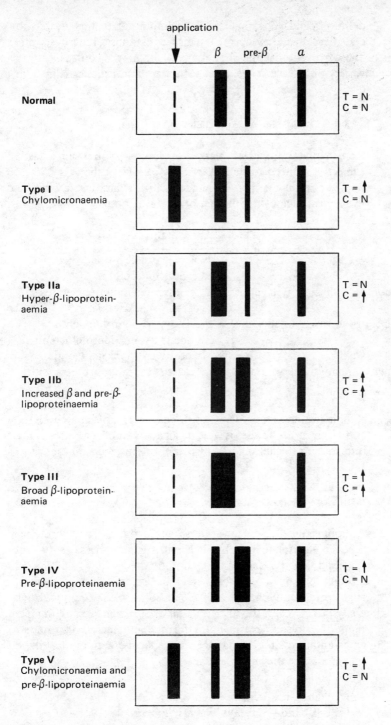

**Fig. 14.2** Diagram of the Fredrickson classification of lipoprotein abnormalities. T triglyceride; C cholesterol.

defining lipid abnormalities. It is important to realize that this classification does not provide a disease diagnosis, but a description of the hyperlipidaemia phenotype, which may of course be due to a variety of disease states, both secondary and primary (Fig. 14.2).

SERUM/PLASMA APPEARANCE

Approximately 4 ml of plasma or serum taken from a fasting subject is allowed to stand vertically in a test tube at 4°C for 18 hours. The tube is then inspected against a dark background and the appearance of the plasma or serum is noted.

| Appearance | Abnormality |
| --- | --- |
| Clear | Normal, raised β-lipoprotein, raised α-lipoprotein |
| Clear subnate with creamy layer at top | Chylomicronaemia (exogenous hypertriglyceridaemia) |
| Even turbidity throughout specimen | Pre-β-hyperlipoproteinaemia<br>Remnant hyperlipoproteinaemia<br>Mixed hyperlipoproteinaemia |
| Turbid with creamy layer at top | Increased pre-β-lipoproteins with chylomicrons |

This test, together with a knowledge of the plasma cholesterol and triglyceride levels, is very informative and provides similar information to that obtained by lipoprotein electrophoresis.

ULTRACENTRIFUGATION

As lipoproteins have a lower density than other plasma proteins it is possible to isolate them by ultracentrifugation in a salt solution of specified density. Using this technique the various fractions containing specific lipoproteins may be isolated and subjected to triglyceride and cholesterol, or in some cases apolipoprotein, analysis. However, the instrumentation is expensive and the technique requires expertise. The information obtained does not play an essential role in the diagnosis and management of most hyperlipidaemias at the present time.

**Indications for plasma lipid investigation**

The manifestations of hyperlipidaemia may include the following.

CLINICAL

The finding of xanthomas is a definite indication, and that of xanthelasma or premature arcus a weaker indication, for plasma lipid examination. Premature vascular disease also requires investigation.

LABORATORY

*Lipaemic plasma* Usually first noted by technical staff, often in samples sent for tests that are unrelated to lipid disorders. A fasting specimen should be requested before further action is taken.

*Screening programmes* It is reasonable to screen sections of the population who are considered to be at risk for hyperlipidaemia. Such groups should include those with:
(1) Vascular disease of early onset, especially premature ischaemic heart disease.
(2) Major risk factors for premature atherosclerosis, e.g. diabetes mellitus, hypertension.
(3) A family history of hyperlipidaemia or premature atherosclerotic disease.
(4) Clinical manifestations suggestive of hyperlipidaemia.

## HYPERLIPIDAEMIA

For the purposes of diagnosis and treatment the most convenient classification is that which is based on the type of lipid abnormality (cholesterol, or triglyceride, or both).

### Hypercholesterolaemia

*Primary* (1) familial hypercholesterolaemia, (2) unclassified.

*Secondary to* (1) hypothyroidism, (2) nephrotic syndrome, (3) obstructive jaundice, (4) dysglobulinaemias.

### Hypertriglyceridaemia

*Primary* (1) hypertriglyceridaemia due to elevated VLDL, (2) hypertriglyceridaemia due to deficient lipoprotein lipase.

*Secondary to* (1) diabetes mellitus, (2) alcoholism, (3) obesity, (4) renal failure, (5) oestrogens (pregnancy, oral contraceptives), (6) glycogen storage disease.

**Combined hypertriglyceridaemia and hypercholesterolaemia**

*Primary* (1) familial combined hyperlipidaemia, (2) broad beta disease (dysbetalipoproteinaemia), (3) unclassified.

*Secondary to* (1) diabetes mellitus, (2) hypothyroidism, (3) renal disease.

**Case examples/pathophysiology**

**Hypercholesterolaemia**

PRIMARY

Patients with increased plasma LDL cholesterol are diagnosed as having a primary disorder when secondary causes are excluded. This familial disorder may be determined by a single mutant gene, or may be more complex in its pathogenesis.

Familial hypercholesterolaemia

This disorder, due to a deficiency or abnormality of cell receptors for LDL, is inherited as an autosomal dominant defect. Hypercholesterolaemia may be noted in infancy. Xanthomas may occur in severely affected members of the family, and ischaemic heart disease may present in some members before the fourth or fifth decade. In heterozygotes, the plasma cholesterol is usually about 8-13 mmol/l, and xanthomas may not occur until the third to fourth decades, if at all. In the rare homozygous state, the plasma cholesterol is of the order of 13-30 mmol/l, and patients rarely survive into their third decade.

Other forms of primary hypercholesterolaemia

In these conditions the pathogenesis of the high LDL is unclear, but probably reflects a combination of genetic effects and dietary factors (high dietary saturated fats and cholesterol). There is an increased risk of ischaemic heart disease.

*Example* Two and a half year old infant whose father has marked hypercholesterolaemia. Mother has a normal plasma cholesterol level.

Plasma appearance  clear
    Chol.          10.9 mmol/l   (<6.5)
    Trig.          1.3 mmol/l    (0.3–1.7)
Electrophoresis    Increased β lipoprotein (LDL, type IIa)

This case is a probable example of heterozygous familial hypercholesterolaemia.

SECONDARY

Hypothyroidism

A 28 year old male with primary hypothyroidism.

Plasma T$_4$    16   nmol/l    (60–160)
    TSH        >40  U/l       (<5.5)
    Chol.      10.3 mmol/l    (<6.5)
    Trig.      1.7  mmol/l    (0.3–1.7)

Hypothyroidism is frequently associated with hypercholesterolaemia due to increased plasma LDL. Lipoprotein turnover studies have indicated that in hypothyroidism there is a decreased rate of LDL catabolism. In some cases the hypothyroidism may also be associated with hypertriglyceridaemia. These lipid abnormalities usually return to normal with the institution of adequate thyroid hormone replacement therapy but if, after treatment, the hyperlipidaemia persists a familial basis for the lipid abnormality should be considered.

Nephrotic syndrome

A 19 year old male with peripheral oedema and proteinuria.

Plasma Alb.    25   g/l       (30–50)    Urine TP  9.6 g/24 h  (<0.15)
    Chol.      17.5 mmol/l    (<6.5)
    Trig.      1.8  mmol/l    (0.3–1.7)
Electrophoresis    increased β lipoprotein (LDL, type IIa)

Nephrosis is usually associated with hypercholesterolaemia; the degree of hypercholesterolaemia tends to be inversely related to the severity of hypoalbuminaemia. The mechanisms involved in the hyperlipidaemia are unclear, but it has been suggested that the loss of albumin and other small proteins in the urine leads to enhanced production of all plasma proteins, including lipoproteins. The lipoproteins are large enough to be retained by the damaged glomeruli.

### Obstructive jaundice

In obstructive jaundice (both intrahepatic and extrahepatic) hyperlipidaemia may be due to the presence of an abnormal lipoprotein that contains albumin, is rich in free cholesterol and phospholipid, and which has the mobility of β-lipoprotein on electrophoresis on paper or agarose. This protein has been called lipoprotein-X (LP-X), and results in an elevated plasma cholesterol, often reaching values greater than 15 mmol/l, a normal triglyceride level and an elevated free cholesterol: esterified cholesterol ratio. The mechanism for this disorder has not been established.

### Dysglobulinaemias

Hyper-γ-globulinaemia (e.g. systemic lupus erythematosis, myeloma) has been associated with a variety of hyperlipidaemias, including hypercholesterolaemia. The mechanism is unknown. Other rare causes of hypercholesterolaemia are: (1) hepatoma, (2) acute intermittent porphyria.

## Hypertriglyceridaemia

PRIMARY

### Hypertriglyceridaemia due to increased VLDL (familial endogenous hypertriglyceridaemia

This is the most common type of hyperlipidaemia and is biochemically characterized by: (1) hypertriglyceridaemia, (2) increased levels of VLDL, (3) normal or moderate increase in plasma cholesterol (VLDL cholesterol), (4) variable LDL levels.

It appears to be transmitted as an autosomal dominant which seems to have a weak penetration. This disorder does not manifest itself until about the fourth decade of life.

Some or all of the following conditions may be associated with the lipoprotein abnormality: (1) obesity, (2) mild glucose intolerance, (3) hyperuricaemia, (4) eruptive xanthoma, (5) lipaemia retinalis.

There is also an associated increased risk of premature ischaemic heart disease. The pathogenesis of the condition in the individual is unclear. In most who have been intensively studied it appears to be due to overproduction of VLDL, but in some it seems to be due to decreased catabolism of VLDL.

### Hypertriglyceridaemia due to lipoprotein lipase deficiency (familial exogenous hypertriglyceridaemia)

This very rare condition is due to lipoprotein lipase deficiency and presents in infants with eruptive xanthoma, abdominal pain and hepatosplenomegaly. The hypertriglyceridaemia is due to chylomicronaemia.

Male, aged 50 years, with eruptive xanthoma, not obese.

Plasma appearance  turbid
    Chol.            8.0 mmol/l  (<6.5)
    Trig.            11.8 mmol/l  (0.3–1.7)
    Urate         0.55 mmol/l  (0.25–0.45)
Electrophoresis    increased pre-$\beta$ (VLDL, Type IV)

This hyperlipidaemia may be primary, or secondary to (1) hypothyroidism, (2) diabetes mellitus, (3) obesity, (4) alcoholism, (5) metabolic stress. These factors must be excluded before a diagnosis of familial endogenous hypertriglyceridaemia is made.

*Example* A 20 year old female admitted with acute abdominal pain and a past history of acute pancreatitis. The plasma was grossly lipaemic.

Plasma appearance  turbid, creamy layer at top after standing at 4°C
    Chol.            9.0 mmol/l  (<6.5)
    Trig.            31 mmol/l   (0.3–1.7)
    Amy.          5425 U/l    (<300)

Hyperchylomicronaemia may be associated with recurrent attacks of acute pancreatitis, but a single attack may result in a temporary hypertriglyceridaemia. This may be due to: (1) pancreatitis inducing some metabolic defect that results in hypertriglyceridaemia, (2) stress uncovering a latent familial hypertriglyceridaemia, (3) alcohol ingestion.

SECONDARY

Causes

(1) Diabetes mellitus, (2) alcoholism, (3) oestrogen excess, (4) renal failure, (5) glycogen storage disease.

Treatment of the primary disorder will result in disappearance of the hyperlipidaemia if there is no underlying familial cause.

Diabetes

Severe insulin deficiency, as may occur in insulin dependent diabetes, is regularly associated with increased circulating levels of VLDL. Chylomicrons may also be present. Hypertriglyceridaemia is a less common finding in patients with insulin independent

diabetes. The pathogenesis of this type of hyperlipidaemia is fairly well understood.
↓ insulin →
(1) ↑ adipose tissue fat mobilization → ↑ FFA release and ↑ FFA uptake by the liver → ↑ triglyceride synthesis → ↑ VLDL
(2) ↓ lipoprotein lipase activity → ↓ clearance of VLDL and chylomicrons

### Alcoholism

This is one of the most common causes of severe hyperlipidaemia. The pathogenesis is unclear, but many of these subjects have a pre-existing familial disorder. The evidence suggests that alcohol decreases the oxidation of FFA in the liver, and induces triglyceride and VLDL synthesis.

### Oestrogen excess

Pregnancy and use of the high oestrogen type of oral contraceptive is accompanied by increased plasma levels of VLDL. If there is a pre-existing endogenous hypertriglyceridaemia, massive increases in the plasma lipids may occur. Oestrogens cause increased synthesis and secretion of VLDL.

### Renal failure

Hypertriglyceridaemia due to increased VLDL is common in renal failure, but hypercholesterolaemia, which occurs in the nephrotic syndrome, is unusual. The plasma triglycerides parallel the plasma insulin levels, which are increased due to insulin resistance. The increased VLDL level is probably due to depressed lipoprotein lipase activity, which may be a consequence of this insulin resistance.

### Glycogen storage disease

In the Type 1 disease, which is due to glucose-6-phosphatase deficiency, hyperlipidaemia is common. The probable cause is postprandial hypoglycaemia (defective glycogenolysis), which results in hypoinsulinaemia. The depressed insulin levels stimulate the release and hepatic uptake of FFA, which is then followed by increased triglyceride and VLDL synthesis. Lipoprotein lipase activity may also be suppressed.

## Combined hypertriglyceridaemia and hypercholesterolaemia

PRIMARY

Increased plasma cholesterol and triglyceride levels are a very common finding in patients with ischaemic heart disease, but they may also occur in apparently healthy subjects. These combined hyperlipidaemias can be due to:
(1) ↑ LDL and ↑ VLDL—familial combined hyperlipidaemia.
(2) ↑ VLDL—increased cholesterol is due to VLDL cholesterol. The triglyceride level is relatively greater than the cholesterol level.
(3) Broad β disease.

Familial combined hyperlipidaemia

This is the commonest type of familial hyperlipidaemia. These patients represent a heterogenous group. Some are members of families with monogenic hyper-β-lipoproteinaemia with increased LDL due to the familial disorder, and the increased VLDL due to either alcoholism, obesity or diabetes mellitus, whilst others may have two lipoprotein abnormalities.

The majority of these subjects, however, have a combined hyperlipidaemia which has many features in common with endogenous hypertriglyceridaemia, i.e. a high incidence of obesity, glucose intolerance and hyperuricaemia. The mechanism of this disorder (↑ LDL and ↑ VLDL) is unknown, but has been ascribed to an increased synthesis of these two lipoproteins. The lipid disorder comes to light in the third to fourth decade, and is associated with an increased risk of ischaemic heart disease.

Broad β disease
(dysbetalipoproteinaemia, remnant hyperlipoproteinaemia)

This uncommon disorder is characterized on electrophoresis by a broad single band in the pre-β region, and represents an intermediate density lipoprotein (IDL) which is rich in both cholesterol and triglyceride. The diagnosis can be suspected on inspection of the EPP, but confirmation requires ultracentrifugation studies. Patients with this disease have a high incidence of vascular (mainly peripheral) disease, and may present with tuberous and planar xanthomata. This disorder appears to be associated with an abnormality of apoprotein E.

*Example* Moderately obese 45 year old man, admitted with a myocardial infarct. Specimen taken 8 weeks after infarct.

Plasma appearance turbid
Chol. 10.5 mmol/l (<6.5)
Trig. 9.5 mmol/l (0.3-1.7)
Electrophoresis broad β band (Type III)

In view of the premature ischaemic heart disease in this patient, it would be recommended to carry out lipid studies on his family. In patients who have had a myocardial infarct it is important to wait 1-2 months before carrying out any plasma lipid studies, as the stress of this situation often results in increased triglyceride and lowered cholesterol levels.

#### SECONDARY

Combined secondary hyperlipidaemias may occur in: (1) hypothyroidism, (2) nephrotic syndrome, (3) diabetes mellitus, (4) renal failure, (5) dysglobulinaemia.
*Example* A 70 year old female with overt myxoedema. Plasma taken for thyroid function tests was found to be turbid.

Plasma appearance turbid
$T_4$ <10 nmol/l (60-160)
TSH >40 U/l (<5.5)
Chol. 9.7 mmol/l (<6.5)
Trig. 9.1 mmol/l (0.3-1.7)

The hyperlipidaemia in this patient resolved when complete thyroid hormone replacement was effected.

## HYPOLIPOPROTEINAEMIA

Low plasma levels of lipoproteins are rare disorders and may be due to the deficient production of apoprotein, malnutrition or occasionally hyperthyroidism.

#### FAMILIAL α-LIPOPROTEINAEMIA DEFICIENCY (TANGIER DISEASE)

A rare disorder inherited as an autosomal recessive. There is a deficient production of an apoprotein called apo-A, and that leads to: (1) decreased or absent HDL, (2) low plasma cholesterol, (3) accumulation of cholesterol in reticuloendothelial cells (enlarged yellowish tonsils and adenoids), (4) hepatosplenomegaly and lymphadenopathy.

## ABETALIPOPROTEINAEMIA

This disorder is due to a deficiency of an apoprotein called apo-B which results in: (1) absence of chylomicrons, VLDL and LDL, (2) low plasma levels of cholesterol and triglyceride, (3) accumulation of fat in intestinal cells, (4) malabsorption and steatorrhoea, (5) progressive ataxia, retinitis pigmentosa, acanthosis.

## MALNUTRITION

Severe malnutrition (starvation, malabsorption, terminal cancer) may be associated with low levels of both cholesterol and triglyceride.

## HYPERTHYROIDISM

In thyrotoxicosis the plasma levels of cholesterol are usually less than 5 mmol/l, but rarely less than 3 mmol/l. Although the exact mechanism is unknown, the low LDL levels are presumably due to increased catabolism. Plasma levels of cholesterol in thyrotoxicosis are of little significance and play no part in diagnosis.

# 15 Steatorrhoea — malabsorption

Steatorrhoea is a clinical term that describes the production of faeces which contain excessive amounts of fat. From the laboratory point of view, steatorrhoea is generally considered to be present in an adult if the faecal excretion of fat exceeds 7 g/day. In infants and children the faecal fat excretion is considered to be increased if it exceeds 10% of the dietary intake. Increased faecal fat excretion is due to defective intestinal absorption, but in the majority of cases the condition is also associated with the malabsorption of other nutrients. This chapter will therefore consider both fat and general malabsorption. For the sake of completeness other syndromes due to defective absorption of specific substances will also be included.

## Gastrointestinal digestion and absorption

### LIPIDS

Digestion and absorption of lipids requires: (1) bile salts, (2) pancreatic lipase, (3) normal intestinal mucosa.

The steps in digestion and absorption of triglycerides are:
(1) Emulsification by bile salts.
(2) Hydrolysis of fatty acid-glycerol bonds at positions 1 and 3 by pancreatic lipase, to produce 2-monoglyceride and free fatty acids (FFA).
(3) Aggregation of monoglycerides, FFA and bile salts to form micelles.
(4) Uptake of monoglycerides and FFA by the mucosal cells of the intestine—bile acid micelles remain in the intestinal lumen.
(5) Absorbed FFA and monoglycerides recombine to produce triglycerides.
(6) Intracellular formation of chylomicrons from triglycerides, cholesterol, phospholipids, fat soluble vitamins and apoproteins.
(7) Extrusion of chylomicrons into the central lacteals of the villi.

## BILE SALTS

*Steatorrhoea-malabsorption*

These salts are formed in the liver from cholesterol, secreted in the bile, reabsorbed in the distal ileum by an active carrier-mediated process and resecreted by the liver (enterohepatic circulation). The total bile pool of the body is around 3.5 g, but with recirculation about 25 g can be secreted into the gut each day; of this less than 0.5 g is lost in the faeces daily. Interruption of this enterohepatic circulation may be caused by: (1) biliary obstruction, (2) deconjugation by bacterial action in the gut (unconjugated bile salts are poorly absorbed), (3) ileal resection or disease—and will result in a reduced amount of bile salt in the intestine, and a consequent malabsorption of lipids.

## CARBOHYDRATES

Dietary carbohydrate comprises a mixture of polysaccharides (starch, glycogen) and disaccharides (sucrose, lactose). The polysaccharides are hydrolysed by salivary and pancreatic amylase to disaccharides (maltose, isomaltose). The disaccharides (maltose (glucose-glucose), lactose (glucose-galactose) and sucrose (glucose-fructose)) are hydrolysed into their constituent monosaccharides by specific disaccharidases that are located in the brush border of the cells lining the intestinal lumen. Most of the monosaccharides are then absorbed by an energy-dependent selective process into the portal blood.

## PROTEINS

Dietary proteins are first degraded to peptones and large polypeptides, by the action of pepsin in the stomach. Further breakdown of large proteins and polypeptides to peptides and amino acids occurs in the small intestine, by the action of trypsin and chymotrypsin. Final hydrolysis of peptides into constituent amino acids is accomplished by dipeptidases and amino-polypeptidase enzymes that are located in the epithelial cells lining the intestinal lumen. The amino acids are then absorbed in the duodenum and jejunum by specific processes that are believed to involve the same sodium cotransport system that effects the absorption of sugars.

## VITAMINS AND MINERALS

Fat soluble vitamins

The vitamins A, D, E and K are solubilized by bile acids, incorporated into micelles and absorbed along with other lipids.

### Water soluble vitamins

The C and B group of vitamins (excluding vitamin $B_{12}$) are absorbed in the upper small intestine.

### Vitamin $B_{12}$

Vitamin $B_{12}$ is absorbed in the lower ileum as a complex with intrinsic factor, a glycoprotein that is secreted by the oxyntic cells in the stomach. If intrinsic factor is absent (pernicious anaemia), vitamin $B_{12}$ will not be absorbed.

### Iron

Absorbed in the upper duodenum.

### Calcium

Absorbed in the upper small intestine, and requires the active metabolite of vitamin D to stimulate the synthesis of a specific calcium-binding protein in the intestinal cells.

### Magnesium

Absorbed in upper small gut.

## LABORATORY INVESTIGATION

In cases of suspected malabsorption biochemical investigations are used firstly to demonstrate that a state of malabsorption exists, and then to assist in defining the abnormality. The final diagnosis will depend on assessment of the clinical picture, in conjunction with the results of biochemical investigations, and perhaps also with those of radiology and jejunal biopsy. Although many tests have been developed for the investigation of malabsorption syndromes, the majority of biochemical laboratories offer only faecal fat estimations and xylose absorption tests. These two tests together with the vitamin $B_{12}$ absorption test generally provide sufficient information for a correct diagnosis to be reached.

# Tests of fat absorption

### FAECAL FAT

A quantitative determination of faecal fat is the most reliable measure of steatorrhoea. In the normal subject on a fat-free diet the total fat output is approximately 3 g/day. This fat is derived from intestinal secretions, sloughing of mucosal cells and the normal bacterial flora. On a normal diet (50-100 g of fat per day) the daily output is less than 7 g/day, i.e. >90% of dietary fat is absorbed. When determining faecal fat excretion a number of conditions should be met:
(1) Diet must contain a significant amount of fat (50-100 g/d) for several days prior to the collection of faeces.
(2) Regular bowel movements.
(3) The collection should be over a period of 3 days to allow for irregular bowel evacuations.
(4) Stool collection must be complete.

### PLASMA CAROTENE LEVELS

The carotenes are lipid soluble dietary precursors of vitamin A. Plasma levels are low in patients who have fat malabsorption. Although a useful screening test for generalized malabsorption, low plasma levels of carotenes are also found in other conditions, particularly poor dietary intake of vegetables, or liver disease.

### PLASMA OPTICAL DENSITY

In this test 60 g of fat (taken as buttered toast is suitable) is given orally, and the plasma optical density is measured (wavelength 620 nm) before the fat load is taken, and then hourly for the next 4 hours. Normal subjects have a plasma optical density >0.1 between 2-4 hours post fat load.

### PLASMA TRIGLYCERIDES

60 g of fat is given as in the above test and plasma triglycerides are measured prior to the test, and at hourly intervals for the next 4 hours. Maximum plasma triglyceride levels normally occur 2-4 hours after fat ingestion.

Both this test and the plasma optical density estimation show great overlap between normal and diseased subjects, and therefore their value is limited to a screening role.

### VITAMIN A ABSORPTION

Vitamin A is a fat soluble vitamin, and therefore it would be expected that its absorption would be impaired in conditions associated with impaired fat uptake. However this test has not gained widespread acceptance, partly because of technical difficulties, and partly because clinical trials have shown that a significant number of false positives and false negatives can occur.

### $^{14}$C-LABELLED TRIGLYCERIDE TEST

After digestion and absorption of an oral dose of $^{14}$C-labelled triglyceride ($^{14}$C in fatty acid component), $^{14}CO_2$ is excreted in the breath. If there is triglyceride malabsorption the excretion of $^{14}CO_2$ is low. The use of this test is becoming popular as it provides a quick and sensitive test for the detection of fat malabsorption.

## Tests of protein absorption

### FAECAL NITROGEN

This determination provides an indirect measure of protein absorption. With a 50–100 g/day protein intake the normal faecal nitrogen excretion is less than 2.5 g/day. This excretion rate will be increased both in protein malabsorption and in protein-losing enteropathy. The test is not commonly performed because of the technical difficulties involved in the analysis of protein nitrogen.

## Tests of carbohydrate absorption

### XYLOSE ABSORPTION TESTS

Xylose is a pentose which is: (1) not normally present in blood in significant amounts, (2) actively absorbed in the upper small gut, (3) not metabolized by the body to a significant extent, (4) filtered by the glomerulus.

Poor intestinal absorption occurs in disease states where there is significant loss of the functional integrity of the jejunum. Absorption of xylose does not depend on liver function (bile salts), or on pancreatic function (amylase, lipase, proteolytic enzymes). In the usual absorption test 5 g is given orally after an overnight fast, and urine is then collected in two aliquots (2 h, 3 h) over the next 5 hours. In the normal subject more than 23% of the dose ($\sim$ 1.2 g) is excreted in the 5 hours, with at least 50% of this being excreted in the first 2 hour aliquot. In mild cases of malabsorption the 5 hour excretion rate may be normal, but the majority of the xylose that is excreted is present in the 3 hour aliquot (i.e. 2 hour:5 hour ratio is less than 40%). Decreased xylose excretion occurs in: (1) malabsorption, (2) delayed gastric emptying, (3) massive bacterial overgrowth in the small intestine, (4) renal insufficiency.

In infants where satisfactory urine collection may be difficult, blood samples for xylose estimations can be collected at 30 and 60 minutes after the oral dose. In the normal infant the blood levels should be greater than 1.0 mmol/l and 2.0 mmol/l at 30 and 60 minutes respectively. The usual xylose dose for infants is 10 g/m$^2$ surface area, up to a maximum of 5 g.


GLUCOSE TOLERANCE TEST

A flat response is classically associated with malabsorption syndromes, but this type also occurs with very efficient absorption and is often seen in normal young adults. A GTT is not worth performing, either as a screening test for malabsorption, or for a differential diagnosis.


LACTOSE TOLERANCE TEST

Lactase deficiency, the most common of the disaccharidase deficiencies, can be diagnosed by the use of the lactose tolerance test. In this test 50 g of lactose (children, 1 g/kg) is given orally, and blood samples are taken as for a glucose tolerance test. In the normal subject the plasma glucose level should rise by at least 1.1 mmol/l. If the test result is abnormal a control test should be performed, using a glucose/galactose mixture equivalent in amount to the lactose given.


FAECAL pH AND SUGARS

In any carbohydrate malabsorption the unabsorbed sugars: (1) cause diarrhoea (osmotic effect), (2) are broken down to acids by bacterial contents of the gut and result in a faecal pH of less than 7, (3) appear in the faeces.

Measurement of faecal pH and sugars (qualitative and quantitative) can be helpful in the diagnosis of carbohydrate malabsorption in infants. These tests have not proved to be useful in adults.

## SMALL INTESTINE DISACCHARIDASE ASSAYS

The activity of the small gut disaccharidases (lactase, sucrase, maltase) can be estimated from tissue obtained by biopsy. These tests provide the definitive diagnosis of the disaccharidase deficiencies.

## Tests of vitamin absorption

### VITAMIN $B_{12}$ ABSORPTION

The absorption of vitamin $B_{12}$ occurs in the lower small intestine, and depends on:
(1) Binding of the vitamin to intrinsic factor released from the parietal (oxyntic) cells of the stomach lining.
(2) Recognition by specific receptors and transport of the complex through the wall of the ileum.
(3) A possible pancreatic factor.

Absorption of the complex is decreased in:
(1) Pernicious anaemia (intrinsic factor deficiency).
(2) Bacterial overgrowth of the gut (metabolize $B_{12}$ locally).
(3) Diseases and resections of the terminal ileum.
(4) Severe pancreatic disease.

In the Schilling test radioactive $B_{12}$ is given by mouth, and a large flushing dose of unlabelled $B_{12}$ is given parenterally to saturate the peripheral vitamin $B_{12}$ binding sites. Urine is collected over the next 24 hours, and the amount of radioactive $B_{12}$ excreted is estimated. Normally the amount excreted is greater than 5–8% of the oral dose. In a case of intrinsic factor deficiency $<2.5\%$ of the dose is excreted. The test should then be repeated by administering the radioactive $B_{12}$ already complexed to intrinsic factor. If the $B_{12}$ deficiency is due to malabsorption the radioactive excretion will remain low in this second test.

## BLOOD FOLATE LEVELS

*Steatorrhoea-malabsorption*

Folic acid is absorbed throughout the small intestine, and so blood levels (plasma and red cell) are usually reduced in malabsorption syndromes that involve the integrity of the gut wall, e.g. coeliac disease, tropical sprue. Estimation of blood folate has been used as a screening test for these disorders, but it is unreliable because:
(1) Coeliac disease can co-exist with normal blood levels.
(2) There are many other causes of low blood folate levels, e.g. (a) folate deficiency in the diet, (b) excessive utilization due to cellular proliferation, e.g. haemolytic anaemia, myeloproliferative disease, carcinoma, (c) drug effects, e.g. phenytoin sodium (dilantin) and other anticonvulsant drugs may decrease folic acid absorption.

## Pancreatic function tests

### SECRETIN-PANCREOZYMIN TEST

Pancreatic function can be assessed by measuring the volume of juice and the amount of bicarbonate and enzymes secreted in response to administered secretin and pancreozymin. Briefly, a double-lumen tube is passed via the mouth so that one tube can collect gastric juices whilst the other tube samples duodenal contents. A 20 minute basal aspiration of duodenal contents is obtained, after which secretin is given by intravenous injection. A further 20 minute sample is obtained, and 1 hour after the secretin injection an injection of pancreozymin is given, followed by a further 20 minute aspiration. The pH of gastric and duodenal samples are measured to ensure that cross contamination, prior or during sampling, has not occurred. The samples are then assayed for volume, bicarbonate and amylase. Trypsin and lipase may also be measured.

In pancreatic disorders (chronic pancreatitits, cystic fibrosis) the volume, bicarbonate content and enzyme activities of the pancreatic juice are reduced. This test is difficult to interpret, and should only be carried out by experienced personnel. The Lundh test is a similar test to the above, whereby a meal containing 19 g of corn oil, 15 g of milk powder and 40 g of glucose is given to stimulate pancreatic secretion. These tests are not widely used for diagnostic purposes.

### FAECAL TRYPSIN ACTIVITY

This test has been used for the investigation of suspected pancreatic dysfunction in children, but it has proved to be very unreliable and should not be used as a screening test.

**Miscellaneous tests**

### BILE SALT $^{14}CO_2$ BREATH TEST

This is a sensitive test for the detection of bacterial overgrowth in the small bowel. $^{14}C$-glycine conjugated bile salts are given orally and the expired air is analysed for $^{14}CO_2$. If there is bacterial overgrowth the bile salt is deconjugated, and the $^{14}C$-glycine is released and converted to $^{14}CO_2$ by bacterial or tissue enzymes. The $^{14}CO_2$ is then excreted in the expired air. In the normal subject the $^{14}C$-bile salt conjugate is absorbed intact and re-excreted by the liver. The lung excretion of $^{14}CO_2$ is also increased if there is disease or resection of the terminal ileum, because unabsorbed bile salt then reaches the colon where it is rapidly metabolized by the normal bacterial flora.

### SWEAT ELECTROLYTES

Cystic fibrosis, a common inborn error of metabolism, is associated with malabsorption due to pancreatic dysfunction in infancy. In these children the sweat sodium and chloride concentrations are high ($>60$ mmol/l). The sweat electrolyte test, if high values are obtained, confirms the diagnosis of cystic fibrosis in children. Results are difficult to interpret in adults as the sweat electrolyte concentration alters with age.

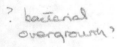

### URINARY INDICANS

Some intestinal bacteria are capable of metabolizing the amino acid tryptophan to indole, which is absorbed and converted by the liver to indoxyl sulphate (indican). The indican is then secreted by the kidney. Urinary indican levels have been suggested as a diagnostic test for steatorrhoea which is due to abnormal bacterial overgrowth in the gut (e.g. blind loop syndrome). The urinary indican level is frequently raised in untreated gluten enteropathy (coeliac disease), Crohn's disease, and in the blind loop syndromes, but is normal in steatorrhoea that is due to either pancreatic deficiency or liver dysfunction. However, there is considerable overlap between the various groups, and therefore this test is of limited use in the differential diagnosis of steatorrhoea.

Syndromes due to generalized malabsorption are often associated with decreased plasma levels of iron, calcium (vitamin D malabsorption) and albumin (amino acid malabsorption). Analysis of these analytes is of little help in the elucidation of the causes of malabsorption, but are useful in the evaluation of the severity of the disorder.

## Selection of tests

The selection of tests will depend on the clinical picture. If steatorrhoea is suspected a faecal fat estimation should be performed to establish the diagnosis. This may be followed by xylose absorption and/or $B_{12}$ absorption tests to allocate the disorder into an appropriate category (Table 15.1). The aetiological diagnosis will finally depend on the results of radiological examination and jejunal biopsy. The more specific tests (pancreatic function, disaccharidase estimations, etc.) are necessary only when indicated by the clinical findings. They are difficult and time consuming, and are not routinely performed by most laboratories.

**Table 15.1** Results of biochemical tests in steatorrhoea

|  | Faecal fat | Xylose absorption | $B_{12}$ absorption | Jejunal biopsy |
|---|---|---|---|---|
| Obstructive jaundice | ↑ | N | N | N |
| Pancreatic dysfunction | ↑ | N | ↓ or N | N |
| Gluten enteropathy | ↑ | ↓ | ↓ or N | Flat villi |
| Tropical sprue | ↑ | ↓ | ↓ or N | Flat villi |
| Terminal ileal disease/resection | ↑ | N | ↓ | N |
| Blind loop syndrome | ↑ | ↓ or N | ↓ | N |

# STEATORRHOEA

The major causes of malabsorption are listed below. These conditions usually result in multiple malabsorption problems, however, fat absorption is often impaired at an early stage and is frequently the presenting feature.

## Classification

Maldigestion

*Lipase deficiency* (1) Chronic pancreatitis, (2) pancreatic resection, (3) carcinoma of pancreas, (4) cystic fibrosis.

*Bile salt deficiency* (1) Biliary obstruction, (2) chronic liver disease, (3) disease/resection of terminal ileum, (4) blind loop syndrome.

Intestinal defects

(1) Gluten enteropathy (coeliac disease), (2) tropical sprue, (3) small bowel resections, (4) small bowel ischaemia, (5) regional enteritis, (6) abetalipoproteinaemia, (7) infiltrations (lymphoma, amyloid, scleroderma).

Miscellaneous

(1) Postgastrectomy, (2) Zollinger-Ellison syndrome, (3) carcinoid syndrome, (4) diabetes mellitus, (5) parasitic infections, (6) Whipple's disease.

**Maldigestion**

LIPASE DEFICIENCY

Defective digestion, and therefore absorption, of triglyceride occurs with a deficiency of exocrine pancreatic secretions. Pancreatic juice also contains proteinases and therefore there is also malabsorption of protein. The most common causes of lipase deficiency are cystic fibrosis and chronic pancreatitis.

Cystic fibrosis

This disease is of unknown pathogenesis and is one of the most common inborn errors of metabolism. It is inherited as an autosomal recessive, and is characterized by exocrine glands that secrete an abnormally viscous mucus and sweat that has a high electrolyte content. It may present with meconium ileus at birth, malabsorption in infancy, or as persistent chronic bacterial infections of the lung. Malabsorption occurs in over 80% of cases and is due to pancreatic insufficiency (pancreatic acini are fibrosed and contain inspissated mucus). Diagnosis is confirmed on the basis of the sweat electrolyte estimations, in which case the chloride and sodium concentrations are both greater than 60 mmol/l. Heterozygotes have normal sweat electrolytes.

## Chronic pancreatitis

Steatorrhoea is a late complication of chronic pancreatitis and may be particularly severe. The diagnosis is usually made on clinical grounds, although in obscure cases pancreatic function tests may be required. In malabsorption due to pancreatic deficiency both the $B_{12}$ and xylose absorption tests are usually normal, although in severe deficiency there may be malabsorption of $B_{12}$ (?deficient pancreatic factor).

### BILE SALT DEFICIENCY

In bile salt deficiency there is a failure to incorporate the dietary fatty acids and monoglycerides into micelles prior to absorption. In this situation the absorption of the fat soluble vitamins, particularly D and K, is also impaired. If prolonged, this may result in metabolic bone disease (osteomalacia, rickets) and bleeding disorders (increased prothrombin time).

The bile salt deficiency syndrome may occur in: (1) obstructive jaundice, (2) chronic liver disease, (3) resection/disease of terminal ileum, (4) bacterial overgrowth (stagnant loop, blind loop syndrome).

### Obstructive jaundice/liver disease

Steatorrhoea due to either of these two disorders usually does not require further investigation as the cause will be clinically obvious. Xylose and $B_{12}$ absorption is normal.

### Resection/disease of terminal ileum

The terminal part of the ileum is the main site for the absorption of vitamin $B_{12}$ and bile salts. Steatorrhoea due to either surgical resections or disease (e.g. Crohn's disease) of this region of the intestine are associated with a decreased absorption of vitamin $B_{12}$, but a normal absorption of xylose (upper small gut absorption).

### Stagnant loop syndrome

The detergent properties of bile salts, which enable them to form micelles, are due to the conjugated amino acids glycine and taurine (conjugated bile salts). Certain bacteria (e.g. E. coli) can deconjugate bile salts to form free bile salts; these do not have detergent properties and are therefore unable to form micelles. Any disorder that alters the motility or emptying time of the small intestine may result in stasis and bacterial overgrowth, e.g. (1) afferent loop after gastrectomy, (2) surgical blind loops, (3) small bowel diverticula, (4) strictures of the small bowel, (5) fistulas of the small bowel.

Bacterial overgrowth may:
(1) Deconjugate bile salts → fat malabsorption and steatorrhoea.
(2) Metabolize tryptophan → increased secretion of urinary indican.
(3) Metabolize dietary viatmin $B_{12}$ → decreased $B_{12}$ absorption.
(4) Metabolize xylose → decreased xylose absorption.

These abnormalities may return to normal after therapy with broad spectrum antibiotics.

## Intestinal defects

MUCOSAL LESIONS

In both coeliac disease and tropical sprue there is extensive involvement of the absorbing surface in the disease process, resulting in malabsorption of protein, carbohydrate and steatorrhoea.

Coeliac disease (gluten-induced enteropathy)

This disease occurs in both adults and children, and is characterized by malabsorption, intolerance to gluten (wheat protein), and abnormal villous structure in the small intestinal mucosa. It is not known how the gluten induces the structural change, but it may be due to an immune mechanism. The tests for malabsorption may reveal abnormalities ranging from minimal alteration to severe changes. Diagnosis requires the demonstration of characteristic histological changes (villous atrophy, flat villi) in a jejunal biopsy.

*Example* Child aged 5 years with failure to thrive, abdominal distension and anaemia. Stools were loose, bulky, pale and offensive. A jejunal biopsy revealed villous atrophy.

| Plasma | Ca | 1.85 | mmol/l | (2.15-2.55) | Plasma Fe | 4 µmol/l (14-29) |
|---|---|---|---|---|---|---|
| | $PO_4$ | 0.62 | mmol/l | (0.60-1.3) | TIBC | 75 µmol/l (45-75) |
| | Alb. | 29 | g/l | (32-50) | | |
| | AP | 400 | U/l | (<250) | | |
| Faecal fat excretion | | 35 | g/3 days | (<21) | | |
| Xylose absorption | | | | | | |
| (5 g dose) | | 5 | hour renal excretion 0.5 g (>1.2) | | | |

The degree of malabsorption in coeliac disease will depend on the extent of the disease process. In this case there has been malabsorption of:

*Steatorrhoea-malabsorption*

fat → steatorrhoea
protein → hypoalbuminaemia
vitamin D → hypocalcaemia (hypoalbuminaemia also → hypocalcaemia)
iron → anaemia and ↓ plasma Fe (anaemia may also be due to deficient absorption of vitamin $B_{12}$ and folic acid)

The xylose absorption test reveals defective intestinal absorption. The diagnosis was confirmed by a jejunal biopsy. After treatment with a gluten-free diet the symptoms in this patient improved within 3 weeks and he began to gain weight.

Tropical sprue

This condition occurs in tropical and subtropical zones and, although the mucosal abnormalities appear to be similar to that seen in coeliac disease, it is not related to gluten ingestion. The cause is unknown, although the usually favourable response to oral broadspectrum antibiotic therapy suggests a bacterial aetiology.

STRUCTURAL LESIONS

Ischaemic lesions of the small intestine (mesenteric artery insufficiency), and resections of this part of the gut will result in varying degrees of malabsorption, but the diagnosis is usually clinically obvious and does not require biochemical investigation.

INFILTRATIONS AND INFLAMMATIONS

Infiltrations (amyloidosis, lymphoma, scleroderma) and inflammatory lesions (regional enteritis, Crohn's disease) of the small intestine may be associated with malabsorption due to:
(1) Involvement of absorptive structures in the gut wall (mucosa, lymphatics, etc.).
(2) Interference with bile salt metabolism, (a) involvement of the absorptive surface of the terminal ileum, (b) bacterial overgrowth due to decreased motility.

BIOCHEMICAL LESIONS

Abetalipoproteinaemia (see p. 333).

**Miscellaneous causes**

POSTGASTRECTOMY

Postgastrectomy steatorrhoea is usually mild and of little clinical significance, however some patients may have gross steatorrhoea that requires careful assessment. Factors that contribute to the steatorrhoea are:
(1) Rapid intestinal transit (insufficient time for emulsification etc.) due to rapid gastric emptying.
(2) Blind loops and afferent loops as part of the operative procedure, resulting in stasis and bacterial proliferation.
(3) Duodenal bypass; bile salts and pancreatic enzymes secreted into the duodenum do not mix with the gastric contents, which are emptied into the jejunum.

*Example* Female, 46 years, gastrectomy for duodenal ulcer 18 months prior to admission. Presented with anaemia, weight loss and complaining that her stools were frequent, yellowish and bulky.

| | | | |
|---|---|---|---|
| Plasma Alb. | 43 | g/l | (32–50) |
| Ca | 2.02 | mmol/l | (2.15–2.55) |
| $PO_4$ | 0.71 | mmol/l | (0.6–1.3) |
| AP | 202 | U/l | (30–120) |
| Faecal fat | 75 | g/3 days | (<21) |

The complications (biochemical, haematological) of gastrectomy are:
(1) Steatorrhoea.
(2) Osteomalacia. Up to 30% of patients in some series have shown some degrees of osteomalacia. This may be due to malabsorption of vitamin D and calcium, or it could reflect an inadequate diet. The earliest biochemical manifestation is a raised plasma alkaline phosphatase. If the deficiency is severe hypocalcaemia and hypophosphataemia may occur.
(3) Postprandial hypoglycaemia.
(4) Anaemia. This occurs in approximately 50% of patients. The most common type is iron deficiency (dietary, malabsorption), but folic acid deficiency is also common, and $B_{12}$ deficiency occurs if the gastrectomy has been total (absent intrinsic factor).

ZOLLINGER-ELLISON SYNDROME

This disease is due to excessive gastric acid secretion stimulated by inappropriate levels of gastrin, a peptide hormone that is normally secreted by the G cells in the pyloric glands. The autonomous production of gastrin originates from tumours (gastrinoma) found in the pancreas (islets) and occasionally in the gut wall. The features of the syndrome are: (1) recurrent or intractable duodenal ulceration (hyperacidity), (2) diarrhoea, (3) high plasma gastrin levels, (4) excessive secretion of HCl by the stomach.

This syndrome is frequently associated with malabsorption, which may be due to: (1) acidification of duodenal contents causing lipase inactivity, (2) precipitation of glycine-conjugated bile salts at low pH.

DIABETES MELLITUS

Steatorrhoea in patients with diabetes, although well documented, is rare. It may be due to autonomic nervous system degeneration, resulting in decreased motility and bacterial overgrowth.

PARASITIC INFECTIONS

Infections with parasites, particularly Giardia, may be associated with mild steatorrhoea and abnormal xylose absorption. The cause of the steatorrhoea is unknown.

WHIPPLE'S DISEASE

This is a rare disorder characterized by diarrhoea, malabsorption and infiltration of the small bowel wall by characteristic cells (macrophages containing glycoprotein) and bacteria. The cause is unknown, but it responds to antibiotic therapy.

CARCINOID SYNDROME

Diarrhoea and flushing are the cardinal features, but malabsorption may occasionally be found. This steatorrhoea may be due to changes in gastrointestinal motility, lymphatic obstruction by the tumour, or to resection of the gut.

**Specific malabsorption syndromes**

VITAMIN $B_{12}$

Pernicious anaemia—lack of intrinsic factor.

## DISACCHARIDASE DEFICIENCY

The most common disaccharidase defect is that of lactase deficiency. In this disorder the ingestion of lactose (milk) is followed by abdominal cramps, distension, borborygmus (↑ abdominal sounds) and frothy frequent diarrhoea. The diarrhoea is due to the increased osmotic load, and the irritation of the colon by lactic acid and short chain fatty acids which are derived from lactose by bacterial degradation. The condition may be primary or secondary to disease (e.g. sprue). The diagnosis is usually made on clinical grounds, and can be confirmed by a lactose tolerance test or intestinal disaccharidase estimations. The stool usually has a pH less than 6 and contains lactose, glucose or galactose (reducing sugars).

## ISOLATED TRANSPORT DEFECTS

### Carbohydrate

*Glucose-galactose transport deficiency.*

### Amino acids

*Cystinuria* Cystine, ornithine, arginine, lysine (transport deficiency in both gut and renal tubules).

*Hartnup disease* Defective intestinal and renal transport of most neutral amino acids.

tryptophan absorption ↓ → nicotinamide (niacin) deficiency
→ dermatitis → pellagra, inflammation of intestinal wall,
　　　　　　　　→CNS lesions → dementia
intestinal tryptophan ↑ → bacterial action → ↑ urinary indoles

## PROTEIN-LOSING ENTEROPATHY

This disease is due to excessive gastrointestinal loss of protein, and although not a malabsorptive condition it is conveniently classified with this group. It may be associated with a wide variety of conditions (see p. 246) and is usually manifested by hypoproteinaemia. Biochemical detection of the disorder involves the intravenous administration of a radioactive macromolecule ($^{125}$I-PVP, $^{51}$Cr-labelled albumin) and the measurement of the radioactivity subsequently appearing in the stool. Faecal $\alpha_1$-antitrypsin analysis provides a useful screening test.

# 16 Iron

Iron is a toxic metal, but this potentially harmful property is avoided because nearly all the iron in the body is associated with protein. Iron plays an essential part in the biological activity of many proteins. Such proteins can be divided into those that contain haem, e.g. haemoglobin, myoglobin, cytochromes, and those that do not, e.g. ferritin, transferrin, flavoproteins, most oxygenases. From the biological function of these proteins it is seen that iron has a central role in both oxygen transport and energy metabolism. The body normally contains 3.5–4.0 g of iron, the approximate distribution is shown below.

| Compartment | | Iron content (mg) | % of total |
|---|---|---|---|
| Haem protein | Haemoglobin | 2500 | 60–70 |
| | Myoglobin | 130 | 3–5 |
| Haem enzymes | Cytochromes etc. | 8 | 0.2 |
| Plasma | Transferrin | 4 | 0.1 |
| Stores | Ferritin, haemosiderin | 1000 (male) 800 (female) | 20–30 |
| Total (approx.) | | 3500–4000 | |

**Homeostasis** (Fig. 16.1).

LOSS

An unusual feature of iron metabolism is the lack of a physiological control mechanism for excretion. Unavoidable loss of iron does occur via normal blood loss into the gut (2–3 ml/day), the secretions of body fluids and the shedding of cells. The iron content of the gut epithelial cells reflect the general body stores, and so the loss of such cells (life-span 2–3 days) does constitute an excretory mechanism, i.e. when body stores are high, the gut cell iron content is high, so iron loss is increased. The amount of iron involved in daily loss and replacement is small compared with that passing through the internal pathways (Fig. 16.1).

## Chapter 16

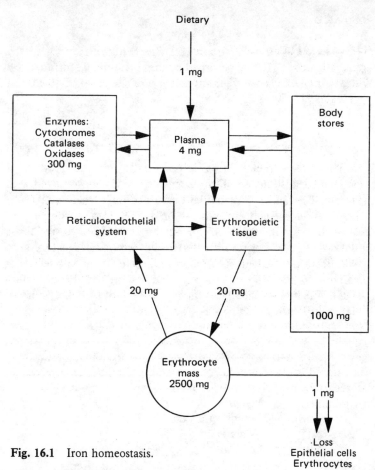

**Fig. 16.1** Iron homeostasis.

| Daily loss (adult) | mg | |
|---|---|---|
| Cells  skin | 0.3 | |
|       intestine | 0.7 | |
| Urine | 0.1 | |
| Faeces blood | 1.0–1.3 | (2–3 ml blood) |

| Other: menstruation | ~30 | (35–40 ml blood) |
|---|---|---|
|       pregnancy | 800–1200 | |
|       lactation(6 m) | 150 | |

The relatively large loss of iron during pregnancy is due to:

| | mg |
|---|---|
| (1) Fetal requirement | 300 |
| (2) Expansion of maternal blood volume | 400 |
| (3) Normal cell loss | 300 |
| (4) Parturition (blood, placenta) | 200 |

## INTAKE

The average Western diet contains 10–30 mg of iron, of which normally only 5–10% is absorbed by the intestine. Recommended dietary intake for non-pregnant adults is of the order of 10–20 mg per day.

## ABSORPTION

There are two different pathways for dietary iron absorption, one for iron attached to haem, and one for uncomplexed ferrous iron. Haem, derived from dietary haemoglobin and myoglobin, is released from its apoprotein by the gastric juice. In the small intestine the haem enters the mucosal cell intact, and the iron is then cleaved from the porphyrin rings. The rest of the dietary iron is mainly in the ferric state ($Fe^{3+}$) and in organic form; it is split from its organic complexes by gastric HCl and then reduced to the ferrous form ($Fe^{2+}$) by reducing substances present in the diet. The ferrous ion is more readily absorbed than the ferric ion. After uptake by the mucosal cells, the iron is either transported across the cell to the plasma (transferrin), or it is incorporated into ferritin and stored in the epithelial cell, depending on the level of tissue stores in the body. The precise mechanism controlling absorption is unknown.

Factors influencing duodenal iron absorption

(1) Quantity of dietary iron present.
(2) Acid gastric juice favours haem release and $Fe^{3+}$ solubilization and reduction.
(3) Presence of solubilizers (sugars, amino acids), chelators (ascorbic acid) and reducing compounds (ascorbate, glutathione).
(4) Presence of precipitating chelators (phytates, phosphates, oxalate, carbonate).
(5) Intestinal mobility.
(6) State of tissue iron stores—iron deficiency increases uptake.
(7) Rate of erythropoiesis.

## IRON TRANSPORT

Iron is transported in the plasma attached to transferrin, a $\beta_1$-glycoprotein (mol. wt. $\sim$ 80 000 Daltons) synthesized in the liver. Each molecule of transferrin is capable of binding two atoms of ferric iron ($Fe^{3+}$). The normal plasma concentration of iron results in about one third of transferrin being saturated. Plasma transferrin can be indirectly quantitated by determining the amount of iron it will carry when all plasma transferrin iron-binding sites are occupied, the 'total iron binding capacity' (TIBC).

## IRON CYCLE

Erythrocytes have an average life span of 120 days, at the end of which they are engulfed by the macrophages of the reticuloendothelial system. During this degradation process 20–30 mg of iron (from 8 to 10 g of haemoglobin) is liberated and reutilized in the synthesis of new haemoglobin. A small proportion of this iron enters the bloodstream, and is transported to other tissues to be stored as ferritin or haemosiderin.

### Iron stores

About 30% of the body iron is in storage form, occurring as complexes with protein within the reticuloendothelial and parenchymal cells of many organs, particularly the liver, bone marrow and spleen.

## FERRITIN

A large spherical molecule that comprises a protein shell (apoferritin) and a core of ferric hydroxide phosphate. The protein has 24 polypeptide subunits, resulting in a total molecular weight of approximately 460 000 Daltons. When fully saturated, the iron contributes to about 50% of the dry weight of the ferritin, but with normal saturation it is about 23%. This storage iron is readily available when needed, as when the iron released from catabolized red cells is insufficient to meet the prevailing requirement for erythropoiesis. A small amount of ferritin is found in the circulating plasma, the level of which reflects the tissue stores.

## HAEMOSIDERIN

This substance is a more stable and less available source of iron than ferritin. Tissues containing haemosiderin show characteristic staining with Prussian Blue. Haemosiderin is derived from ferritin, and exists in the tissues as water insoluble granules that comprise iron and apo-ferritin-like protein.

# LABORATORY INVESTIGATION

The iron stores of the body may be evaluated by a variety of biochemical and non-biochemical procedures.

Non-biochemical

*Peripheral blood picture*   Iron deficiency is reflected by the appearance of hypochromic microcytic anaemia. With decreasing iron availability the following sequence occurs: microcytic → hypochromic → anaemia.

*Biopsy*   A rough assessment of iron stores can be made by estimating the iron content of biopsies taken from bone marrow or liver. Prussian Blue is used to detect tissue haemosiderin.

Biochemical (excludes required haematological tests)

(1) plasma iron, (2) plasma total iron binding capacity (TIBC), (3) plasma ferritin, (4) chelation.

## PLASMA IRON

Care is required in interpreting a plasma iron result because it is influenced by a variety of physiological factors.

*Sex*   Male levels are 10–20% higher than female levels.

*Pregnancy*   Increased levels, due to increased transferrin synthesis, are often found during pregnancy and the use of oral contraceptives (oestrogen effect). However, the increased demand for iron that occurs in pregnancy may overshadow the effect of the increased transferrin, and be reflected by a low plasma iron.

*Circadian rhythm*   Plasma levels are highest in the morning, and then decline to their lowest point in the evening.
(1) Day-to-day—variations from day to day may be two to three fold.
(2) Menstrual cycle—very low levels may be seen immediately prior to, and during, menstruation.

Pathological variation

*Chronic disease*   Many chronic diseases, particularly infections, malignancy, and renal failure, may be associated with low plasma levels of iron in the presence of normal or increased body stores.

*Acute liver disease* During acute liver disease enough ferritin may be released into the plasma to give a transient rise in the iron concentration.

*Haemolytic anaemia* During a haemolytic crisis the release of iron from stores may result in a temporarily increased plasma iron level.

In isolation, plasma iron levels have limited diagnostic value because they do not reflect body iron stores unless the stores are severely depleted. The diagnostic value of a plasma iron estimation can be improved if it is performed in conjunction with a plasma total iron binding capacity (TIBC).

## TOTAL IRON BINDING CAPACITY (TIBC)

The TIBC is estimated by saturating an aliquot of plasma with iron, and then measuring the total protein-bound iron level, thus giving a measure of the circulating transferrin. In the normal subject there is sufficient transferrin to bind 40–70 μmol/l, but normally about one third (20–50%) of the transferrin carries iron (plasma iron: 15–30 μmol/l). The 'percentage saturation' of the plasma transferrin (Serum iron/TIBC × 100%) is a useful parameter if properly interpreted.

In the normal subject the percentage saturation lies in the range 20–50%, with values towards the upper end of the range occurring in the early part of the day. If transferrin saturation falls below 16% the basal level of erythropoiesis is depressed due to a lack of available iron, whilst saturation of less than 10% is indicative of iron deficiency (whole body), i.e. if iron stores are markedly depleted the plasma iron is usually low, and there is a high TIBC (increased apotransferrin synthesis), causing a low percentage saturation. The percentage saturation may also be low in certain conditions where iron stores are normal, but the release of iron from them is blocked, e.g. (1) acute infections (↓plasma iron), (2) premenstrual period (↓plasma iron).

In iron overload the plasma iron rises and the TIBC falls, indicating a high percentage saturation, usually 80–100%. High percentage saturations without iron overload may occur in: (1) hypoplastic anaemia (↑plasma iron), (2) liver cirrhosis (↓TIBC), (3) protein-losing states (↓TIBC).

Increased TIBC may occur in: (1) iron deficiency, (2) childhood, (3) pregnancy, (4) oral contraceptives.

Decreased TIBC may occur in: (1) iron overload, (2) cirrhosis of liver, (3) chronic inflammatory disease, (4) protein-losing states.

## TRANSFERRIN

The plasma levels of this iron transport protein can be measured by immunological techniques. However, because of the ease with which TIBC can be estimated along with plasma iron, the direct measurement of plasma transferrin has not gained wide acceptance.

## PLASMA FERRITIN

Ferritin is the major iron storage protein present in the reticuloendothelial cells of the liver, bone marrow and spleen. A small amount is present in plasma, all in the apo form, its concentration normally varying directly with the body stores of iron (10–200 µg/l). Decreased levels are found in iron deficiency. Increased plasma ferritin levels are found with increased iron stores and with iron overload (e.g. haemochromatosis), but levels can also be increased in a number of conditions that are not usually associated with iron overload, e.g.
(1) Acute liver disease—release from liver cells to plasma.
(2) Neoplasms (a) leukaemias, (b) lymphomas, (c) carcinomas (? formation of tumour ferritin).
(3) Acute/chronic infections—release of iron from ferritin is blocked (unknown mechanism).
(4) Haemolysis.

The main clinical application of measuring plasma ferritin is to aid in the diagnosis of iron deficiency, and so it is important to realize that in this condition the plasma levels of ferritin may be normal or even increased if there is concurrent disease, e.g. infection.

## CHELATION

Iron stores can be assessed by measuring the urinary excretion of iron following the injection of an iron-chelating agent such as desferrioxamine. This test has been used in the assessment of iron stores in haemochromatosis, but otherwise it has limited clinical value.

## IRON DEFICIENCY

### Causes

*Inadequate intake* (1) Poor nutrition—very rare (compare stores vs loss Fig. 16.1), (2) malabsorption—very rare.

*Excessive loss* Blood loss—menorrhagia, intestinal malignancy, intestinal parasites.

*Increased requirement* Pregnancy.

**Case examples/pathophysiology**

Female, aged 40 years, with severe menorrhagia. Patient was pale, had a haemoglobin of 50 g/l (120–160) and the red cell picture was hypochromic and microcytic.

| Plasma | Fe | 4 µmol/l | (14–29) |
|---|---|---|---|
| | TIBC | 86 µmol/l | (45–72) |
| | Saturation | 4.6% | (20–50) |
| | Ferritin | 6 µg/l | (10–200) |

Females during the reproductive period are in precarious iron balance, so that any increase above the normal menstrual blood loss (menorrhagia) can produce iron deficiency. A low plasma ferritin level and/or a low percentage saturation associated with a hypochromic anaemia is strong evidence for iron deficiency. The definitive diagnosis requires an estimation of the iron stores (e.g. bone marrow biopsy), but if the anaemia responds to iron therapy this is sufficient to prove the diagnosis of iron deficiency. Hypochromic anaemia may also occur in chronic diseases (e.g. rheumatoid arthritis, renal failure, chronic infections), in which case the anaemia is probably due to a failure of iron release from the stores, resulting in a low plasma iron, reduced TIBC (but normal % saturation) and a normal or increased plasma ferritin. However, there may be iron deficiency in patients with rheumatoid arthritis in whom long term treatment with aspirin can cause gastric bleeding.

# IRON OVERLOAD

**Causes**

*Increased absorption* (1) Idiopathic haemochromatosis, (2) increased red cell turnover—chronic haemolytic anaemia, (3) increased dietary intake—e.g. Bantu siderosis.

*Parenteral administration* Multiple transfusions (especially treatment for thalassaemia).

In iron overload the excess iron is deposited in the tissues as ferritin and haemosiderin. Haemosiderosis is a histological diagnosis that describes an increased stainable iron (haemosiderin), being detectable in reticuloendothelial cells and to a lesser extent in parenchymal cells. If there is associated parenchymal damage the condition is referred to as 'haemochromatosis'. Haemosiderosis is not necessarily associated with increased body iron stores since it can occur as a local phenomenon (e.g. local haemorrhage). Haemochromatosis is associated with increased body iron stores and often functional disturbance of the involved organ.

### Case examples/pathophysiology

A 42 year old male with epigastric pain and nausea. He had slate grey skin and hepatomegaly.

Plasma
| | | | | | | | |
|---|---|---|---|---|---|---|---|
| Fe | 38 | µmol/l | (14–29) | Plasma ALT | 270 | U/l | (<35) |
| TIBC | 43 | µmol/l | (45–72) | AP | 400 | U/l | (30–120) |
| Saturation | 88% | | (20–50) | Bili. | 20 | µmol/l | (<20) |
| Ferritin | 1700 | µg/l | (10–200) | | | | |

OGTT
| | | |
|---|---|---|
| fasting glucose | 5.7 mmol/l | (3.0–5.5) |
| 60 minute glucose | 14.7 mmol/l | |
| 120 minute glucose | 11.4 mmol/l | |

*Liver biopsy*  Features of cirrhosis, and a large amount of dark brown granular material staining a bright blue with Prussian Blue.

*Diagnosis*  Idiopathic haemochromatosis. This is an inherited disorder (? autosomal recessive), characterized by excessive iron deposition and associated parenchymal damage in a variety of organs. The main tissues involved are the liver (cirrhosis), the pancreas (diabetes mellitus) and the heart (cardiac failure). Other tissues (e.g. adrenal) are involved to a lesser extent. There is a characteristic grey-brown pigmentation of the skin due to increased deposition of melanin, hence the name 'bronze diabetes' that is often used to describe this condition. In the above case there is severe liver involvement and diabetes. The basis of the disorder is an increased absorption of iron over a period of many years (iron stores can be of the order of 25 g). Studies have shown that in the intestinal mucosal cell the ratio of iron transported across the cell to that taken up is increased, but the precise mechanism of the defect is unknown. The diagnosis is made on the clinical findings, the biochemical evidence of increased iron stores (ferritin, % saturation), and the results of a liver biopsy. When diagnosed the patient's relatives should also be investigated. The best single test in this regard is the measurement of plasma ferritin.

# 17 Porphyrins

Haem, the iron-containing prosthetic group of both haemoglobin and the cytochromes, is synthesized from glycine and succinate by many tissues, particularly the liver and bone marrow. Important intermediates in the haem biosynthetic pathway are the porphyrinogens, a group of tetrapyrroles that differ from each other in their side-chain composition.

**Porphyrin structure**

The porphyrin molecule comprises four rings linked by methene bridges (tetrapyrrole), and provides the basic structural unit that is common to all the porphyrins (Fig. 17.1). Reduction of the methene bridges forms the corresponding porphyrinogen (Fig. 17.1). The attachment of various side chains (methyl, vinyl, propionic acid, acetic acid) to the pyrrole rings give rise to the different porphyrins. A series of isomers arise when the tetrapyrroles have the same side chains attached at different positions. The naturally occurring isomers are designated Type I and Type III (Fig. 17.2). The porphyrin molecule is planar and hydrophobic, but the water solubility of porphyrins is increased by the presence

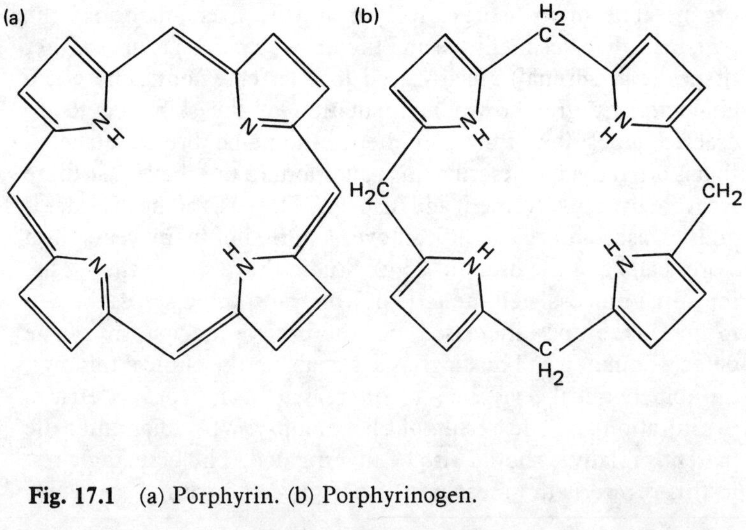

**Fig. 17.1** (a) Porphyrin. (b) Porphyrinogen.

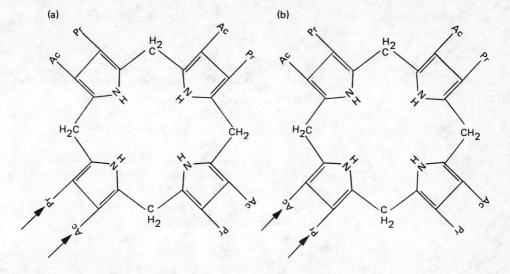

**Fig. 17.2** (a) Uroporphyrinogen I. (b) Uroporphyrinogen III. Ac, acetate substituent ($-CH_2COOH$); Pr, propionate substituent ($-CH_2CH_2COOH$).

of hydroxyl or carboxyl containing side chains. The degree of water solubility is an important determinant of the route of excretion taken by porphyrins.

## Haem biosynthesis

A number of enzyme defects, mostly hereditary, can occur in the pathway for haem synthesis, giving rise to a variety of distinct clinical disorders that result from the harmful accumulation of intermediate metabolites. Important points to note in the normal haem pathway (Fig. 17.3) are as follows.

### ALA SYNTHASE

The condensation of glycine and succinyl CoA to form δ-aminolaevulinic acid (ALA) requires vitamin $B_6$ (pyridoxal phosphate) as cofactor. ALA synthase is subject to feedback repression by haem and is the rate-limiting enzyme, all subsequent steps in the pathway to haem being virtually irreversible. ALA synthase can be induced by a wide variety of compounds, particularly drugs that are microsomal enzyme inducers (e.g. phenobarbital) and which probably act by depleting free haem. ALA synthase activity is repressed by glucose loading or administration of haematin.

## Chapter 17

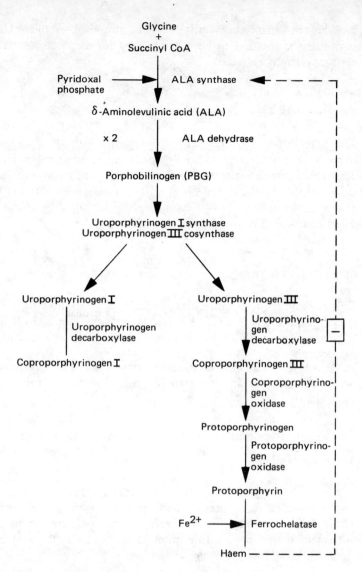

**Fig. 17.3** Biosynthesis of haem.

ALA DEHYDRASE

Two moles of ALA are condensed to form a monopyrrole, porphobilinogen (PBG). This enzyme is inhibited by lead.

UROPORPHYRINOGEN FORMATION

Two enzymes, uroporphyrinogen I synthase and uroporphyrinogen III cosynthase associate together to convert PBG to uroporphyrinogen III. Normally very little PBG is converted to uroporphyrinogen I. Uroporphyrinogen I synthase is defective in acute intermittent porphyria, whilst deficient uroporphyrinogen III cosynthase is thought to be the inherited lesion in congenital erythropoietic porphyria.

## UROPORPHYRINOGEN DECARBOXYLASE

The four acetate side-chains of uroporphyrinogen III undergo stepwise decarboxylation, to finally yield coproporphyrinogen III. The activity of this enzyme is decreased in porphyria cutanea tarda, whilst the acquired form (often associated with liver damage) may arise from this enzyme being inhibited by chlorinated hydrocarbons (insecticides).

## COPROPORPHYRINOGEN OXIDASE

This enzyme requires molecular oxygen to oxidize the two propionate side chains to yield protoporphyrinogen. The activity of this enzyme is inhibited by lead, and is defective in hereditary coproporphyria.

## PROTOPORPHYRINOGEN OXIDASE

Oxidation of the ring-linking bridges to give protoporphyrin.

## FERROCHELATASE

This enzyme catalyses the chelation of ferrous iron ($Fe^{2+}$) by protoporphyrin to yield haem, and is thought to be the enzyme defect underlying both protoporphyria and variegate porphyria.

## PORPHYRINS

The pathway from PBG to protoporphyrin is only via isomer Type III porphyrinogens. The corresponding porphyrins do not take part in the pathway, but arise as a result of oxidation of the various porphyrinogen intermediates.

## Excretion

Normal urine contains small amounts of ALA, PBG and porphyrin, whilst faeces contain some porphyrin. Enzyme defects in the haem biosynthetic pathway can cause abnormal levels of various intermediates. The method of excretion of these compounds depends upon their solubility, which in turn is partly governed by the carboxyl-group ($-COOH$) content of their side chains.

### Uroporphyrinogen, uroporphyrin

With eight COOH groups these molecules are very water-soluble, and are almost exclusively excreted in the urine.

### Coproporphyrin

Four COOH groups, allowing excretion by both urinary and faecal routes.

### Protoporphyrin

With two COOH groups this molecule is insoluble in water. The lipophilic properties of bile allow excretion via the faecal route.

## LABORATORY INVESTIGATION

Diagnosis of disorders of porphyrin metabolism depends in the first instance on simple qualitative screening tests (urinary PBG and porphyrins, faecal porphyrins, red cell porphyrins). A definitive diagnosis may require porphyrin fractionation, and in some cases the measurement of appropriate enzyme activities.

Porphyrinogens are colourless, and are very easily oxidized to their corresponding porphyrin by weak oxidizing agents, or by light in the presence of air. This instability means that uroporphyrinogens cannot be tested for in patient specimens. Porphyrins are coloured, stable compounds that have characteristic absorption spectra, and produce an intense red fluorescence when irradiated with UV light; both these properties form the basis for useful qualitative tests. In acid solutions porphyrins are red-purple in colour, whilst in alkaline conditions the colour tends towards red-brown.

### Urine

The colourless porphyrinogens and PBG are unstable in an acid medium and will oxidize to the corresponding coloured porphyrins when exposed to sunlight. Urine that contains large amounts of PBG or porphyrins will gradually darken (port wine urine) if left to stand for a period in the light, thus providing one of the simplest tests for these compounds.

URINARY δ-AMINOLEVULINIC ACID (ALA)

This assay requires an acidic urine (tartaric acid added to collection). A preliminary chromatographic step is needed if PBG is to be separately measured. The estimation is not essential for the diagnosis of porphyria, but is of value as a sensitive test for detecting exposure to lead.

URINARY PORPHOBILINOGEN (PBG)

PBG is unstable, colourless and oxidizes under acidic conditions to uroporphyrin and porphobilin, causing urine to gradually darken if present in large amounts. Urine should therefore be analysed as soon as possible after collection, and if 24 hour estimations are required the pH of the specimen should be maintained near neutrality during the collection period (by the addition of 5 g sodium carbonate to the container) and kept in the dark. Screening procedures (Watson-Schwartz test) are suitable for most diagnostic purposes, and are best performed on early morning specimens. Normal subjects excrete less than 15 μmol of PBG per day, whilst the screening tests become positive at a minimum excretion rate of approximately 30 μmol/day. Positive tests are found during the acute phases of acute intermittent porphyria (AIP), variegate porphyria (VP) and hereditary coproporphyria (HC). A positive test may also be seen during the latent phase of AIP.

Positive screening tests, and all suspected carriers of AIP, should have a quantitative estimation performed. Not all carriers of AIP, VP and HC will show increased PBG levels.

**Porphyrin screening tests**

URINE

The porphyrins are extracted into organic solvent and examined under UV light for red fluorescence. This procedure detects porphyrins down to a level of $\sim 1.0$ μmol/l (normal excretion $\lesssim 0.5$ μmol/l).

FAECES

Porphyrins are extracted from faeces as for urine and examined under UV light. Dietary chlorophyll also gives a red fluorescence, and is excluded by first extracting the porphyrins into hydrochloric acid. This test is sufficiently sensitive as to detect porphyrins in some normal subjects, and therefore all positive tests should be followed by a quantitative examination.

RED CELLS

Nearly all blood porphyrin is contained within erythrocytes. The red cell content of porphyrins may be qualitatively examined by solvent extraction, followed by UV inspection for fluorescence in the acid extract. Alternatively, fluorescence microscopy allows an unfixed blood smear to be scanned for red-fluorescing erythrocytes. The fluorescence due to protoporphyrins fades quickly due to destruction of the molecule by UV light. This screening test gives a positive result for the porphyrias which cause photosensitivity and are of the erythropoietic type.

PORPHYRIN QUANTITATION

Ideally all positive screening tests for porphyrins should be followed by quantitation and fractionation. This is necessary in some cases, but usually the diagnosis becomes evident from the clinical features, family history and the results of screening tests. The quantitative procedures are usually only carried out by laboratories that have a special interest in this area.

ENZYME ASSAYS

Enzyme assays are readily performed on both red cells (ALA dehydrase, uroporphyrinogen synthase, uroporphyrinogen decarboxylase) and leucocytes (ALA synthase, coproporphyrinogen oxidase, protoporphyrinogen oxidase, ferrochelatase). These measurements are required in some instances for definitive diagnosis and for the detection of carriers.

## Selection of tests

Normal urine contains small amounts of ALA, PBG and porphyrins that are generally not detected by the screening tests. For routine diagnostic purposes a rapid classification of the porphyrias can be made from the clinical findings and the results of the simple screening tests (Figs 17.4 and 17.5, and Table 17.1).

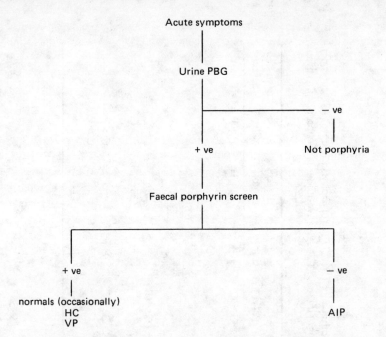

**Fig. 17.4** Acute symptom screen. AIP, acute intermittent porphyria; HC, hereditary coproporphyria; VC, variegate porphyria.

Positive urine PBG screen

(1) Acute intermittent porphyria (AIP)—acute and latent, (2) hereditary coproporphyria (HC)—acute, (3) variegate porphyria (VP)—acute.

Positive urine porphyrin screen

(1) Primary porphyrias, (2) secondary porphyrias (mainly coproporphyrin), (a) liver disease, (b) lead poisoning, (c) heavy metals other than lead (gold, arsenic), (d) haematological disorders (occasional)—haemolytic anaemia, pernicious anaemia, aplastic anaemia.

Positive faecal porphyrin screen

(1) Some normals, (2) porphyrias, (3) gastrointestinal haemorrhage.

Positive red cell screen

(1) Erythropoietic porphyrias, (2) iron deficiency anaemia.

**Table 17.1** Biochemical findings in the porphyrias. Abbreviations: PBG, porphobilinogen; Uro. (U), uroporphyrin; Copro. (C), coproporphyrin; Proto. (P), protoporphyrin; +–+++, degree of abnormality; ±, normal to slight increase; C/U, relative amounts of coproporphyrin and uroporphyrin; C/P, relative amounts of coproporphyrin and protoporphyrin

| | | Urine | | | | Faeces | | Red cell |
|---|---|---|---|---|---|---|---|---|
| | PBG | Uro. | Copro. | C/U | Copro. | Proto. | C/P | Screen |
| Congenital erythropoietic porphyria | | ++ | ++ | U>C | +++ | | C>P | + |
| Erythropoietic protoporphyria | | | | | | ++ | P>C | + |
| Acute intermittent porphyria | | | | | | | | |
|   Acute | ++ | ++ | ++ | | | | | |
|   Latent | ± | ± | ± | | | | | |
| Variegate porphyria | | | | | | | | |
|   Acute | ++ | ++ | ++ | C>U | + | + | P>C | |
|   Latent | ± | ± | ± | | | + | P>C | |
| Hereditary coproporphyria | | | | | | | | |
|   Acute | ++ | ± | + | C>U | ++ | | C>P | |
|   Latent | ± | ± | ± | | + | | C>P | |
| Porphyria cutanea tarda | | ++ | + | U>C | ± | ± | | |

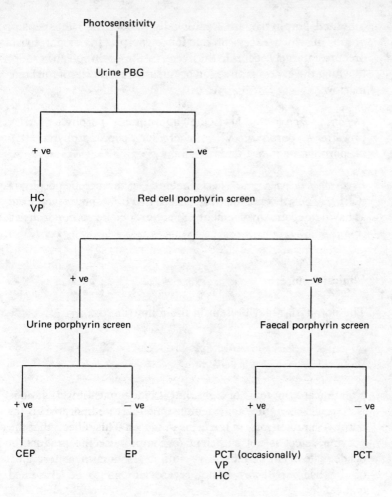

**Fig. 17.5** Photosensitivity screen. HC, hereditary coproporphyria; VP, variegate porphyria; PCT, porphyria cutanea tarda; CEP, congenital erythropoietic porphyria; EP, erythropoietic protoporphyria.

## PORPHYRIAS

The porphyrias are a group of rare disorders caused by enzyme defects in the haem biosynthetic pathway. Although most of the porphyrias are inherited, porphyria can be induced by chemical insult in the normal individual. It is important to note that in the acute hepatic porphyrias the enzyme defects generally do not express themselves clinically unless the haem biosynthetic pathway is stressed by an event that induces ALA synthase.

### Classification

Haem is probably synthesized by most tissues of the body but, based on the tissue most obviously affected by a biosynthetic defect, the inherited porphyrias are traditionally divided into the hepatic group, in which excess production of porphyrins or porphyrin precursors largely occurs in the liver, and the erythropoietic group, in which the excess production occurs in the red cells of the bone marrow.

*Hepatic porphyrias* (1) Acute intermittent porphyria (AIP), (2) variegate porphyria (VP), (3) hereditary coproporphyria (HC), (4) porphyria cutanea tarda (PCT).

*Erythropoietic porphyria* (1) Congenital erythropoietic porphyria (CEP), (2) erythropoietic protoporphyria (EP)—now recognized as having hepatic involvement, (3) erythropoietic coproporphyria (EC).

**Clinical features**

The porphyrias may present in the following forms.

LATENT

No clinical symptoms or signs, but if biochemical investigations were undertaken they may demonstrate increased porphyrins (or porphyrin precursors) in the urine or faeces. This reflects that the enzyme defect is not sufficient to compromise the production of adequate haem, and that any resulting biochemical abnormality of the haem pathway is not severe enough to be expressed clinically.

ACUTE

An acute attack describes the appearance of the signs and symptoms of porphyria, these being due to an accumulation of one or more intermediate substances in the haem pathway. These abnormal levels arise because:
(1) ALA synthase, the rate-limiting enzyme, has been induced, thereby speeding up the haem biosynthetic pathway.
(2) The defective enzyme present cannot cope causing (a) accumulation of intermediates behind the block (b) inadequate haem production and thus induction of ALA synthase.
   A wide variety of apparently unrelated substances, many of them therapeutic drugs (barbiturates, sulphonamides, oestrogens), or acute illness, can precipitate acute attacks in subjects with hepatic porphyria. It is thought that some of the drugs induce increased

production of cytochrome P-450, the synthesis of which requires haem, whilst other compounds deplete haem by other mechanisms, resulting in induction of ALA synthase.

Acute attacks occur in AIP, VP and HC and can be manifested by: (1) abdominal symptoms—pain, vomiting, dehydration, (2) peripheral neuritis—limb pain, muscle weakness, (3) psychogenic symptoms—depression, anxiety, psychosis, (4) red urine ('port wine' urine).

The neurological symptoms are related to abnormally high levels of ALA and PBG.

PHOTOSENSITIVITY

Skin lesions may occur in those areas where the high circulating levels of porphyrins are exposed to sunlight. Porphyrins absorb energy intensely at around 400 nm, and it is thought that the resulting 'excited' porphyrins damage lysosome membranes in the skin tissues, releasing lysozymes and proteases that then cause further local tissue damage.

## Case examples/pathophysiology

HEPATIC PORPHYRIAS

The four hepatic porphyrias are acute intermittent porphyria (AIP), variegate porphyria (VP), hereditary coproporphyria (HC) and porphyria cutanea tarda (PCT). These disorders clinically manifest themselves as intermittent acute attacks of neurological dysfunction (abdominal pain, mental disturbance, etc.), and/or photosensitivity. These attacks are generally precipitated by drugs or an acute illness that induces ALA synthase, probably by depleting haem levels. The resulting increased activity of the haem pathway causes intermediates to pile up behind the specific enzyme block. The liver is the most obvious site of this overproduction.

Thus the biochemical abnormalities seen in each of the hepatic porphyrias can be explained by the presence of both: (1) a genetic defect causing reduced activity of a specific enzyme in the haem pathway, (2) increased ALA synthase activity.

## ACUTE INTERMITTENT PORPHYRIA (AIP) (Fig. 17.6)

Transmission

Autosomal dominant.

Defect

Reduced uroporphyrinogen I synthase activity.

Chemical pathology

*Urine* ↑ ALA and ↑ PBG, but PBG readily oxidizes in acid conditions and sunlight to porphobilin, so that urine collected during an attack will darken on standing.

*Faeces* No significant increase in porphyrins.

Clinical

Usually occurs in early adulthood, and is characterized by acute attacks of abdominal pain, peripheral neuritis and psychotic illness; often brought on by drugs, many of which induce microsomal enzymes (barbiturates, oestrogens, sulfonamides), or acute illness. AIP is not associated with photosensitivity.

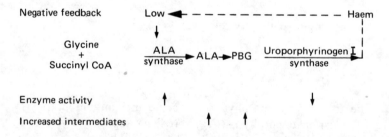

Fig. 17.6 Haem pathway in acute intermittent prophyria.

Diagnosis

*Acute phase* ↑ urine PBG, normal faecal porphyrin screen.

*Latent phase* May be increased urinary PBG, but definitive diagnosis depends on demonstrating a lack of red cell uroporphyrinogen synthase I activity.

*Example* A 21 year old female admitted to hospital with a colicky abdominal pain and diarrhoea of 4 weeks duration. As a child she had a history of recurrent episodes of abdominal pain. This admission followed her commencing oral contraceptive therapy.

Urine screen  PBG         Present in excess
              Porphyrins  Negative
Faecal screen Porphyrins  Negative
Family screen Mother and a brother were found to excrete increased amounts of PBG, but they had no history of acute symptoms

VARIEGATE PORPHYRIA (VP) (Fig. 17.7)

Transmission

Autosomal dominant.

Defect

?ferrochelatase and ?protoporphyrinogen oxidase deficiency.

Chemical pathology

*Urine* ↑ALA and ↑PBG (may be normal or slightly raised during latent periods), ↑ uroporphyrin and ↑ coproporphyrin (coproporphyrin > uroporphyrin).

*Faeces* ↑ protoporphyrin and ↑ coproporphyrin (protoporphyrin > coproporphyrin) at all times.

Clinical

This disorder is associated with acute attacks as in AIP. Photosensitive skin lesions also occur (unlike AIP), but they can arise independently of the acute attack. Acute attacks are often associated with certain types of drug therapy, particularly barbiturates, sulphonamides and alcohol.

Diagnosis

*Acute phase*  Urine (1) ↑ PBG, (2) ↑ uro- and coproporphyrins (copro- > uroporphyrins). Faeces—↑ proto- and coproporphyrins (proto- > coproporphyrins).

*Latent phase*  Increase coproporphyrins and protoporphyrins in the faeces.

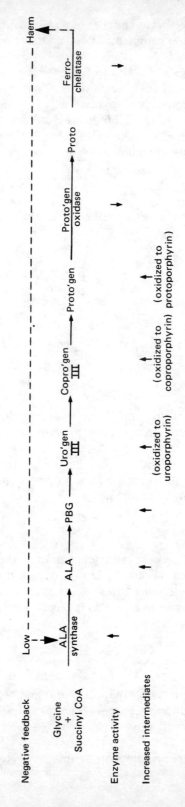

**Fig. 17.7** Haem pathway in variegate porphyria.

# HEREDITARY COPROPORPHYRIA (HC) (Fig. 17.8)

### Transmission
Autosomal dominant.

### Defect
Reduced coproporphyrinogen oxidase activity.

### Chemical pathology

*Urine*  (1) ↑ ALA and ↑ PBG (during acute attacks), (2) ↑ uroporphyrin and ↑ coproporphyrin III (copro- > uroporphyrin).

*Faeces*  ↑ coproporphyrin III, protoporphyrin may be normal or slightly raised.

### Clinical

Presents with acute attacks that are similar to those seen in AIP, and in one third of cases there may be associated photosensitivity. Unlike VP, the skin lesions in HC are always associated with acute attacks.

### Diagnosis

*Acute phase*  Urine (1) ↑ PBG, (2) ↑ uro- and ↑ coproporphyrin (copro- > uroporphyrin). Faeces—↑ coproporphyrins with little or no increase in protoporphyrins.

*Latent phase*  Urinary PBG usually normal but faecal coproporphyrins increased.

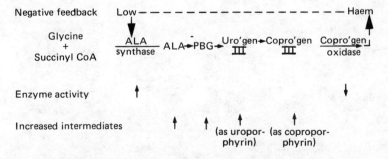

**Fig. 17.8**  Haem pathway in hereditary coproporphyria.

## PORPHYRIA CUTANEA TARDA (PCT) (Fig. 17.9)

### Transmission

Usually acquired (liver disease), but many cases appear to have a familial basis.

### Defect

Reduced uroporphyrinogen decarboxylase activity.

### Pathophysiology

PCT is the most commonly occurring porphyria, and is often associated with liver dysfunction. Predisposing factors are often chronic alcoholism or siderosis, or at least increased hepatic iron stores. The exact mechanism of the defect is unknown, but it has been suggested that the reduced uroporphyrinogen decarboxylase activity is sufficient for normal haem production, but that the enzyme is further inhibited by the presence of excess iron. Phlebotomy, by reducing iron levels, effects long-term remissions.

### Chemical pathology

*Urine* PBG is normal. ↑ uroporphyrins (uro- > coproporphyrins).

*Faeces* Occasionally increased porphyrins.

*Plasma* Iron levels often increased. Body stores of iron are increased.

### Clinical

This disorder differs from the other hepatic porphyrias in that clinical manifestations are restricted to the skin lesions.

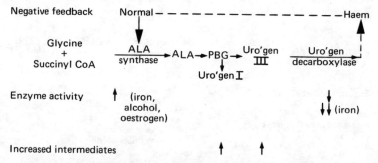

**Fig. 17.9** Haem biosynthesis in porphyria cutanea tarda.

Diagnosis

*Urine* Normal PBG with ↑ uroporphyrins (uro- > coproporphyrins).

*Faeces* Porphyrins may be increased.

*Red cells* Negative screen.

Definitive diagnosis may require porphyrin fractionation and the separate estimation of urinary coproporphyrins and uroporphyrins.

Differential diagnosis (Table 17.2)

(1) Congenital erythropoietic porphyria (CEP), (2) variegate porphyria (VP), (3) hereditary coproporphyria (HC).

It is important to distinguish PCT from HC and VP because PCT may respond to phlebotomy, which, by reducing body iron stores, can bring about clinical remission lasting several years.

**Table 17.2** Differentiation of porphyria cutanea tarda (PCT) from variegate porphyria (VP) and hereditary coproporphyria (HC) on the basis of urine porphyrin fractionation

|     | Urine coproporphyrins | Uroporphyrins | Copro-/uroporphyrins |
| --- | --- | --- | --- |
| PCT | ± | +++ | U>C |
| VP  | ++ | ++ | C>U |
| HC  | +++ | ++ | C>U |

*Example* A 50 year old male with a history of heavy alcohol intake, resulting in a previous diagnosis of alcoholic liver disease. He had a 4 month history of redness and blistering of the back of the hand, face and scalp. He had noticed red urine on five occasions.

| | | |
|---|---|---|
| Urine screen | PBG | Negative |
| | Porphyrins | Positive |
| Faecal screen | Porphyrins | Positive |
| Red cell screen | Negative | |

| | | | |
|---|---|---|---|
| Porphyrin fractionation | Urine coproporphyrins | 40 nmol/day | (<380) |
| | uroporphyrins | 900 nmol/day | (<190) |
| | Faecal coproporphyrins | 300 nmol/g | (<25) |
| | protoporphyrins | 320 nmol/g | (<90) |

## CONGENITAL ERYTHROPOIETIC PORPHYRIA (CEP)

Transmission

Autosomal recessive.

Defect

Unclear, but probably an imbalance of uroporphyrinogen III cosynthase and uroporphyrinogen I synthase activity in the erythrocytes.

Pathophysiology

↑ production of uroporphyrinogen I and coproporphyrinogen I.

Chemical pathology

*Red cells*  ↑ uroporphyrinogen. Premature red cell destruction.

*Urine*  ↑ uroporphyrin I and coproporphyrin I (uro- > copropor-phyrins).

*Faeces*  ↑ uroporphyrin I and coproporphyrin I.

A small increase in the levels of respective III isomers by both routes.

Clinical

Onset with bullous skin lesions (photosensitive) and red urine in the first few years of life. Teeth are brown-pink in colour (fluoresce orange under UV light), and there is often an associated haemolytic anaemia and hirsutism. There are no neurological symptoms (normal ALA and PBG).

Diagnosis

*Red cells*  +ve screen for porphyrins.

*Urine*  PBG is normal. ↑ uro- and coproporphyrins.

*Faeces*  ↑ uro- and coproporphyrins.

Definitive diagnosis depends on the demonstration of increased red cell content of uroporphyrinogen.

## PROTOPORPHYRIA
## (ERYTHROPOIETIC PROTOPORPHYRIA)

### Transmission
Autosomal dominant.

### Defect
Reduced ferrochelatase activity in red cells and the liver.

### Pathophysiology
↓ ferrochelatase → ↑ protoporphyrin accumulation in erythrocytes.

### Chemical pathology

*Red cells* ↑ protoporphyrinogen.

*Urine* Normal (protoporphyrins are not excreted in urine).

*Faeces* ↑ protoporphyrins.

### Clinical
Onset in childhood with photosensitivity and often liver damage, but often a mild disorder.

### Diagnosis

*Red cells* +ve porphyrin screen.

*Faeces* protoporphyrins.

Definitive diagnosis depends on demonstrating increased red cell protoporphyrin.

## ERYTHROPOIETIC COPROPORPHYRIA (EC)

Extremely rare disorder, resembling erythropoietic protoporphyria but associated with increased red cell coproporphyrin.

Chapter 17    **Lead poisoning**

Lead inhibits certain of the haem biosynthetic enzymes which may result in: (1) red cell abnormalities and anaemia, (2) increased urinary excretion of ALA and porphyrins.

The enzymes suppressed by toxic levels of lead are: (1) ALA dehydrase, (2) ferrochelatase, (3) coproporphyrinogen oxidase.

There is also increased activity of ALA synthase resulting from the decreased haem synthesis. These enzyme abnormalities result in: (1) decreased haem synthesis (anaemia), (2) increased ALA synthesis, (3) accumulation of protoporphyrin in red cells, (4) accumulation and increased excretion of coproporphyrin.

These alterations in porphyrin metabolism provide a useful means of detecting and assessing lead intoxication. The available tests are: (1) red cells—ALA dehydrase activity, protoporphyrin concentration, (2) urine—ALA excretion, coproporphyrin excretion.

# 18 Uric acid

Uric acid is the end product of purine metabolism (adenine, guanine) in the human, and is the causative agent of gout. With a normal diet some 5–6 mmol of urate are produced daily, and of this about 3–4 mmol is produced from purines synthesized within the body (*de novo* synthesis), whilst the remaining 1–2 mmol are contributed from purines present in the diet.

**Purine metabolism**

### SYNTHESIS OF PURINES AND URATE PRODUCTION

The first, and also rate-limiting, step is the formation of phosphoribosylamine from phosphoribosyl pyrophosphate (PRPP) and glutamine, by the enzyme phosphoribosyl pyrophosphate amidotransferase (PRPP-AT). This enzyme is subject to negative feedback control by adenylic and guanylic acids. After a series of intermediate steps phosphoribosylamine is converted to inosinic acid, which is then degraded through inosine to hypoxanthine. Hypoxanthine is oxidized by xanthine oxidase through xanthine to uric acid. The free purine bases hypoxanthine, guanine and adenine, can be converted back to purine nucleotides by the enzymes hypoxanthine-guanine phosphoribosyl transferase (HGPRT) and adenine phosphoribosyl transferase (APRT). These reactions play an important regulatory role in purine metabolism (salvage process) (Fig. 18.1).

### EXCRETION OF URIC ACID

Of the 5–6 mmol of urate produced daily, approximately 25–30% is eliminated via the gastrointestinal tract, where it is degraded by bacterial uricases. The remainder of the urate is excreted by the kidney. In renal failure the intestinal excretion can be markedly increased.

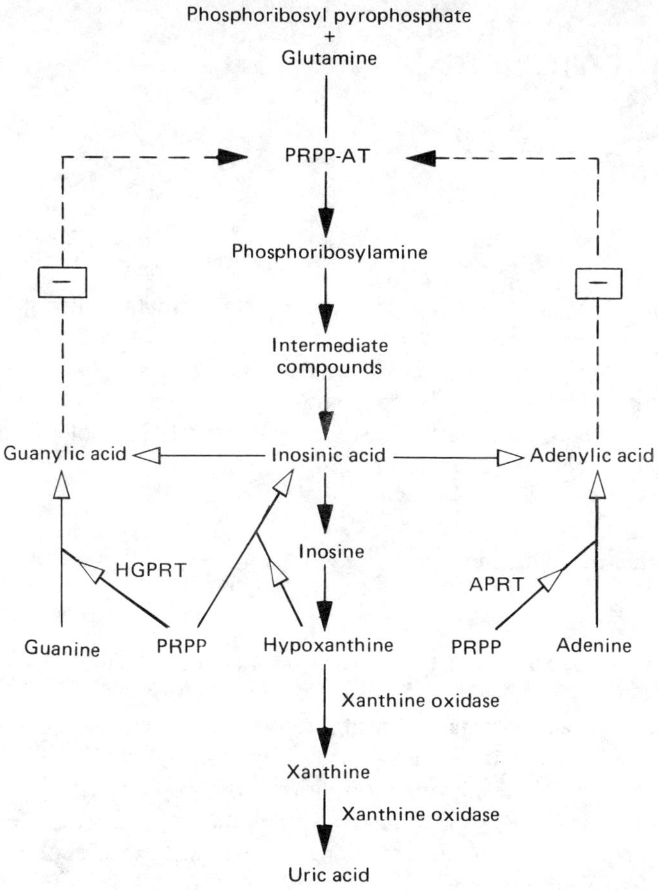

## Chapter 18

```
        Phosphoribosyl pyrophosphate
                     +
                  Glutamine
                     ↓
              →   PRPP-AT   ←
              |      ↓      |
              |Phosphoribosylamine
              |      ↓      |
            [ − ]           [ − ]
              |      ↓      |
              |Intermediate |
              |compounds    |
              |      ↓      |
  Guanylic acid ◁── Inosinic acid ──▷ Adenylic acid
         △              △                  △
         │ ◁ HGPRT      │     Inosine   APRT ▷
         │              │        ↓           │
      Guanine  PRPP   Hypoxanthine   PRPP  Adenine
                         ↓
                   Xanthine oxidase
                         ↓
                      Xanthine
                         ↓
                   Xanthine oxidase
                         ↓
                     Uric acid
```

**Fig. 18.1** Purine metabolism. PRPP-AT, phosphoribosyl pyrophosphate amidotransferase; PRPP, phosphoribosyl pyrophosphate; APRT, adenine phosphoribosyl transferase; HGPRT, hypoxanthine-guanine phosphoribosyl transferase: ----, negative feed back; ──▶, major pathway; ──▷, resynthesis of purine nucleotides from free purine bases.

### RENAL EXCRETION OF URATE

The renal clearance of urate is of the order of 5–10% of inulin clearance. Some 98% of the filtered urate is reabsorbed in the proximal tubule. The 2% that escapes reabsorption contributes about 20% of the total that is excreted, proximal tubular secretion accounting for the remainder. The rate of renal excretion is influenced by various metabolic products and drugs (Fig. 18.2).

*Decreased excretion*   (1) Lactic acid, ketone bodies (inhibit tubule secretion), (2) drugs—thiazides, frusemide, salicylates (low dose inhibits tubule secretion).

*Increased excretion*   (1) Salicylates (large dose inhibits tubule reabsorption), (2) uricosuric agents—probenecid.

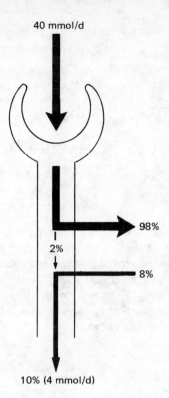

**Fig. 18.2** Renal (proximal tubule) handling of uric acid.

There is good evidence to suggest that the rate of proximal reabsorption is influenced by the extracellular volume (↓ extracellular volume → ↑ reabsorption, ↓ excretion).

## Therapeutic agents used in gout and hyperuricaemia

Three groups of drugs are available for the management of gout and hyperuricaemia: (1) allopurinol, (2) uricosuric agents, (3) anti-inflammatory agents.

### Allopurinol

Allopurinol, and its major metabolite oxypurinol, inhibit the enzyme xanthine oxidase, producing a decrease in the plasma and urine concentrations of urate. This agent is likely to precipitate an acute attack of gout when the initial dose is given, and therefore it should be covered in the early stages of treatment with an anti-inflammatory agent.

### Uricosuric agents

These drugs (e.g. probenecid) increase the urinary excretion of urate by inhibiting its tubular reabsorption.

Chapter 18    Anti-inflammatory agents

The anti-inflammatory agents, used to relieve the pain of an acute gout attack are colchicine and indomethacin. They have no effect on plasma urate levels.

## LABORATORY INVESTIGATION

### PLASMA URATE

Plasma urate concentration is influenced by sex, race and diet. The upper reference limit is difficult to determine, but probably lies somewhere in the area of 0.45 mmol/l for adult males (females are approximately 0.05 mmol/l lower). Reported levels in gouty subjects show a range of 0.35–0.8 mmol/l. During an acute attack of gouty arthritis the plasma level is often high, but it may also be within the reference range.

### PLASMA CREATININE

Hyperuricaemia may cause renal failure, whilst renal failure itself may result in hyperuricaemia. In renal insufficiency the plasma [urate] tends not to rise until the GFR falls below 0.3 ml/s (plasma creatinine of $\sim 0.4$ mmol/l). A useful test to differentiate between urate nephropathy and hyperuricaemia due to renal failure is to determine the urate:creatinine ratio on a 'spot' urine. A ratio of less than 0.7 suggests renal failure as the cause of the hyperuricaemia; values greater than this suggest that the hyperuricaemia is the cause of the renal failure.

### URINARY URATE

Hyperuricaemia may be due to either overproduction or decreased renal excretion, or a combination of both. Although the kidney disposes of 70–75% of the total urate produced, the rate of renal excretion provides a rough index of the production rate. On a purine-free diet of 5–7 days duration the normal renal excretion rate should be less than $\sim 3.5$ mmol/24 hours. Without dietary control the level should be less than $\sim 5.0$ mmol/24 hours. Values in excess of these figures are presumptive evidence of overproduction of urate.

# HYPERURICAEMIA

Hyperuricaemia may be due to:

Primary

*Overproduction of urate* (1) Idiopathic, (2) associated with (a) glucose-6-phosphatase deficiency, (b) HGPRT deficiency (Lesch-Nyhan syndrome).

*Reduced secretion of urate* Idiopathic.

Secondary

*Overproduction* Increased nucleic acid turnover, (1) myeloproliferative disease, e.g. polycythaemia vera, (2) lymphoma, leukaemia, (3) multiple myeloma, (4) treatment of malignant disease, (5) psoriasis.

*Reduced excretion* (1) Renal disease—failure, lead nephropathy, (2) drugs—diuretics, aspirin, (3) lactic acidosis—alcohol, (4) ketoacidosis—diabetes mellitus.

**Case examples/pathophysiology**

PRIMARY GOUT

Male, 50 years, with acutely swollen and painful right knee.

| | | | |
|---|---|---|---|
| Plasma | Urate | 0.71 mmol/l | (0.15–0.45) |
| | Creat. | 0.12 mmol/l | (0.06–0.12) |
| Urine | Urate (purine-free diet) | 10.5 mmol/day | (<3.6) |

20% of patients with idiopathic gout secrete excessive amounts of urate in their urine. These cases are usually best treated with allopurinol, a drug that inhibits xanthine oxidase by acting as a competitive inhibitor (see Fig. 18.1).

Gout is a disease caused by the precipitation of crystals of monosodium urate monohydrate in body tissues, giving rise to: (1) acute arthritis, (2) tophaceous deposits, (3) renal disease (urate nephropathy), (4) urolithiasis.

The diagnosis is usually made on clinical grounds (with evidence of hyperuricaemia), but only the demonstration of urate crystals within synovial fluid leucocytes, or in tophi, establishes a definitive diagnosis.

Chapter 18    PRIMARY GOUT

Acute arthritis of both knees and left metacarpo-phalangeal joint; urate crystals present in knee synovial fluid leucocytes.

| | | | | |
|---|---|---|---|---|
| Plasma | Urate | 0.75 | mmol/l | (0.15–0.45) |
| | Creat. | 0.10 | mmol/l | (0.06–0.12) |
| Urine | Urate (purine-free diet) | 3.2 | mmol/day | (<3.6) |

80% of cases of primary gout are associated with normal or decreased urinary excretion of urate. Providing renal function is adequate, these cases may be best treated in the first instance by uricosuric drugs (probenecid), which reduce the tubular reabsorption of urate.

### SECONDARY HYPERURICAEMIA (OVERPRODUCTION)

Acute myeloid leukaemia in a 26 year old female.

| Date | | 23/4 | 26/4 | | |
|---|---|---|---|---|---|
| Plasma | Urate | 0.92 | 0.18 | mmol/l | (0.20–0.45) |
| | Creat. | 0.09 | 0.11 | mmol/l | (0.06–0.12) |
| Urine | Urate (non purine-free diet) | 10.5 | — | mmol/24 h | (<4.8) |

Allopurinol therapy begun on 23/4.

With increased turnover of nucleic acids in the tumour cells the plasma urate may reach very high levels and, with the institution of chemotherapy, urate levels will increase further as cells are damaged. To lessen the danger of renal damage, these patients should have their plasma urate levels lowered by allopurinol therapy.

### SECONDARY HYPERURICAEMIA (DECREASED EXCRETION)

Chronic renal failure in a 39 year old male.

| | | | | |
|---|---|---|---|---|
| Plasma | Urate. | 0.78 | mmol/l | (0.20–0.45) |
| | Creat. | 0.90 | mmol/l | (0.06–0.12) |
| Creatinine clearance | | 0.1 | ml/s | (1.5–2.0) |

In renal insufficiency the plasma urate does not begin to rise until the GFR has fallen to approximately 0.3 ml/s. Until then the urate level is kept within normal limits by: (1) increased excretion by remaining healthy nephrons, (2) increased elimination via the gut.

## LESCH-NYHAN SYNDROME

This is a rare inborn error of purine metabolism, due to deficiency of the enzyme HGPRT. There is an inability to recycle hypoxanthine and other purines. This condition manifests itself as severe hyperuricaemia in male children. There is also mental deficiency, self mutilation and aggressive behaviour.

## HYPOURICAEMIA

**Causes**

*Decreased production* (1) Xanthine oxidase deficiency (xanthinuria), (2) allopurinol therapy.

*Increased excretion* (1) Fanconi syndrome, (2) uricosuric drugs (high dose salicylates), (3) SIADH (occasionally).

Hypouricaemia is an uncommon occurrence, except in the treatment of hyperuricaemia (especially allopurinol). (See above case of acute myeloid leukaemia.)

# 19 Hypothalamic and anterior pituitary hormones

## HYPOTHALAMUS

The hypothalamus forms part of the floor of the third ventricle and receives neural input from many parts of the brain. Despite its small size ($\sim 10$ g) the hypothalamus is a major centre for the control of many vital body functions, i.e.
(1) Vegetative—body temperature, weight, hunger, thirst, cardiovascular function.
(2) Hormonal—control of anterior pituitary secretions.

**Control of anterior pituitary**

Nuclei within the hypothalamus elaborate and secrete substances which then pass down the axons to the median eminence of the hypothalamus, where they are released into a blood system that flows via the pituitary stalk to the highly vascular anterior pituitary (hypothalamic-hypophyseal portal system). These substances have stimulating and inhibiting effects on the hormone secretions of the anterior pituitary, and are generally released from the hypothalamus in response to the balance of: (1) signals from the brain, CNS and periphery (adrenergic, catecholamines, prostaglandins), (2) feedback effects by target organ hormones (anterior pituitary and its peripheral targets).

The neural signals are complex, and appear to modify hypothalamic function in response to almost any environmental, physiological, biochemical or emotional event. The hypothalamic hormones are generally released in a pulsatile manner and often with a circadian or similar rhythm. These aspects of the control of hypothalamic hormones are imperfectly understood and are outside the scope of this chapter.

Some of the hypothalamic substances have been fully characterized and can be referred to as hormones, whilst others remain structurally ill-defined (factors), although their existence and function seems definite; those of physiological importance in the human are shown in Table 19.1. Many of the hypothalamic hormones have been detected in areas outside the hypothalamus, particularly in the brain and in gastrointestinal tissue, but their physiological functions and control mechanisms remain to be elucidated.

**Table 19.1** The major hormones and factors released from the hypothalamus

| Hormone/factor | Structure | Effect on anterior pituitary |
| --- | --- | --- |
| Corticotropin-releasing factor (CRF, corticoliberin) | Peptide | Stimulates ACTH and β-lipotropin release |
| Growth hormone-releasing factor (GHRF, somatoliberin) | Not known (peptide?) | Stimulates growth hormone release |
| Somatostatin | Tetradecapeptide | Inhibits GH release |
| Prolactin-releasing factor (PRF, prolactoliberin) | Not known (?TRH) | Promotes prolactin secretion (TRH in pharmacological dose) |
| Prolactin-inhibiting factor (PIF, prolactostatin) | Dopamine | Inhibits prolactin release |
| Gonadotropin-releasing hormone (GnRH) also known as Luteinizing hormone releasing hormone (LHRH, gonadoliberin) | Decapeptide | Stimulates LH and FSH release |
| Thyrotropin-releasing hormone (TRH, thyroliberin) | Tripeptide | Release of TSH (prolactin) |

The evidence for the existence or physiological importance of other factors remains contradictory; such factors include MSH inhibiting and stimulating factors, FSH stimulating factor, and LH inhibiting factor. Most of the hypothalamic hormones have releasing or inhibiting effects on more than one anterior pituitary hormone, but these relationships are complex and their physiological significance remains unknown.

Chapter 19  **Hypothalamic disease**

Functional damage to the hypothalamus may have profound effects on many aspects of body physiology and metabolism. Hypothalamic dysfunction can be caused by a variety of conditions, particularly infarction, tumours, infections and mechanical injury, and is often reflected by an inability to secrete one or more hormones. However, the hypothalamic hormones and factors are present in plasma in extremely low concentration and are usually rapidly inactivated; consequently assays have proved difficult to develop and are not available for diagnostic purposes. Diagnosis of hypothalamic disease is made on clinical grounds, and therefore biochemical evidence of dysfunction must be sought from its effects on the anterior pituitary and its peripheral targets.

## ANTERIOR PITUITARY

The pituitary gland lies in a bony pocket, the sella turcica, at the base of the brain and is connected to the hypothalamus by the pituitary stalk. The posterior pituitary derives from an outgrowth of hypothalamus, whilst the anterior pituitary is of a different embryological origin, arising from the buccal cavity. The anterior pituitary contains several histologically distinct types of cell that can produce a number of polypeptide hormones (Table 19.2). These hormones fall into one of three groups, based on structural features and physiological properties; the glycoproteins (FSH, LH, TSH), the somatomammotropins (GH, PRL) and the pro-opiocortins (ACTH, β-lipotropin, endorphins). These hormones are synthesized and released in response to the balance between the appropriate stimulating or inhibiting hormones from the hypothalamus, and the feedback from both the target organ hormones (long loop) and their own concentration (short loop). The hormones exert their initial biological activity by stimulating the adenyl cyclase system of their target tissue cells. Many of these hormones have been found to be present in other tissues, particularly in the CNS and gastrointestinal tract, but their function and control is unknown.

### ACTH and related peptides

Recent research in animals has shown that a large polypeptide (variously termed pro-opiocortin, pro-opiomelanocortin, prolipocortin) is synthesized by cells in the anterior pituitary. The primary structure of pro-opiocortin suggests that it acts as a precursor for a number of biologically active peptides that can be released into the circulation. Experimental evidence suggests that the anterior pituitary and *pars intermedia* (indistinct in man) may cleave the

**Table 19.2** Hormones synthesized and released by the anterior pituitary

| Common name | Systematic name | Abbreviation | Major target |
|---|---|---|---|
| Adrenocorticotropic hormone | Corticotropin | ACTH | Adrenal cortex |
| Growth hormone | Somatotropin | GH | Liver |
| Prolactin | Prolactin | PRL | Breast |
| Follicle-stimulating hormone | Follitropin | FSH | Gonad |
| Luteinizing hormone | Lutropin | LH | Gonad |
| Thyroid stimulating hormone | Thryotropin | TSH | Thyroid |

pro-opiocortin differently, resulting in the secretion of peptides that are characteristic of each region (Fig. 19.1). The available evidence suggests that a similar precursor system will be revealed for the human pituitary.

Except for ACTH, little is known about the physiological activity of most of the peptides released from the precursor, and so ACTH currently remains the peptide from this group that is of diagnostic interest.

### ACTH (ADRENOCORTICOTROPIC HORMONE, CORTICOTROPIN) (Figs 19.1, 19.2)

ACTH is a 39 residue polypeptide that stimulates, via adenyl cyclase, the synthesis and release of glucocorticoid (cortisol) by the adrenal cortex. ACTH also has some effect on adrenal androgen production, and plays a permissive role in the release of aldosterone from the adrenal cortex. The synthesis and episodic release of ACTH is:

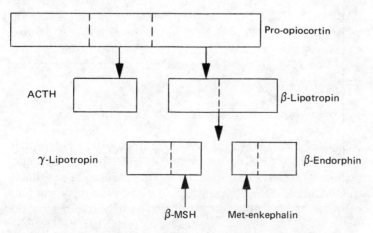

**Fig. 19.1** Major peptides derived from pro-opiocortin.

(1) Stimulated by corticotropin-releasing factor (CRF), which is secreted episodically from the hypothalamic median eminence.
(2) Probably directly inhibited by cortisol (negative feedback).

The release of ACTH must also reflect the activity of those factors that control CRF secretion. These are:
(1) Circadian secretion, with the highest levels occurring in the early morning—entrained by sleep-wake pattern and activity (nyctohemeral).
(2) Negative feedback by cortisol.
(3) Stress—emotional, physical, biochemical.

It is important to note that stress (physical, emotional, biochemical) can override the other two control mechanisms.

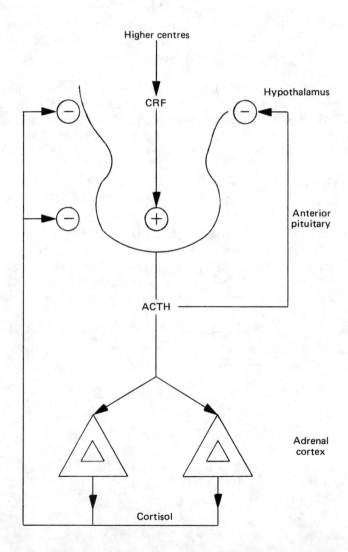

**Fig. 19.2** Hypothalamic-pituitary-adrenal axis in the normal adult.

## Growth hormone (GH) (Fig. 19.3)

*Hypothalamic and anterior pituitary hormones*

Growth hormone (somatotropin) is secreted episodically by specific cells of the anterior pituitary and has a polypeptide structure (mol. wt. 22 000 Daltons) that contains portions that are identical to both human placental lactogen (HPL) and prolactin. GH has a number of metabolic effects: (1) stimulates epiphyseal bone growth, (2) increases cellular protein synthesis, (3) decreases

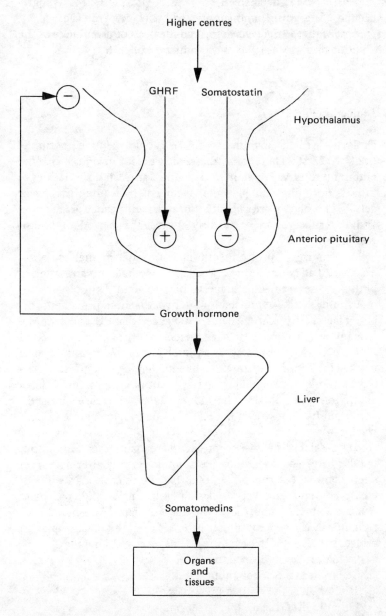

**Fig. 19.3** Major pathways controlling growth hormone secretion.

glucose utilization by opposing insulin action, (4) increases fat mobilization.

Some of these effects may be directly due to the action of GH, whilst others are due to polypeptide growth factors (somatomedins) that are released from the liver by GH. GH release is under the control of two hypothalamic hormones, GH-releasing hormone (not characterized) and GH-release inhibiting hormone (somatostatin, 14 amino acid residues). The secretion of GH is increased by: (1) deep sleep, (2) stress, (3) exercise, (4) hypoglycaemia, (5) arginine; and suppressed by: (1) hyperglycaemia.

Most of these effects have formed the basis of useful tests of the anterior pituitary reserve and production of GH.

**Prolactin** (Fig. 19.4)

Prolactin (PRL, mammotropin) is a polypeptide hormone (mol. wt. 22 500 Daltons) synthesized by the lactotroph cells of the anterior pituitary. Prolactin bears marked structural similarities to growth hormone and placental lactogen. In women the major actions of prolactin are the initiation and maintenance of lactation, and the suppression of fertility during the period of breast-feeding.

The control of prolactin output is unusual in that secretion increases if all connection between the hypothalamus and anterior pituitary is severed. This suggests that prolactin output is autonomous unless it is suppressed by the hypothalamic prolactin-inhibiting factor (PIF). Strong evidence suggests that PIF is dopamine. Dopamine antagonists such as metoclopramide cause increased PRL secretion, whilst agonists such as bromocryptine cause marked inhibition of release. No hypothalamic prolactin-stimulating hormone has been characterized, but pharmacological doses of thyrotropin-releasing hormone (TRH) increase prolactin secretion. It is not known if TRH has a physiological role in PRL release.

Low basal levels of plasma prolactin are maintained throughout the day by episodic secretion ($\sim 60$ min). Shortly after nocturnal sleep commences the pulses of prolactin release markedly increase, so that high levels of plasma prolactin (3–4 fold basal) are reached some 5–8 hours after the onset of sleep. This broad nocturnal peak is sleep-entrained. After waking, the level of plasma prolactin falls rapidly, reaching the basal level by midday.

Plasma prolactin levels are:
(1) Stable in males throughout adulthood.
(2) Higher after the female menopause.
(3) Often higher during the luteal phase.
(4) Increased throughout pregnancy (up to 20 fold).

*Hypothalamic and anterior pituitary hormones*

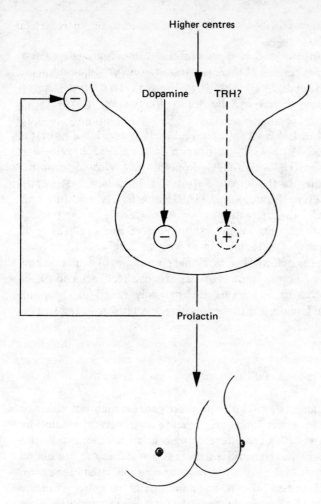

**Fig. 19.4** Major pathways controlling prolactin release.

(5) Increased by stress—emotional or physical.
(6) Sometimes mildly increased by nipple stimulation in males and non-lactating females.
(7) May remain elevated for a year or more in nursing mothers.

### Gonadotropins—follicle-stimulating hormone (FSH) and luteinizing hormone (LH) (Fig. 19.5)

FSH and LH are glycoprotein hormones secreted by the gonadotroph cells of the anterior pituitary. Both hormones stimulate the maturation and function of the gonads. They both comprise two polypeptide subunits, designated α and β. The same α chain is common to LH, FSH, TSH and hCG, whilst the β chains differ and invest the hormonal activity that is unique to each hormone.

LH (mol.wt. ~28 000 Daltons)

A major function of LH is to stimulate steroidogenesis by the gonads. In the testes LH increases testosterone production by the interstitial (Leydig) cells, whilst in the ovary androgen (androstenedione) synthesis by the theca interna of the ovary is stimulated.

In the adult LH is synthesized and then secreted in a pulsatile manner (1–2 h) under the influence of gonadotropin-releasing hormone (GnRH), which is itself episodically released from the hypothalamus. In the female during the follicular phase of the menstrual cycle the output of GnRH and LH is inhibited and modulated by the increasing amounts of oestrogen (oestradiol) secreted by the developing ovarian follicles. At mid-cycle the high level of plasma oestradiol has a positive feedback effect that releases a surge of GnRH and LH that culminates in ovulation and luteinization. Testosterone provides negative feedback control in the male. Plasma LH levels are markedly raised in the postmenopausal female, due to the loss of ovarian function and steroid feedback.

FSH (mol.wt. ~33 000 Daltons)

A major function of FSH is to promote gonadal maturation and to stimulate the conversion of androgen to oestrogen (oestradiol) by the granulosa cells of the developing ovarian follicle, and by the Sertoli cells of the testes. FSH secretion is pulsatile and stimulated by GnRH, negative feedback probably being mediated in the male by a protein named inhibin, and additionally by oestrogen in the female. In ovulating females the pattern of FSH secretion is fairly similar to that of LH, but the midcycle surge is much less marked. In the postmenopausal female plasma FSH levels are markedly elevated due to lack of ovarian oestrogen production and consequent loss of feedback inhibition.

**Thyroid stimulating hormone (TSH, thyrotropin)** (Fig. 19.6)

TSH (mol.wt. ~28 000 Daltons) is a two-chain glycoprotein hormone that contains the same α-polypeptide subunit as FSH, LH and hCG. TSH is secreted by the thyrotroph cells of the anterior pituitary, and interacts with specific cell surface receptors on the follicular cells of the thyroid gland, resulting in increased synthesis and release of thyroid hormones. TSH release is stimulated by thyrotropin-releasing hormone (TRH), and subject to negative feedback inhibition by the thyroid hormones. Somatostatin also blocks the TRH-stimulated release of TSH.

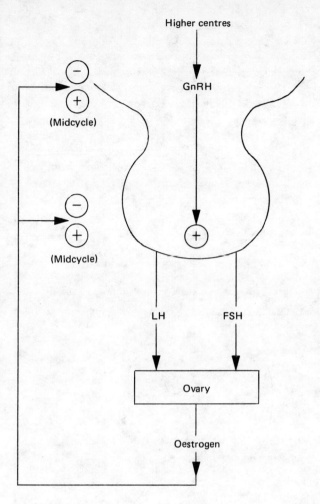

**Fig. 19.5** Major pathways controlling the secretion of gonadotropins in the fertile female.

## LABORATORY INVESTIGATIONS

*Note* The test protocols given below are in outline only and do not list all contraindications and side-effects. As protocols can vary in detail it is recommended that the appropriate local specialist medical unit or laboratory always be consulted for information.

### Hypothalamic hormones

As noted earlier, no assays are routinely available for these hormones.

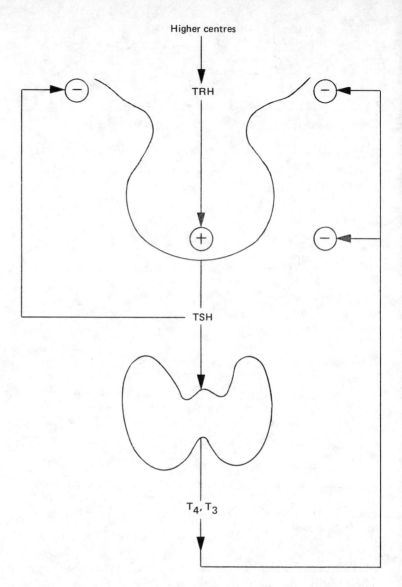

**Fig. 19.6** Major pathways controlling the secretion of TSH.

### Anterior pituitary hormones

Any biochemical investigation of anterior pituitary function poses three questions:
(1) Is the anterior pituitary output of hormone(s) normal?
(2) Does the anterior pituitary have a normal reserve capacity of hormone(s)?
(3) Does hormone output respond to normal control mechanisms?

It is important to note that a normal output of hormone(s) does not necessarily reflect a normally functioning anterior pituitary and so single plasma measurements of hormones can, in cases of suspected hypofunction or situations of stress, be of limited or misleading value. Generally, the anterior pituitary must be investigated by assessing its response to specific stimulation or suppression by applying an appropriate dynamic function test.

Polypeptide hormones are difficult to purify, often share similar regions of structure, and may not be stable. However, all the anterior pituitary hormones can be measured in plasma using the radioimmunoassay technique with varying degrees of acceptability in terms of imprecision, specificity and sensitivity. It should also be recalled that *immunoreactivity* does not necessarily reflect *biological activity*.

Major problems that beset the assay of these hormones remain cross-reactivity, non-specific interference and matrix effects, and the calibration of kit standards against different primary standards. These problems are often clearly reflected in the markedly different reference ranges quoted by laboratories that use different commercial or home-made kits. The magnitude of the problem is largely unknown at present because widely-based quality assurrance programmes for these hormones are only now becoming available, thus allowing direct interlaboratory assay comparisons. For the present it is unwise to compare plasma peptide hormone values quoted here with those obtained by your own laboratory, but a helpful comparison can be obtained if the values are each related to their own reference range.

## ACTH

Although RIA assays for ACTH have been available for a number of years they have often proved to be unreliable. ACTH adheres to glass, and is also rapidly degraded by proteolytic enzymes in plasma. The wide range of plasma peptides that are able to cross-react have caused problems with obtaining suitable antisera to ACTH, and also led to additional preassay purification steps. These difficulties have decided most laboratories to measure plasma cortisol instead of ACTH. If adrenal function is normal this approach is acceptable because:
(1) Plasma cortisol levels reflect the circulating level of biologically active, as opposed to immunoreactive, ACTH.
(2) The hormone cascade effect ensures that the small quantities of ACTH are amplified to relatively large amounts of easily detected cortisol.
(3) The assay of cortisol is quick and simple.

Therefore, the use of plasma cortisol as an indirect measure of plasma ACTH will be described below, and largely repeated in Chapter 21.

Despite the technical difficulties mentioned above, a plasma ACTH measurement can be helpful in differentiating ectopic

ACTH production from excessive pituitary secretion of ACTH as a cause of Cushing's syndrome, or for locating the site of an ACTH-secreting tumour. In such cases it is advisable to send the patient sample to a laboratory that has expertise in the ACTH assay.

PLASMA CORTISOL

Cortisol is stable and is readily measured in plasma or serum by RIA. Drug intake by the patient, particularly steroids, should be noted; prednisolone may cross-react heavily with the cortisol antibody, whilst oestrogens raise the total plasma cortisol level by increasing the liver synthesis of cortisol-binding globulin. Prior to RIA the fluorimetric estimation of cortisol as an 11-hydroxycorticosteroid was widely used, but lack of specificity and technical problems relative to RIA make this approach unattractive.

The circadian rhythm of CRF is shadowed by plasma cortisol, and therefore blood samples for cortisol estimation should be taken at a defined time, generally between 07.00 and 09.00 hours, when the level can normally be expected to be high. Samples may also be taken, if clinically relevant, around 22.00–24.00 hours, when the cortisol concentration is normally low. The circadian rhythm of cortisol is reflected in the wide reference range quoted by most laboratories, few bothering to obtain a range for the early morning period alone.

An isolated plasma cortisol estimation is of limited value and may even be misleading, for the following reasons:
(1) Suspected inadequacy of ACTH output by the anterior pituitary. The plasma cortisol may fall well within the reference range, but this does not give any indication as to whether the anterior pituitary can respond to a stress (biochemical, physical, emotional) situation, i.e. despite pituitary dysfunction, enough ACTH can be produced to maintain a basal level of circulating cortisol.
(2) An elevated plasma cortisol can equally reflect a pathological level of circulating ACTH or the patient's anxiety about the immediate clinical examination and venepuncture.

Generally, the estimation of a single plasma cortisol for diagnostic purposes should be discouraged, and an appropriate dynamic function test used.

PLASMA ELECTROLYTES

A fall in extracellular osmolality normally results in the inhibition of ADH release, followed by water excretion. Cortisol plays an important role in the inhibition of ADH. ACTH deficiency causes a loss of cortisol secretion, which in turn is reflected by the inability of the kidney to rapidly excrete an acute water load. Therefore hyponatraemia, due to water retention, is often associated with cortisol deficiency syndromes.

PLASMA GLUCOSE

Cortisol opposes the action of insulin by promoting the conservation of glucose and hyperglycaemia. Reduced ACTH secretion will lower cortisol output, so that the unopposed action of insulin may be reflected by hypoglycaemia.

URINARY CORTISOL

Cortisol bound to cortisol-binding globulin (CBG) does not cross the renal glomerular membrane, and so urinary cortisol reflects the free fraction of the hormone present in plasma. A lower reference limit for cortisol excretion is difficult to define, and therefore urinary estimations have generally been used as a screen only for situations of excess ACTH or cortisol production. The use of urinary cortisol estimations is declining, partly because of the usual practical drawbacks of 24 hour collections and partly because of the simplicity of performing a more informative suppression test.

## Dynamic function tests

### Stimulation

Anterior pituitary dysfunction with respect to ACTH production will disrupt the hypothalamic-pituitary adrenal (HPA) axis. Therefore a partial or complete loss of ACTH reserve and secretion can be investigated by stimulating the HPA axis, and assessing its integrity by sampling the adrenal response in terms of plasma cortisol, provided adrenal cortex function is normal. Normally:

stress → ↑ CRF → ↑ ACTH → ↑ cortisol

A standardized biochemical stress can be achieved by creating a hypoglycaemic episode with insulin.

INSULIN HYPOGLYCAEMIA TEST

Precautions

The test must be performed under medical supervision and should be avoided if possible in those with cardiac or cerebrovascular disease. A 50% glucose solution must be available for IV administration if the test requires immediate termination.

*Insulin*   0.05–0.15 units/kg body weight depending also on age and clinical findings (i.e. cortisol deficient patients are more sensitive to insulin).

*Samples*   Blood at 0, 30, 60, 90, 120 minutes for plasma or serum cortisol and glucose estimations.

*Notes*   Adequate stress must occur for the test to be valid. There should be clinical symptoms of hypoglycaemia—sweating, light-headed, tachycardia, tremor, etc. The plasma glucose should fall to less than 50% of the basal level, and in one sample be below 2.2 mmol/l. If the test is abandoned by injecting glucose because of a severe hypoglycaemia, the timed sample-taking for cortisol assay should be continued as the stress has obviously been adequate.

Interpretations

For a normal response by the HPA axis the plasma cortisol concentration in at least one sample should both: (1) increase over the basal value by at least 200 nmol/l, (2) exceed an absolute value of 550 nmol/l.

A lack of response can indicate either hypothalamic or anterior pituitary dysfunction. Obviously this method of assessing the HPA axis is dependent on a normal adrenal cortex. Therefore it is wise to test adrenal function as an integral part of the insulin hypoglycaemia test of the HPA axis, but this is often omitted if there is no clinical or biochemical evidence of primary adrenal disease. Discussion of this test is deferred to Chapter 21. (See also triple function test.)

**Suppression**

ACTH secretion by the anterior pituitary is stimulated by CRF, and subject to negative feedback by cortisol. CRF secretion by the hypothalamus is also inhibited by high cortisol levels. The ability of the hypothalamus/anterior pituitary to respond normally to feedback inhibition can be tested using dexamethasone, a powerful glucocorticoid that behaves as cortisol in its negative feedback effects.

DEXAMETHASONE SUPPRESSION TEST

Basal blood sample taken for cortisol estimation at 08.00–09.00 hours, and 2 mg dexamethasone given to patient at 22.00–23.00

hours. Blood sample taken at 08.00–09.00 hours the following morning for plasma cortisol estimation.

A normal response is shown when the postdexamethasone plasma cortisol level has suppressed by at least 70% (should be <150 nmol/l) of the basal level. A failure to suppress suggests that the normal pituitary 'set' that cuts off ACTH release in response to a rising plasma cortisol is disturbed (Cushing's disease), or that the ACTH secretion is autonomous from non-pituitary sites. These two conditions can be differentiated by increasing the dexamethasone dose to 8 mg (2 mg every 6 h for 2 days); the higher dose suppresses ACTH in the case of Cushing's disease (see Chapter 21 for other aspects of the dexamethasone test).

## GROWTH HORMONE (GH)

GH is a fairly stable hormone, and is readily estimated in plasma or serum by RIA techniques. Cross-reactivity of the antiserum with hCG should be noted when estimating GH in samples taken from subjects in late pregnancy.

### PLASMA GH

Basal levels of plasma GH are markedly affected by a variety of factors, particularly food, exercise, sleep and anxiety; levels reached in these situations can often exceed those found in acromegaly. In addition, basal GH concentrations are often undetectable in many normal subjects by the assay methods presently available, and therefore it is usually not possible to distinguish GH deficiency from normal. For these reasons the measurement of GH in random blood samples should be abandoned in favour of stimulation or suppression tests that are appropriate to the clinical question. Since a basal sample is generally taken as part of a dynamic function test it should be noted that oestrogens enhance GH release, and therefore plasma GH levels are generally higher in females than in males, particularly at midcycle. Plasma GH is raised in infants, and usually falls to adult levels by 2–3 months.

### Dynamic function tests

**Stimulation**

The following tests are designed to demonstrate the pituitary reserve of GH. The difficulty of ensuring a reliable standard stimulation and defining a normal response is reflected in the variety of tests available. The more commonly used tests are given below.

*Note* National bodies responsible for the provision of GH for treatment purposes generally have their own numerical definitions of what constitutes a normal response to GH stimulation tests.

SLEEP

Deep sleep (Stage III or IV) causes a reproducible marked elevation of plasma GH, particularly in children and adolescents. The response may be reduced or absent by middle age. This sleep-related rise in GH is not abolished by hyperglycaemia.

Blood samples are obtained via an in-dwelling catheter, whilst the sleeping subject is monitored by EEG. Samples are collected every 30 minutes for 3–4 hours after the onset of sleep. The spurt of GH generally occurs within 30–90 minutes of sleep onset and lasts for 1–2 hours (Fig. 19.7). The sleep peak is absent in both GH deficiency states and acromegaly. This test provides a reliable method for detecting GH deficiency, but it is under-used because of the facilities required for its performance, it generally being reserved for those subjects for whom the other stimulation tests are unsuitable.

EXERCISE

Hard physical exercise will stimulate a marked increase in plasma GH concentration in the normal individual. The patient can be subjected to a standardized workload by using a bicycle ergometer.

Protocol

*Patient* No special preparation if exercise is medically safe for the subject.

*Exercise* Appropriate workload for 20–30 minutes.

*Blood sample* Basal, and 2 and 20 minutes after cessation of exercise.

Interpretation

Laboratories should determine their own reference value for a minimal normal response, usually approximately 10x basal ($\sim 20$ mU/l).

Comment

The exercise test is an ideal screening test for pituitary GH reserve in children. However, maximal effort during exercise cannot be

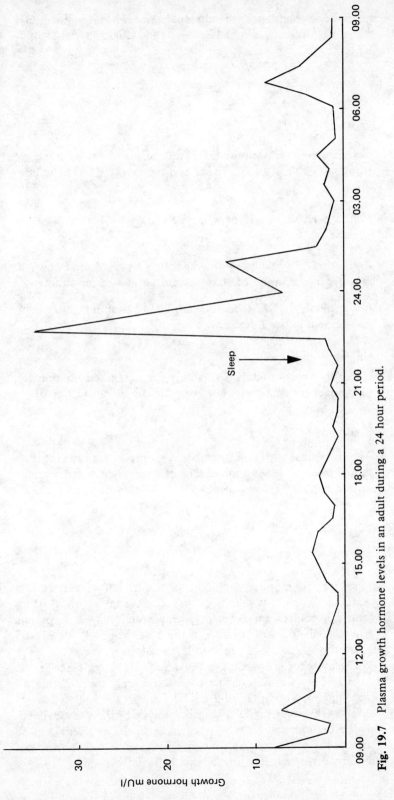

Fig. 19.7 Plasma growth hormone levels in an adult during a 24 hour period.

guaranteed, and so any individual with an inadequate GH response to exercise should be subjected to another type of stimulation test.

### ARGININE LOAD

Certain amino acids stimulate GH release, probably via adrenergic stimulation. This stimulation test, if used, is well suited to children.

*Patient preparation*   Hyperkalaemia can occur if renal disease is present.

*Amino acid load*   0.5 g arginine/kg body weight infused IV over 30 minutes (a maximum of 30 g for adults and 20 g for children).

*Blood samples*   Basal, then 30, 60, 90, 120 minutes after commencement of the infusion.

Interpretation

The response is variable, particularly due to sex, stage of puberty, phase of menstrual cycle, but normally should be >20 mU/l.

Comment

Any individual with an inadequate response to arginine infusion should be subjected to another type of stimulation test, preferably insulin hypoglycaemia provocation.

### INSULIN HYPOGLYCAEMIA

A severe fall in plasma glucose concentration stimulates GH secretion. The test is valid if: (1) the subject shows clinical symptoms of hypoglycaemia (sweating, tachycardia, etc.), (2) the plasma glucose falls to less than 50% of the basal level, (3) the plasma glucose is <2.2 mmol/l in at least one sample.

The protocol and precautions are identical to those described on p. 401. This test is probably the most widely used method of investigating pituitary reserve of GH. Normal individuals demonstrate a GH peak of >20 mU/l, whilst in GH deficiency, depending on the severity, little (<10 mU/l in partial deficiency) or no response is seen (see triple function test). The GH response is diminished in subjects with obesity or hypothyroidism.

## Suppression

The suppression test is designed to differentiate those patients with high circulating levels of GH due to a pathological process from those with high levels due to stress or other non-pathological factors. Hyperglycaemia will suppress GH secretion in the normal individual.

GLUCOSE SUPPRESSION TEST

Protocol

*Patient* Preparation and precautions as for an oral glucose tolerance test (see p. 213).

*Load* 75 g glucose in 250–350 ml chilled water drunk in <15 minutes.

*Blood samples* 0, then 30, 60, 90, 120 minutes postload for GH and plasma glucose estimation. Samples should be obtained via an in-dwelling needle inserted 30 minutes prior to the test.

Interpretation

The suppression of GH to within the reference limit, or to undetectable levels, indicates a normal response. In acromegaly and gigantism there is generally a failure to suppress from the high basal value, and in some cases there is even a rise in GH levels in the later samples, possibly in response to the fall in glucose levels as the load is cleared from the plasma. In some cases of acromegaly there is a small but clearly inadequate suppression. A failure to suppress GH levels may occur in subjects with liver cirrhosis or uraemia.

# PROLACTIN (PRL)

PRL is a stable hormone that is readily assayed in plasma or serum by the RIA technique. There is no detectable cross-reactivity with GH or hCG if the antisera is raised against pure PRL.

PLASMA PRL

Basal plasma PRL levels are higher in women than men, this difference first appearing at puberty and disappearing again in old age. The sex difference and the changes in PRL levels across the menstrual cycle are not significant enough to affect the interpretation of assay values. Circulating PRL increases during pregnancy, reflecting the expanding pituitary caused by the increased number of PRL-secreting cells. Plasma PRL levels fall towards normal postpartum, with breast-feeding causing large fluctuations (up to $\sim 20$ fold) for 30–60 minutes. Plasma PRL is markedly elevated (four to five fold) during non-REM sleep, this increase being unrelated to that seen with GH. PRL is also raised, but to a lesser extent, by either physical or emotional stress. With an awareness of these factors a random PRL estimation provides a useful and adequate screening test for most diagnostic purposes. Equivocal values or clinical findings may require further investigation with dynamic function tests.

## Dynamic function tests

### Stimulation

#### TRH

Thyrotropin-releasing hormone (TRH) stimulates the release of PRL, acting directly on the anterior pituitary by a mechanism that is independent of that for TSH release. PRL rises more rapidly in response to TRH than does TSH, and also requires a smaller TRH dose to achieve maximal secretion. For protocol see p. 439.

Interpretation

Peak PRL secretion (10 fold) normally occurs within 30 minutes of TRH injection and returns to basal levels in 2–3 hours. The magnitude of response is greater in females, being further enhanced during the luteal phase, and declines with age in both sexes.

Low basal plasma PRL, and a failure to respond to TRH stimulation, is seen in panhypopituitarism, whilst high basal PRL and a response, often blunted, to TRH is seen in hyperprolactinaemia due to hypothalamic or pituitary disease.

#### OTHER STIMULATION TESTS

Metoclopramide is a dopamine antagonist that increases PRL secretion ($\sim 10$ fold) within 30 minutes of oral administration, presumably by opposing the dopamine-mediated hypothalamic PIF. Phenothiazines (e.g. chlorpromazine) are psychotropic drugs which lower the level of hypothalamic dopamine, causing a marked increased in PRL secretion in the normal individual.

**Suppression**

Dopamine agonists such as L-dopa and the ergot alkaloid called bromocryptine cause suppression of PRL in both normal and hyperprolactinaemic subjects. Tests have been based on these compounds, but the responses are similar in all hyperprolactinaemic patients regardless of the aetiology. Therefore these tests are of little diagnostic value.

## GONADOTROPINS (FSH, LH)

RIA methods are widely available for the measurement of both FSH and LH in plasma and serum. Possible cross-reactivity of hCG with LH antiserum should be noted in any assay used.

### PLASMA FSH AND LH

Both hormones are low or undetectable in children, rising to comparable plasma levels in both sexes in the adult. In the female both hormones rise during the follicular phase of the menstrual cycle prior to a sharp preovulatory peak (FSH $\sim$ 2-3 fold, LH $\sim$ 10 fold), followed by a return to basal levels during the luteal phase. In old age, plasma levels of both hormones increase as gonadal function declines and target organ feedback is lost, by up to 10-20 fold in the postmenopausal female, and by two to three fold in the male. Measurement of basal levels of plasma gonadotropins provide a useful screening test, but it should be restated that adequate basal levels give no evidence of normal pituitary reserve, for which a dynamic function test is required.

**Dynamic function tests**

**Stimulation**

### GnRH (LHRH) TEST

Gonadotropin-releasing hormone (GnRH) stimulates the release of both FSH and LH from the anterior pituitary (see triple function test).

*Patient preparation*   Overnight fast.

*Stimulation*   100 µg synthetic GnRH, usually IV.

*Blood samples*   0, then 30, 60, 90 minutes poststimulation.

Interpretation

In the normal adult an LH surge (5-10 fold) is seen within 45 minutes, followed by a smaller FSH peak (2-3 fold) at 60-90 minutes. No response by either hormone is seen in panhypopituitarism, whilst the response in hypothalamic disease is variable.

CLOMIPHENE TEST

Clomiphene is a synthetic oestrogen that prevents endogenous oestrogen from exerting a negative feedback effect by blocking hypothalamic oestrogen receptors, resulting in increased GnRH secretion and a consequent increase of LH and FSH output. Protocols are variable and should be sought from a specialist centre. Clomiphene also provides a valuable means of initiating ovulation in certain types of infertility.

**Suppression**

50 µg ethinyl oestradiol given daily will suppress the gonadotropins in the female, whilst 10-15 mg testosterone propionate will do so in males. A suppression test is rarely required for diagnostic purposes.

# THYROID STIMULATING HORMONE (TSH)

Assays for TSH in plasma or serum by RIA are widely available. Many methods currently employed are insensitive to the extent that with some kits up to 30% of normal individuals have undetectable levels of TSH in plasma. Commercial reagent kits also markedly vary in the upper reference range limit they quote (2-3 fold) probably due to specific and non-specific interferences deriving from the various base sera (horse, human, sheep, etc.) that are used as a matrix for calibration standards.

PLASMA TSH

Basal plasma TSH is two to three fold higher at birth relative to adulthood, probably as a response to body cooling postpartum; levels decline to those of the adult within 2-3 days. Plasma TSH levels are comparable in both sexes, tend to be higher at night, and are not affected by physical or emotional stress. Therefore basal plasma TSH can be measured in randomly taken samples and used for diagnostic purposes. However, as noted above, most assays cannot detect low normal levels, and thus a basal plasma TSH is presently useless as a diagnostic test for pituitary deficiency of TSH.

## Dynamic function test

### Stimulation

The anterior pituitary reserve of TSH can be demonstrated by administering synthetic TRH, and monitoring the change in plasma TSH.

#### TRH TEST

For the use and interpretation of this test in thyroid disease see Chapter 20.

*Patient preparation*   No specific requirement. Note drug history and age.

*Stimulation*   Adult: 200 µg TRH given IV. Transient side effects include nausea, metallic taste, light headedness.

*Blood samples*   0, then 20, 60 minutes, post TRH. Some protocols advocate more frequent sampling to locate the peak response for both TSH and PRL, but this should be rarely necessary.

Interpretation

In normal individuals a marked rise in TSH (3-5 fold of basal) occurs by 20-30 minutes. A failure to respond, in conjunction with a low or undetectable basal TSH, and a low plasma level of thyroid hormone, is typical of pituitary deficiency. A sluggish TSH response (60 min value $>$ 20 min value), in association with a low basal level of TSH and thyroid hormone, is suggestive of hypothalamic dysfunction. The normal TSH response is greater in females, and declines with age, particularly in males (see triple function test and Chapter 20).

#### TRIPLE FUNCTION TEST

An isolated deficiency of a pituitary hormone is extremely rare, and therefore a clinical suggestion of pituitary dysfunction generally requires that the pituitary reserve of all hormones be evalu-

*Hypothalamic and anterior pituitary hormones*

ated. This is most simply accomplished by combining several of the stimulatory dynamic function tests in a modified form.

*Patient preparation*  Contraindications and medical supervision as for insulin hypoglycaemia.

*Protocol*  See Table below. Blood samples should be taken via an in-dwelling needle.

|  | Plasma/serum for | | | | | | |
|---|---|---|---|---|---|---|---|
|  | Glucose | Cortisol | GH | TSH | LH | FSH | PRL |
| Basal | √ | √ | √ | √ | √ | √ | √ |
| 0.05–0.15 units insulin | | | | | | | |
| 200 µg TRH, 100 µg GnRH IV simultaneously | | | | | | | |
| 30, 60, 90, 120 min | √ | √ | √ | √ | √ | √ | √ |
| 250 µg Synacthen | | √ | | | | | |
| 30, 60 min | | | | | | | |

Note that a synacthen test should be routinely appended at the end of the pituitary function tests so as to demonstrate that the adrenal cortex is functional, and able to reflect a surge of endogenous ACTH. However, this is often not done in cases when there is no clinical or other evidence of primary adrenal disease.

*Interpretation*  As for the individual tests.

*Note*  The peptide hormone assays used in the case examples were calibrated against the following primary standards:

| | | |
|---|---|---|
| GH | First International Reference Preparation (IRP) | 66/217 |
| TSH | Medical Research Council WHO Standard | |
| PRL | First IRP | 75/504 |
| LH | First IRP | 69/104 |
| FSH | First IRP | 69/104 |

## HYPOPITUITARISM

Hypopituitarism describes the partial or complete failure of the anterior pituitary to secrete one, several or all (panhypopituitarism) of its normal complement of hormones. These situations can arise in several ways:
(1) Loss of appropriate stimulation may result from hypothalamic disease or section of the pituitary stalk. Typically the anterior pituitary hormones secretions are low, except prolactin which rises as a result of the loss of hypothalamic inhibition.

(2) Loss of hormone-secreting cells may be caused iatrogenically, or as a result of tumour expansion. If the tumour is anterior pituitary in origin, it may well autonomously secrete large quantities of its own hormone product.
(3) Isolated hormone deficiency. Inability to secrete a specific hormone can be due to loss of the appropriate hypothalamic stimulation, or dysfunction of the hormone secreting cell itself.

Some of the major causes of hypopituitarism are:

Hypothalamic disease

*Tumours* (1) Craniopharyngioma, (2) chromophobe adenomas, (3) optic gliomas, (4) astrocytomas.

*Infections* (1) Encephalitis, (2) syphylis, (3) meningitis, (4) tuberculosis.

*Injury* (1) Head injury, (2) aneurysm.

*Iatrogenic* (1) Surgery, (2) irradiation therapy.

*Genetic* (1) Cysts, (2) hyperplasia.

*Other* Malnutrition, anorexia nervosa.

Anterior pituitary disease

*Space-occupying lesions* (1) Chromophobe, acidophil, basophil adenomas, (2) craniopharyngioma, (3) secondary carcinoma—particularly from breast, lung.

*Necrosis* Postpartum (Sheehan's syndrome).

*Infection* Encephalitis.

*Iatrogenic* (1) Irradiation, (2) surgery—hypophysectomy.

*Idiopathic* Single hormone deficiency, i.e. GH.

*Miscellaneous* (1) Embolus, (2) trauma to head.

**Case examples/pathophysiology**

NORMAL HYPOTHALAMIC-PITUITARY FUNCTION

A 35 year old married woman with a history of migraine type

headaches since the age of 22 years. The headaches had increased in frequency over the last few months, including one prolonged severe episode. She had a good appetite, no tiredness, no abnormal lactation or cold intolerance, nor any other clinical evidence of hypopituitarism. A plain skull X-ray suggested a calcified pituitary lesion. A triple pituitary function test was performed to exclude craniopharyngioma, pituitary apoplexy or the rupture of a pituitary cyst.

| Time | Glu. mmol/l (3.0–5.5) | Cort. nmol/l (140–690) | GH mU/l | TSH mU/l ($<$5.5) | PRL µg/l ($<$30) | LH IU/l (1.6–9.0) | FSH IU/l (1.1–6.4) |
|---|---|---|---|---|---|---|---|
| 09.00 | 3.9 | 220 | $<$1.0 | 4.3 | 7.0 | 4.6 | 1.5 |
| | Insulin: 9.0 units; TRH: 200 µg; GnRH: 100 µg | | | | | | |
| 09.30 | 1.3 | 210 | $<$1.0 | 12.0 | 67 | 36.0 | 2.7 |
| 09.50 | 50 g oral dextrose, 12.5 g dextrose IV | | | | | | |
| 10.00 | 8.3 | 435 | $>$40 | 9.7 | $>$200 | 36.0 | 2.7 |
| 11.00 | 4.6 | 625 | $>$40 | 7.1 | 150 | 36.0 | 3.6 |

Interpretation

*Test conditions* Plasma glucose levels declined by more than 50% of the basal level, and also fell below 2.2 mmol/l, whilst clinically the hypoglycaemia was severe enough to require dextrose administration. Note that although the hypoglycaemia was terminated early, samples should continue to be taken.

*Cort.* Two test samples exceeded the basal value by more than 200 nmol/l, whilst one sample also rose to more than 550 nmol/l, indicating a normal response by the hypothalamic-pituitary axis.

*GH* A normal response to the hypoglycaemia episode was indicated by a sample exceeding 20 mU/l.

*TSH* Normal response to TRH.

*PRL* Normal response to TRH. Sample times are insufficient to locate the peaks of PRL and TSH, and although PRL usually peaks first it may not have done so in the above case.

*Gonadotropins* Normal response by both hormones, with the characteristically greater rise shown by LH.

The overall response suggests the hypothalamic-pituitary axis is intact and that the anterior pituitary reserve of hormones is normal.

## HYPOTHALAMIC DISEASE

*Hypothalamic and anterior pituitary hormones*

A 20 year old man who was clinically prepubertal was referred by his medical practitioner to an endocrinologist to investigate the possibility of Klinefelter's syndrome. Low basal gonadotropins suggested the need (Klinefelter's syndrome has ↑FSH, +N or ↑LH) for a full evaluation of pituitary function.

| Time | Glu. mmol/l (3.0–5.5) | Cort. nmol/l (140–690) | GH mU/l | TSH mU/l (<5.5) | PRL µg/l (<30) | LH IU/l (0.7–2.0) | FSH IU/l (2.0–5.0) |
|---|---|---|---|---|---|---|---|
| 09.30 | 4.9 | 156 | <0.5 | 4.5 | 115 | 0.5 | nd |
| 09.31 | Insulin: 7 U; TRH: 200 µg; GnRH: 100 µg | | | | | | |
| 10.00 | 2.0 | 169 | <0.5 | 18 | 109 | 0.4 | 0.1 |
| 10.30 | 3.1 | 210 | <0.5 | 29 | 273 | 2.5 | 0.6 |
| 11.00 | 4.4 | 128 | <0.5 | 36 | 252 | 2.4 | 0.3 |
| 11.30 | 4.7 | 146 | <0.5 | — | 216 | 2.5 | 0.7 |
| | Synacthen: 250 µg | | | | | | |
| 12.00 | | 183 | | | | | |
| 12.30 | | 251 | | | | | |

Interpretation

*Test conditions* An adequate biochemical stress was achieved (Glu. <2.2 mmol/l), which was confirmed clinically with the patient sweating and light-headed.

*Cort.* Although the basal level was normal, there was no response to hypoglycaemia. There was also no response to a short Synacthen stimulation; however, a subsequent 3 days depot test (see p. 472) was normal. This slow response suggests that the adrenal glands have not been stimulated for some time.

*GH* No response to hypoglycaemia.

*TSH* TSH normally shows a peak response by 40 minutes. In the above case the sluggish increase over 90 minutes is suggestive of a hypothalamic lesion, the pattern indicating that the thyrotroph cells have been somewhat dormant due to lack of stimulation.

*PRL* Basal PRL high, followed by a blunted response to stimulation. This suggests either increased PRL production by the anterior pituitary, or loss of hypothalamic inhibition.

*Gonadotropins* The basal levels of both LH (0.7–2.0 IU/l) and FSH (2.0–5.0 IU/l) are abnormally low, with a subsequently poor response to GnRH.

The results are consistent with poor anterior pituitary function as a result of hypothalamic dysfunction. CAT scans and surgery revealed a hypothalamic astrocytoma.

PITUITARY TUMOUR

An 18 year old man was investigated for slow onset of puberty. He was of short stature, slightly obese and complained of frequent headaches. Basal plasma gonadotropins were inappropriately low for the low plasma testosterone (3 nmol/l; ref. range: 8–30 nmol/l). A skull X-ray indicated an enlarged pituitary fossa. Pituitary function was investigated by means of a triple function test.

| Time | Glu. mmol/l (3.0–5.5) | Cort. nmol/l (140–690) | GH mU/l | TSH mU/l (<5.5) | PRL µg/l (<30) | LH IU/l (0.7–2.0) | FSH IU/l (2.0–5.0) |
|---|---|---|---|---|---|---|---|
| 09.15 | 4.8 | 212 | <1 | <2.5 | 296 | 0.8 | 1.6 |
| | Insulin: 8 units; TRH: 200 µg; GnRH: 100 µg | | | | | | |
| 09.45 | 1.8 | 209 | <1 | <2.5 | 350 | 1.9 | 1.8 |
| 10.15 | 2.9 | 335 | <1 | <2.5 | 390 | 1.3 | 1.7 |
| 11.15 | 4.6 | 188 | <1 | 2.5 | 300 | 1.1 | 1.7 |
| | Synacthen: 250 µg | | | | | | |
| 11.45 | | 195 | | | | | |
| 12.15 | | 220 | | | | | |

Interpretation

*Test condition*   The achievement of an adequate hypoglycaemia was suggested by the blood glucose level (<2.2 nmol/l), and confirmed clinically by sweating and a pulse-rate of 96/minute.

*Cort.*   No response to hypoglycaemia. The short Synacthen test showed no adrenal response. A long stimulation later demonstrated normal adrenal function, suggesting the glands had been dormant for some time.

*GH*   No response.

*TSH*   Flat response to TRH.

*PRL*   Basal plasma prolactin approximately 10 fold above the reference range upper limit, with a subsequently poor response to TRH stimulation.

*Gonadotropins*   Failure to rise following GnRH stimulation.

An abnormally high basal plasma PRL level, in conjunction with subnormal values for the other hormones and their failure to rise in response to appropriate stimulation, suggests the presence of a prolactin-secreting pituitary tumour. The growth of such a tumour would progressively destroy hormone-producing tissue, whilst autonomously secreting increasing amounts of its own product. Transphenoidal surgery to remove a tumour and subsequent histology confirmed a PRL-secreting adenoma.

HYPOPHYSECTOMY

A 56 year old previously fit man complained of headaches for approximately a month. He also noted occasional visual disturbance in his left eye. Basal hormone levels and a triple function test did not reveal any abnormalities of endocrine function. A transphenoidal hypophysectomy was performed to remove a benign chromophobe adenoma. A triple function test was subsequently done to assess the postoperative hormone capacity of the anterior pituitary.

| Time | Glu. mmol/l (3.0–5.5) | Cort. nmol/l (140–690) | GH mU/l | TSH mU/l (<5.5) | PRL µg/l (<30) | LH IU/l (0.7–2.0) | FSH IU/l (2.0–5.0) |
|---|---|---|---|---|---|---|---|
| 10.00 | 4.7 | 440 | <1.0 | 3.1 | 23 | 0.7 | 0.9 |
| | Insulin: 9 units; TRH: 200 µg; GnRH: 100 µg | | | | | | |
| 10.30 | 1.4 | 595 | <1.0 | 5.7 | 75 | 1.6 | 1.0 |
| 11.00 | 2.0 | 620 | 2.5 | 5.8 | 69 | 1.4 | 1.2 |
| 11.30 | 2.5 | 605 | 1.0 | 5.1 | 58 | 1.3 | 1.0 |
| 12.00 | 2.9 | 460 | <1.0 | 4.8 | 41 | 1.0 | 0.9 |

Interpretation

*Test condition* Sufficient clinical and biochemical hypoglycaemia was achieved.

*Cort.* A barely adequate response to hypoglycaemia, in that no sample exceeded the basal level by more than 200 nmol/l.

*GH* Insufficient response.

*TSH* Partial response, with some suggestion of a reduced pituitary reserve.

*PRL* Normal response to TRH.

*Gonadotropins* Poor response, suggesting a reduced reserve.

Hypophysectomies are generally incomplete; in the above case, the pituitary reserve of cortisol, PRL and, to some extent, TSH were preserved. On the above results the patient required only a precautionary low daily replacement dose of steroid. The loss of GH in an adult seems to be without deleterious effect and does not require replacement, whilst in the above case TSH and the gonadotropins may do so at some time. Therefore, this patient will have a triple function test performed at intervals to assess any changes.

PITUITARY TUMOUR

A 53 year old man was fully investigated following an isolated fit. A CT scan revealed an irregular pituitary fossa, whilst tomograms indicated the presence of a pituitary mass. Bitemporal upper quadrant visual field defects were also detected. A modified pituitary triple function test was performed.

| Time | Glu. mmol/l (3.0–5.5) | Cort. nmol/l (140–690) | GH mU/l | TSH mU/l (<5.5) | PRL μg/l (<30) | LH IU/l (0.7–2.0) | FSH IU/l (2.0–5.0) |
|---|---|---|---|---|---|---|---|
| 10.00 | 4.6 | 325 | 1.0 | 4.4 | >200 | 1.2 | 1.3 |
| | Insulin: 8 units; TRH: 200 μg; GnRH: 100 μg | | | | | | |
| 10.30 | 1.3 | 520 | 2.0 | 18.0 | >200 | 4.5 | 1.9 |
| 11.00 | 3.2 | 710 | 5.5 | 15.0 | >200 | 5.2 | 2.2 |
| 12.00 | 3.3 | 545 | 4.0 | 9.8 | >200 | 5.4 | 2.3 |

Interpretation

*Test condition*  Satisfactory clinical and biochemical evidence of hypoglycaemia.

*Cort.*  Normal response.

*GH*  Inadequate response to hypoglycaemia, indicating poor pituitary reserve of GH.

*TSH*  Normal response to TRH stimulation.

*PRL*  Grossly elevated basal level.

*Gonadotropins*  Partial response by LH, but a low basal FSH failing to respond.

These results are consistent with a prolactin-secreting tumour that has diminished the capacity of the anterior pituitary to synthesize and secrete GH and gonadotropins.

# HYPOGONADOTROPIC HYPOGONADISM

A 25 year old man thought by his local medical practitioner to have Klinefelter's syndrome. Past history revealed undescended testes at age 12, followed by poor sexual development. Examination revealed obesity, eunuchoid stature, poor body hair, small firm testes and a normal sense of smell. A basal plasma testosterone value was 1.7 nmol/l (8–30), but basal gonadotropins were low, thereby excluding a diagnosis of Klinefelter's syndrome. A triple function test was performed.

| Time | Glu. mmol/l (3.0–5.0) | Cort. nmol/l (140–690) | GH mU/l | TSH mU/l (<5.5) | PRL µg/l (<30) | LH IU/l (0.7–2.0) | FSH IU/l (2.0–5.0) |
|---|---|---|---|---|---|---|---|
| 09.00 | 5.2 | 650 | 2.2 | 5.0 | 7 | 0.9 | 1.3 |
| Insulin: 7 units; TRH: 200 µg; GnRH: 100 µg ||||||||
| 09.30 | 2.3 | 580 | 2.2 | 19.0 | 27 | 2.9 | 2.7 |
| 10.00 | 1.9 | 500 | 2.5 | 15.0 | 24 | 3.3 | 3.2 |
| 10.30 | 3.1 | 610 | 5.4 | 11.0 | 26 | 2.7 | 3.3 |
| 11.00 | 4.9 | 520 | 2.1 | 6.2 | 12 | 2.1 | 3.3 |

Interpretation

*Test condition*  Adequate hypoglycaemic stress obtained.

*Cort.*  Adequate basal level, with subsequent samples failing to rise further, suggesting the patient was already stressed at the start of the test.

*GH*  Partial, but poor response. The rate of fall of glucose may have been too slow to give a good stimulation. Requires a re-test.

*TSH*  Response indicates a good pituitary reserve of TSH.

*PRL*  Adequate response to TRH.

*Gonadotropins*  Basal LH and FSH were inappropriately low for the testosterone value and the provisional diagnosis of Klinefelter's syndrome. Both hormones showed a poor response to exogenous GnRH.

X-ray studies showed no evidence of a hypothalamic lesion. The clinical and biochemical findings were consistent with an isolated deficiency of hypothalamic GnRH. A chronic administration of GnRH would probably demonstrate a normal pituitary reserve of gonadotropins.

*Example* A 13 year old girl, thought by her parents to be very small for her age. The local medical practitioner took a blood sample for growth hormone analysis.

|  | mU/l | Reference range |
|---|---|---|
| Plasma GH | <1 | not quoted |

Some 30–50% of normal subjects have plasma GH levels undetectable by presently used assays, and therefore the above result could represent hormone deficiency, or a normal pituitary reserve. The child subsequently had several blood samples taken as she slept during a one-night hospital admission.

| Time | GH (mU/l) |
|---|---|
| 19.30 | <1 |
| 22.00 | 1.9 |
| 24.00 | 46.0 |
| 08.00 | 2.0 |

A normal response was seen in this case (>20 mU/l). The increase in GH can persist for up to 2 hours, but the above test may have been a little fortunate in catching the rise with so few samples taken around the critical period. A failure to detect a good peak (<20 mU/l) is suggestive of poor pituitary reserve, but because of the possibility of missing the peak a different stimulation test should be performed as a check.

*Example* An 11 year old boy was investigated for short stature. An exercise test was performed to assess the pituitary reserve of GH.

| Time (min) | GH (mU/l) |
|---|---|
| 0 | 2.5 |
| 18 min pedalling against a load | |
| 2 | 9.5 |
| 20 | 15 |

The peak plasma GH value reached demonstrates at least a partial GH reserve. Careful clinical assessment did not point to GH deficiency, and therefore an arginine stimulation test was performed at a later date.

| Time (min) | GH (mU/l) |
|---|---|
| 0 | <1 |
|  | Arginine (0.5 g/kg body weight) as 10% solution IV over 30 min |
| 45 | 16 |
| 60 | 36 |

A normal GH response (>20 mU/l) was obtained. In the exercise test the exercise may have been inadequate, or the sample times could have missed the peak response of GH.

*Example* A 15 year old boy investigated for short stature. A bicycle ergometer was used for an exercise test.

| Time (min) | GH (mU/l) |
|---|---|
| 0 | 1.0 |
| 20 min pedalling against appropriate load | |
| 2 | 4.5 |
| 20 | 9.2 |

This partial response was investigated further by performing an insulin hypoglycaemia test.

| Time (min) | Glu. (mmol/l) | GH (mU/l) |
|---|---|---|
| 0 | 4.8 | <1 |
|  | Insulin: 5 units | |
| 30 | 2.0 | 7 |
| 60 | 3.6 | 11 |
| 90 | 4.7 | 7 |
| 120 | 4.6 | 2 |

An adequate clinical and biochemical hypoglycaemia was obtained, and therefore a partial but inadequate pituitary reserve of GH was demonstrated.

## HYPERPITUITARISM

Hyperpituitarism is the term generally used to describe an inappropriate overproduction of a hormone by the anterior pituitary.

The hormones most commonly secreted in excess are: (1) GH—acromegaly, gigantism. (2) PRL—prolactin-secreting adenoma. (3) ACTH—Cushing's disease (see Chapter 21).

## ACROMEGALY

A 54 year old farmer consulted his new general practitioner for a prescription renewal. The doctor noted the man had classic acromegalic facies, and arranged investigations. Acromegaly was overt on clinical examination, but there was no history of headaches or visual disturbance. A glucose suppression test was performed.

| Time | Glu. (mmol/l) (3.0–5.5) | GH (mU/l) |
|---|---|---|
| 08.40 | 10.2 | 36 |
| | 100 g glucose (recommended dose now 75 g) | |
| 09.10 | 9.9 | 28 |
| 09.40 | 9.5 | 32 |
| 10.10 | 9.8 | 30 |
| 10.40 | 9.1 | 30 |
| 11.10 | 7.8 | 38 |
| 11.40 | 5.9 | 56 |

The basal GH level is markedly elevated, and fails to be suppressed by a large glucose load. The test was extended, and demonstrates the rise in GH level that is often seen 2–3 hours post-load, probably in response to the insulin-mediated clearance of the glucose. The basal plasma glucose is raised, possibly reflecting the anti-insulin effect of excess GH. The high basal GH and its non-suppressibility supports the clinical diagnosis of active acromegaly.


## ACROMEGALY

A 32 year old obese woman complained of headaches and slight visual disturbance. Her medical practitioner took a single blood sample for GH estimation and other tests.

Plasma GH    11 mU/l

This result could be interpreted in two ways:
(1) If the GH value is typical of the circulating level at any time, then it is suggestive of acromegaly.
(2) If the patient felt anxious during the medical consultation and venepuncture then this is sufficient to markedly elevate the plasma GH level.

The random GH estimation was therefore valueless as a diagnostic investigation. At a later date a glucose suppression test was performed.

| Time | Glu. (mmol/l) (3.0-5.5) | GH (mU/l) |
|---|---|---|
| 09.30 | 4.9 | 18 |
| | 50 g glucose (75 g now recommended) | |
| 10.00 | 4.8 | 5 |
| 10.30 | 3.5 | 2 |
| 11.00 | 4.7 | <1 |
| 11.30 | 3.8 | <1 |

Note the high basal level again. At least one sample was suppressed to <1 mU/l, indicating a normal GH response to a glucose load.

HYPERPROLACTINAEMIA

A 25 year old woman ceased taking the oral contraceptive pill 18 months previously, and had been amenorrhoeic since. Prolactin was found to be elevated in a random blood sample, whilst gonadotropins were normal. Radiological studies and a triple function test was performed to assess the pituitary.

| Time | Glu. mmol/l (3.0-5.5) | Cort. nmol/l (140-690) | GH mU/l | TSH mU/l (<5.5) | PRL µg/l (<30) | LH IU/l (1.6-9.0) | FSH IU/l (1.1-6.4) |
|---|---|---|---|---|---|---|---|
| 08.35 | 5.2 | 420 | 27.0 | 4.9 | 125 | 3.2 | 3.4 |
| | Insulin: 7 units; TRH: 200 µg; GnRH: 100 µg | | | | | | |
| 09.05 | 0.7 | 405 | 61.0 | 42.0 | 175 | 11.5 | 6.2 |
| 09.35 | 3.1 | 635 | 55.0 | 24.0 | 150 | 8.6 | 5.2 |
| 10.05 | 4.1 | 640 | 41.0 | 18.0 | 125 | 6.4 | 4.8 |
| 10.35 | 5.0 | 670 | 32.0 | 12.0 | 125 | 4.9 | 4.6 |

Interpretation

*Test condition* Adequate hypoglycaemia by both clinical and biochemical criteria.

*Cort.* Normal response, i.e. basal + >200 nmol/l and one sample >550 nmol/l.

*GH* Note high basal value, presumably due to some emotional stress regarding the test. A normal response to the hypoglycaemia.

*TSH* A normal response. Note the higher values often achieved in the younger age group (cf. previous cases).

*PRL*  Such a high basal value is unlikely to be caused by stress. A high basal level with a poor response to TRH stimulation is typically seen in hyperprolactinaemia.

*Gonadotropins*  In the basal sample both hormones are within the follicular phase reference range, and following stimulation they respond normally to GnRH stimulation.

The triple function test suggests hyperprolactinaemia, with all other hormone systems intact. The final diagnosis was that of a prolactin-secreting microadenoma. High levels of prolactin appear to abolish the pulsatile release of LH and FSH by interfering with the release of GnRH, thereby causing loss of the ovarian cycle, with consequent amenorrhoea and infertility. Normoprolactinaemia with consequent restoration of the menstrual cycle is often rapidly restored following treatment with the dopamine agonist bromocyptine.

# 20 Thyroid hormones

The thyroid hormones are synthesized and secreted by the thyroid gland under the hormonal control of the anterior pituitary and hypothalamus.

**Hypothalamic-pituitary-thyroid axis**

Thyrotropin-releasing hormone (TRH) is a tripeptide that is synthesized by neurosecretory cells in the hypothalamus and released into the hypothalamic-hypophyseal portal system. TRH is carried to the anterior pituitary where it binds to thyrotroph cell receptors and stimulates, via adenyl cyclase, the synthesis and secretion of thyroid-stimulating hormone (TSH, thyrotropin). TRH release is inhibited to a minor extent by both TSH and thyroid hormone.

TSH is a glycoprotein hormone (mol. wt. ~28 000 Daltons) that comprises two polypeptide chains; the α chain is similar to that found in LH, FSH and hCG, whilst the β chain embodies the TSH hormone activity. After secretion TSH binds to specific receptors on the follicular cells of the thyroid gland and stimulates, via adenyl cyclase, the synthesis and release of the thyroid hormones. TSH secretion is subject to feedback inhibition by thyroid hormone. Evidence suggests that this negative feedback is governed by plasma non-bound (free) thyroxine and triiodothyronine ($fT_4$, $fT_3$) levels, the $T_4$ first being converted to triiodothyronine ($T_3$) within the pituitary cells. The intrapituitary $T_3$ binds to nuclear receptors and stimulates the synthesis of a protein that reduces the responsiveness of the TSH-secreting cells to TRH (Fig. 20.1).

**Thyroid hormone production**

The main structural feature of the highly vascular thyroid gland is the follicles, each comprising epithelial cells enclosing a colloid filled space. The function of the follicles is the synthesis, storage and secretion of thyroid hormones, the major steps of which are described below, and shown diagrammatically in Fig. 20.2.

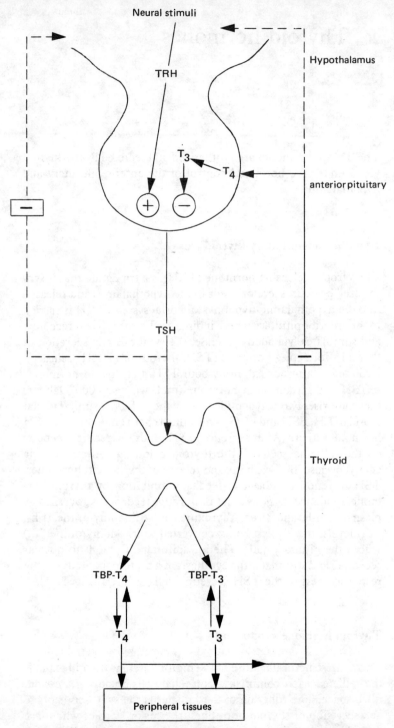

**Fig. 20.1** The normal control mechanisms of thyroid hormone secretion. ⊖, negative feedback; ⊕, positive feedback; ——, major feedback loop; ---, minor feedback loop.

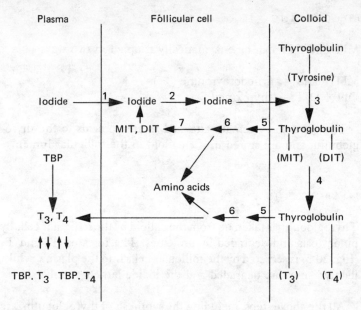

**Fig. 20.2** Biosynthesis of thyroid hormones. 1, trapping of dietary iodide; 2, oxidation by a peroxidase enzyme; 3, organification close to cell-colloid interface; 4, intra- and intermolecular coupling; 5, pinocytosis; 6, proteolysis; 7, salvage of iodine-containing intermediates; ( ), variable number of residues attached to thyroglobulin; TBP, thyroid binding proteins.

(1) TRAPPING IODIDE

The thyroid contains 90–95% of body iodide, and can trap and concentrate it within the follicular cells against a high concentration gradient. This energy-dependent uptake process is competitively inhibited by both thiocyanate and perchlorate. Various other tissues can also trap iodide, particularly the salivary and mammary glands.

(2) ORGANIFICATION

The trapped iodide is enzymatically oxidized to inorganic iodine, which is then able to iodinate some of the tyrosyl amino acid residues of thyroglobulin. Thyroglobulin is a very large protein (mol. wt. ~ 660 000 Daltons) that is stored in the colloid space. One or two iodine atoms can be attached to the tyrosyl groups, forming monoiodotyrosine (MIT) and diiodotyrosine (DIT) respectively. Iodination is blocked by carbimazole, thioureas and excess iodide.

## (3) COUPLING

MIT and DIT can be enzymatically coupled in two ways:

MIT + DIT → triiodothyronine ($T_3$)
DIT + DIT → thyroxine ($T_4$)

The $T_4$ and $T_3$ remain attached as amino acids to the thyroglobulin, and are stored in the colloid in the follicular lumen.

## (4) SECRETION

Thyroglobulin is taken up from the colloid by the follicular cells by pinocytosis and degraded by proteases. The released $T_4$ and $T_3$ (Fig. 20.3) is secreted by the follicular cells into the plasma, whilst iodotyrosines are degraded and the iodide largely reutilized.

All the above steps, including the synthesis of thyroglobulin, are stimulated by TSH.

### Plasma thyroid hormones

Nearly all of the plasma thyroid hormones ($\sim$99.95% $T_4$, $\sim$99.5% $T_3$) are bound to plasma proteins, thereby providing a readily available pool of hormone, and at the same time protecting the body cells from the full physiological effect of the total hormone present in plasma. The proteins involved, sometimes referred to as thyroid binding proteins (TBP) are:

|  | % bound | |
| --- | --- | --- |
|  | $T_4$ | $T_3$ |
| Thyroxine-binding globulin (TBG) | 70–75 | 70–75 |
| Thyroxine-binding prealbumin (TBPA) | 15–20 | Trace |
| Albumin | 5–10 | 25–30 |

The binding of both hormones may be shown as:

$T_4$ + TBP $\rightleftharpoons$ TBP.$T_4$,    $T_3$ + TBP $\rightleftharpoons$ TBP.$T_3$
<0.05%         >99.95%    <0.5%          >99.5%

The unbound or 'free' hormones ($fT_4$, $fT_3$) are widely considered to be the fractions that have the potential to enter cells, determine the thyroid status of the body and provide the feedback inhibition at the pituitary, whilst the protein-bound hormones act as a metabolically inactive reservoir. (In the following sections: $T_4$ = (bound

$T_4$ + free $T_4$); $T_3$ = (bound $T_3$ + free $T_3$); unbound (free) *Thyroid hormones* $T_4 = fT_4$; unbound (free) $T_3 = fT_3$.)

The approximate plasma concentrations of the total thyroid hormones, and their half-lives in plasma are:

|  |  | Half-life |
| --- | --- | --- |
| $T_4$ | 60–160 nmol/l | ~9 days |
| $T_3$ | 1.2–2.8 nmol/l | ~1.5 days |

$T_4$ is the principal hormone secreted by the thyroid gland. Approximately 25–30% of plasma $T_4$ is β deiodinated by peripheral tissues (especially the liver and kidney) to form $T_3$; the majority (~80%) of plasma $T_3$ is derived from plasma $T_4$. $T_3$ is metabolically more potent than $T_4$ and is believed to exert the majority of thyroid hormone activity.

Plasma $T_4$ (30–35%) is also α deiodinated to form a biologically inactive product, *reverse $T_3$* ($rT_3$) (Fig. 20.3).

Many forms of illness, metabolic stress or trauma alter the peripheral metabolism of thyroid hormones. Evidence suggests that these alterations reflect a reduced activity of the enzyme that

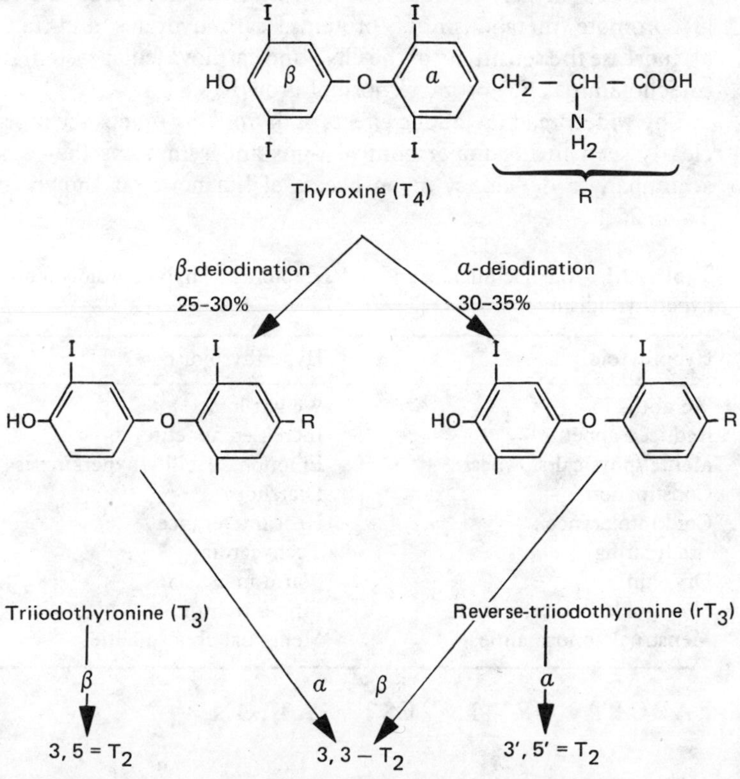

**Fig. 20.3** Peripheral metabolism of thyroxine.

deiodinates the β (outer) thyronine ring, thereby lowering both $T_3$ production and $rT_3$ catabolism (Fig. 20.3). These changes principally have the effect of lowering plasma $T_3$ and raising $rT_3$ levels, whilst $T_4$ often remains little affected unless the illness is severe. Despite the reduction in circulating $T_3$ patients appear both clinically and biochemically (normal TSH) euthyroid, suggesting that free hormone levels are normal. This 'sick euthyroid' or 'low $T_3$ syndrome' can be seen with almost any non-trivial illness or trauma, e.g. acute or chronic illness such as myocardial infarct, uncontrolled diabetes mellitus, renal failure, respiratory disease, febrile illness, trauma such as surgery or isolated anaesthesia, starvation, carbohydrate deprivation.

The physiological advantages of lowering $T_3$ production during illness are not clear, but possibly there is a need to reduce the potential metabolic effect of the circulating $T_3$ at such times. With a severe non-thyroidal illness the plasma $T_4$ may also be markedly reduced, and is generally a poor prognostic sign. Some evidence suggests that in this situation the low $T_4$ is due to the presence of an inhibitor that leaks from tissues and interferes with the binding of $T_4$ to its plasma transport proteins.

The thyroid hormones influence the metabolic rate of cells to: (1) increase oxygen consumption, (2) increase heat production, (3) promote metabolism of proteins, carbohydrates and fats, (4) increase the sensitivity of the CNS and cardiovascular system to catecholamines, (5) promote normal growth.

The widespread metabolic effects of thyroid hormones are most clearly seen in the major clinical signs and symptoms that can accompany a deficiency or an excess of hormone, as shown in Table 20.1.

**Table 20.1** Major clinical signs and symptoms of hypothyroidism and hyperthyroidism.

| Hypothyroidism | Hyperthyroidism |
|---|---|
| Weight gain | Weight loss |
| Reduced appetite | Increased appetite |
| Mental/physical slowness | Emotional lability/hyperkinesis |
| Constipation | Diarrhoea |
| Cold intolerance | Heat intolerance |
| Bradycardia | Tachycardia |
| Dry skin | Warm moist skin |
| Slow ankle jerk relaxation | Finger tremor |
| Menstrual abnormalities | Menstrual abnormalities |

## LABORATORY INVESTIGATION

### PLASMA TOTAL THYROXINE ($T_4$)

A wide variety of RIA techniques are available for the direct assay of total $T_4$ in serum or plasma. Other techniques, particularly those employing natural binding protein (competitive protein binding) and antibodies attached to enzymes or fluorescent tags, are also widely used. Most methods are direct and do not require prior extraction or concentration of the hormone. All these methods have overcome the interference and contamination problems experienced with chemical techniques such as the protein-bound iodine determination.

*Thyroid hormones*

Plasma total $T_4$ methods measure both protein-bound and free $T_4$. Since approximately 99.95% of $T_4$ is protein-bound a $T_4$ estimation will be affected by any alteration in the concentration of thyroid-binding proteins, principally TBG. If, for example, the plasma concentration of TBG is gradually raised, the free $T_4$ ($fT_4$) will be taken up and bound until the TBG is again carrying its normal proportion of $T_4$; the reduced $fT_4$ causes increased output of TSH (and therefore $T_4$) until the $fT_4$ level is restored to normal, and the euthyroid state is maintained (Fig. 20.4).

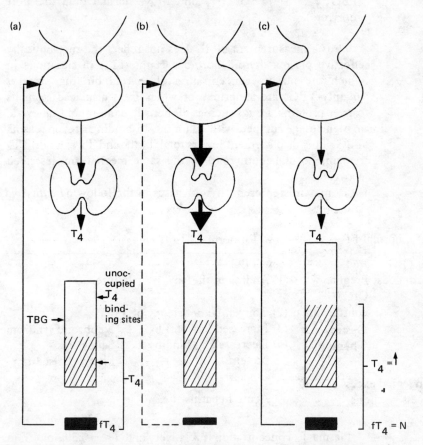

**Fig. 20.4** The effect of increased TBG on plasma thyroid hormones. (a) Normal hypothalamic-pituitary-thyroid axis. (b) Increasing plasma TBG binds more $T_4$; the reduced $fT_4$ removes negative feedback to allow increased $T_4$ secretion. (c) A new equilibrium between TBG and $T_4$ is reached ($T_4$ raised) and $fT_4$ is restored to a euthyroid level. Hatching denotes bound $T_4$.

The equilibrium of $fT_4$ and TBG in plasma may be shown as:

$$fT_4 + TBG \rightleftharpoons TBG.T_4$$

therefore:

$$[fT_4] \propto \frac{[TBG.T_4]}{[TBG]}$$

so, if

$[TBG] \uparrow$ then $[fT_4] = N$ if $[TBG.T_4] \uparrow$

In the new state of equilibrium the total $T_4$ is elevated $[TBG.T_4]$ $\uparrow$ therefore $[T_4] \uparrow$, but $[fT_4] =$ normal (and therefore euthryoid). If TBG is reduced the reverse occurs, i.e.

$[TBG] \downarrow$, $\therefore [TBG.T_4] \downarrow$ but $[fT_4] =$ normal (and therefore euthyroid)

Thus a measurement of $T_4$ does not reflect the metabolically effective plasma thyroid hormone status ($fT_4$) in situations in which the plasma concentration of thyroid binding proteins (mainly TBG) are abnormal, or when $T_4$ is displaced from its carrier proteins. Drugs such as diphenylhydantoin and salicylate in high dosage compete with $T_4$ for the $T_4$-binding sites on plasma TBG. This displacement lowers total $T_4$ (bound $T_4 \downarrow$) whilst $fT_4$ remains normal (euthyroid) by the same mechanism described above.

$T_4$ may misrepresent thyroid status in the following *euthyroid* situations:

| High $T_4$, normal $fT_4$ | Low $T_4$, normal $fT_4$ | |
|---|---|---|
| Raised TBG | Low TBG | |
| (1) Oestrogens Pregnancy Oral contraceptive Oestrogen Therapy | (1) Androgen therapy, liver failure | |
| | (2) Congenital deficiency of TBG | |
| | (3) Increased loss of TBG | Nephrotic syndrome |
| | (4) Decreased $T_4$ binding to protein | Salicylates Phenytoin sodium |
| (2) Congenital excess of TBG | | |
| (3) Excess synthesis | Acute hepatitis | |

Plasma $T_4$ concentration is relatively high ($< \sim 230$ nmol/l) in the full-term newborn infant ($\uparrow$ TBG), and falls over the following 2–3 years towards adult levels. There is no significant sex difference for plasma $T_4$ levels, nor any diurnal or seasonal variation. A random blood sample may therefore be taken for the estimation of $T_4$.

A plasma $T_4$ provides a good assessment of thyroid function provided (1) there are no binding protein abnormalities, (2) there is no drug interference with $T_4$ binding to TBG, (3) there is no endogenous or exogenous excess of $T_3$.

## FREE THYROXINE (fT$_4$) AND FREE TRIIODOTHYRONINE (fT$_3$)

It is presently believed that the non protein-bound plasma hormones ($fT_4$, $fT_3$) are the fractions that have the potential to enter cells, and that they therefore determine the thyroid status of the body. (The alternative, whereby the TBG.$T_4$ interacts with specific cell receptors in order to release $T_4$ or $T_3$ into the cell remains a possibility.) For this reason much effort has been devoted to developing assays to measure the extremely low levels of the free hormones that are present in plasma. Generally, these methods have been based on equilibrium and other dialysis techniques unsuited to routine diagnostic purposes. However, methods based on direct RIA have recently become commercially available, but until more experience is gained it is difficult to assess the additional diagnostic value of free hormone estimations. However, clearly one advantage is that a free hormone estimation is unaffected by alterations in the concentration of TBG, and would therefore obviate the problems that can render total $T_4$ values misleading.

## $T_3$ UPTAKE TESTS

The effect that alterations in plasma TBG concentration can have on total $T_4$ and $T_3$ estimations led to the need for measuring TBG itself. Until very recently TBG assays were unavailable, and so indirect techniques of assessing TBG levels have been employed. A wide variety of techniques are used, but the principle is similar in all. The method used in the authors' laboratory is a $T_3$-uptake test, which is briefly outlined below.

(1) TBG is normally about one third saturated with $T_4$ (bound $T_3$ is relatively insignificant).
(2) In hypothyroidism and in states of increased plasma TBG there is an increased number of unoccupied $T_4$-binding sites per unit volume of plasma.
(3) In hyperthyroidism or states of reduced TBG there are fewer unoccupied $T_4$-binding sites per unit volume of plasma.
(4) Because $T_3$ has a lower affinity than $T_4$ for the binding sites, $T_3$ can bind to these unoccupied $T_4$-binding sites without displacing already bound $T_4$.

(5) Therefore, addition of an excess of radioactive ($^{125}$I) $T_3$ to an aliquot of plasma will ensure that $^{125}$I-$T_3$ will occupy all empty $T_4$-binding sites.

(6) A resin, or similar substance, is added that adsorbs unbound $^{125}$I-$T_3$.

(7) An aliquot of a pool of normal plasma (reference pool) is subjected to the same procedure.

(8) An aliquot of test sample and of reference pool is taken and the radioactivity measured in each.

(9) The amount of radioactivity bound by the reference pool of serum is expressed as a ratio with that bound by the test sample (see Note below). (A $T_3$ resin uptake method ($T_3$RU) uses the ratio of the radioactivity bound by the resin in the test sample to that bound by the reference pool resin.)

$$T_3 \text{ uptake } (T_3U) = \frac{\text{Reference pool}}{\text{Test sample}} \quad \text{(reference range: 0.85--1.05)}$$

If unoccupied sites are increased (TBG ↑ or $T_4$ ↓), then the test sample binds more radioactivity, i.e.

$$T_3U = \frac{\text{Reference pool}}{\uparrow} = \downarrow$$

If TBG is reduced, or $T_4$ ↑ (i.e. fewer unoccupied binding sites), then the $T_3U$ increases.

These types of test can therefore be used as an aid to interpreting a $T_4$ estimation. An elevated $T_4$ associated with a reduced $T_3U$ (above method), suggests that unoccupied binding sites are increased (TBG ↑), and therefore the total $T_4$ is raised because of an increase in thyroid-binding proteins. A high $T_4$ and a high $T_3U$ suggests fewer unoccupied binding sites, i.e. the thyroid binding proteins are more saturated with $T_4$ than is normal (hyperthyroid state). The reverse situations apply for low $T_4$.

*Note* To directly reflect the unoccupied $T_4$-binding sites the uptake ratio should be calculated as:

$$\frac{\text{Test sample}}{\text{Reference pool}}$$

so that a high value is obtained when there are many unoccupied binding sites (↑ TBG or ↓ $T_4$), and conversely for low values. However, many laboratories prefer a method to yield low values for hypothyroidism (↑ sites) and high values for hyperthyroidism and so, as in the above method, the reciprocal calculation is made. Note that if $T_3U$ values are used for further calculation (e.g. FTI) either ratio may be used providing it is appropriately applied (see FTI, p. 436).

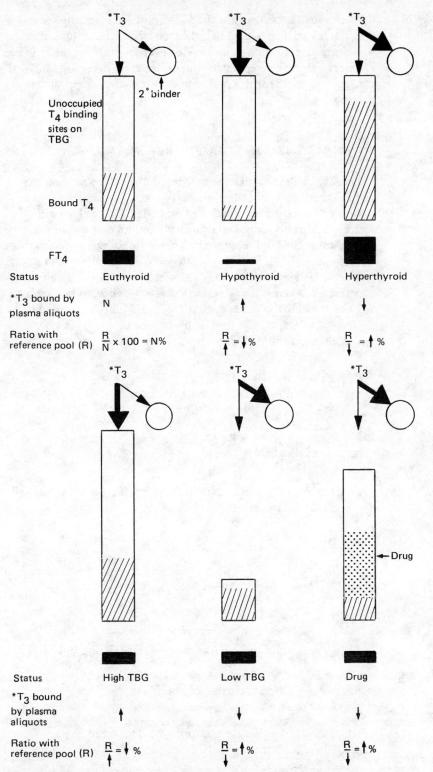

**Fig. 20.5** Diagramatic representation of the mechanism of the $T_3$-uptake test in various thyrometabolic and binding abnormalities. *$T_3$, radioactive ($^{125}$I) $T_3$; 2° binder, Sephadex; Drug, a drug that competes with $T_4$ for binding sites on TBG (e.g. diphenylhydantoin, salicylates).

Although these tests are often insensitive to markedly decreased or increased levels of TBG they do give a clear indication of binding protein abnormalities. However, laboratories often do not report $T_3U$ results separately, but utilize them in conjunction with a $T_4$ measurement to provide an indirect assessment of free $T_4$ levels.

### INDIRECT ESTIMATION OF FREE HORMONE

Because of the technical problems of directly measuring $fT_4$ or TBG, various methods of indirectly assessing the level of circulating $fT_4$ have been developed and widely used during the last decade. Various commercial kits have employed methods that take account of the binding proteins when estimating $T_4$ (i.e. Effective thyroxine ratio (ETR), Mallinckrodt; Normalized $T_4$, Amersham; Normalized serum $T_4$ (T-4N), Abbott Laboratories). The *free thyroxine index* (FTI) is a parameter that is widely used by diagnostic laboratories to indirectly assess $fT_4$. An FTI method is used in the case examples, and its basis is therefore outlined below.

The equilibrium of $T_4$ with its plasma binding proteins may be shown as:

$$fT_4 + TBP \rightleftharpoons TBP \cdot T_4$$
$$\text{unoccupied binding sites} \qquad \text{bound hormone (99.95\%)}$$

$$[fT_4] \propto \frac{[TBP \cdot T_4]}{[TBP]}$$

A $T_4$ assay estimates both bound and free $T_4$, but since bound $T_4$ represents approximately 99.95% of total, it is valid to assume

$$[T_4] = [TBP \cdot T_4]$$

and since TBP can be indirectly estimated by assessing unoccupied binding sites using a $T_3U$ method, then:

$$[TBP] \propto T_3U$$

By substituting these terms the original relationship becomes:

$$fT_4 \propto \frac{[T_4]}{T_3U} = \text{free thyroxine index (FTI)}$$

(when $T_3U$ is calculated as test sample/reference pool. When $T_3U$ is calculated as reference pool/test sample then FTI $= T_4 \times T_3U$).

Therefore two estimations on plasma, $T_4$ and $T_3U$, are required to derive the FTI. The FTI, as with the other methods mentioned, provides a correction to $T_4$ for binding abnormalities and correlates well with clinical findings and diagnosis, but the general inability of $T_3U$ methods to adequately assess very abnormal levels of TBG is also reflected by a poor FTI performance in the same situations. In most cases however, the FTI provides a very good biochemical correlation with clinical thyroid status.

Example of an FTI calculation

Pregnant subject:

$T_4$ 195 nmol/l (60–160)

This suggests hyperthyroxinaemia.

$T_3U = 0.71$ (0.85–1.05)

This suggests increased $T_4$ binding sites, and therefore either TBG ↑ or $T_4$ ↓ (clearly the former fits with the measured $T_4$).

FTI = 195 × 0.71 = 138 units (50–150)

This correctly suggests that the $fT_4$ is normal and that the subject is euthyroid.

## THYROXINE-BINDING GLOBULIN (TBG)

TBG is a glycoprotein (mol. wt. ~60 000 Daltons) that has a plasma concentration of approximately 10–30 mg/l. Only recently have commercial kits for the assay of plasma TBG become available. Most laboratories continue to use the often cheaper $T_3U$ methods which provide an adequate assessment of binding abnormalities, and so it is rather early to assess whether the direct measurement of TBG will find a place in the routine assessment of thyroid function. However, some laboratories are using a $T_4$/TBG ratio as an indirect indication of free hormone status.

## TRIIODOTHYRONINE ($T_3$)

$T_3$ is readily estimated in unextracted serum or plasma by a wide range of techniques that are similar to those used for $T_4$. The $T_4$:$T_3$ molar ratio in plasma is approximately 50:1, so therefore a $T_3$-antiserum used for an assay system must have negligible cross-reactivity with $T_4$. All $T_3$ methods measure both protein-bound

(TBG, albumin) and free $T_3$ ($fT_3$), and therefore $T_3$ estimations are subject to the same effects of binding abnormalities as was discussed earlier for $T_4$. However, laboratories rarely use a FTI type of approach to formally correct a plasma $T_3$ value for binding protein changes.

Plasma $T_3$ is low at birth ($\sim$ half adult concentration), but rises markedly within 24 hours to often double the level found in adults, and then returns to normal in a day or so. As there is no significant difference in plasma $T_3$ levels between males and females, nor any significant diurnal or seasonal variation, a random blood sample is satisfactory for $T_3$ estimation.

Plasma $T_3$ concentrations generally change in association with, and in the same direction as, changes in plasma $T_4$ levels. Thus $T_3$ is usually elevated in hyperthyroidism and reduced in hypothyroidism. In the former situation $T_3$ rises relatively more than $T_4$, and, in the latter situation, it falls relatively less than $T_4$. Generally:

$$\text{Hyperthyroidism: } \frac{T_4}{T_3} = \frac{\uparrow \text{ or N}}{\uparrow \uparrow} \qquad \text{Hypothyroidism: } \frac{T_4}{T_3} = \frac{\downarrow \downarrow}{\downarrow \text{ or N}}$$

Thus the $T_4:T_3$ ratio falls in both situations, and explains why $T_3$ compared to $T_4$ is a more sensitive indicator of hyperthyroidism, and a less sensitive one for hypothyroidism; in fact in the latter situation $T_3$ often remains within the reference range until the primary hypothyroidism is severe.

As discussed earlier plasma $T_3$ is largely derived from the peripheral deiodination of $T_4$, the alternative product being the metabolically inactive reverse $T_3$ ($rT_3$). In many states of acute or chronic illness, or metabolic shock (surgery, myocardial infarct, etc.), the conversion of $T_4$ to $T_3$ is reduced (sick euthyroid). It is therefore advisable to defer thyroid function tests, particularly $T_3$ measurement, if clinical considerations can allow the investigation of thyroid status to await an improvement in the general health of the patient.

### REVERSE $T_3$ ($rT_3$)

Although RIA assays are now available, the measurement of $rT_3$ has not yet found a useful place in the routine investigation of thyroid disease.

### THYROID STIMULATING HORMONE (TSH, THYROTROPIN)

TSH, a glycoprotein of mol. wt. 28 000 Daltons, comprises two subunits, an α subunit that is common to LH, FSH and hCG, and a β subunit that expresses TSH hormone bioactivity. TSH is estimated by RIA techniques. Antisera against TSH are generally preabsorbed with hCG to ensure low cross-reactivity in samples with high levels of hCG (pregnancy) or gonadotropins (postmenopausal). Most routinely available methods are not sensitive enough to detect low plasma levels of the hormone, and so few laboratories are able to provide a lower reference limit. TSH estimations are therefore valueless for clinical situations where plasma TSH may be expected to be low (hypothalamic or pituitary disease), or suppressed (hyperthyroidism).

TSH levels rise markedly (3–4 fold) in the infant immediately postpartum, a response it is thought to extrauterine cooling, and return to normal by the second or third day. Plasma TSH levels are similar in both sexes and are not significantly affected by stress, time of day or season.

The TRH stimulated release of TSH from the pituitary thyrotrophs is inhibited by the feedback effect of the thyroid hormones. Reduction or loss of circulating hormones results in removal of this inhibition and an increase in TSH secretion. TSH assays are therefore a sensitive test for primary hypothyroidism. The inability of present TSH assays to detect the hormone in a proportion of normal subjects means that a dynamic function test is required to investigate the pituitary reserve and feedback mechanism. The major diagnostic value of single sample TSH assays is in the confirmation of primary hypothyroidism, and in the monitoring of thyroid hormone replacement therapy.

THYROTROPIN-RELEASING HORMONE (TRH) TEST

In the normal subject the administration of synthetic TRH results in a marked release of TSH from the anterior pituitary. Plasma TSH rises to a peak (up to 5 fold basal) by 20–30 minutes, and then falls over the next 3–4 hours. The TSH response to a standard exogenous dose of TRH is greater in females than in males, is greater in the follicular than luteal phase, declines with age in males and is exaggerated in primary hypothyroidism. In cases of hypothalamic disease the TSH response to exogenous TRH can be sluggish, presumably reflecting the lack of recent endogenous TRH stimulation to the TSH-secreting cells.

The response of the thyrotrophs to TRH is inhibited by thyroid hormones, and therefore the presence of hyperthyroidism is reflected in a failure to secrete TSH in response to a standard exogenous dose of TRH.

The assessment of the TSH response to administered TRH can therefore be used to investigate both the pituitary reserve of TSH and the negative feedback mechanism (Fig. 20.6).

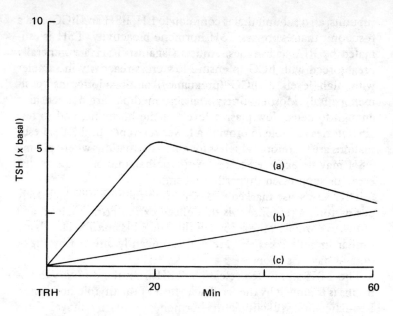

**Fig. 20.6** Typical plasma TSH responses to 200 μg IV TRH in (a) normal adult, (b) hypothalamic (tertiary) hypothyroidism, (c) pituitary (secondary) hypothyroidism or hyperthyroidism.

Protocol

No specific patient preparation. Note drug history, sex and age of patient.

*Stimulation*   200 μg TRH given IV. Transient side-effects may include nausea, a metallic taste, lightheadedness.

*Blood samples*   0, then 20 and 60 minutes post TRH for TSH assay.

*Interpretation*
(1) A three to five fold increase over the basal level indicates an adequate pituitary reserve of TSH and a normal level of thyroid hormone feedback.
(2) A lack of TSH response, in conjunction with a low basal TSH level and a low basal concentration of thyroid hormone, is compatible with anterior pituitary disease.
(3) A lack of TSH response in conjunction with a high or high normal level of plasma thyroid hormones ($T_4 + T_3$ or $T_3$ or $T_4$) is suggestive of hyperthyroidism. A flat response may also be seen in euthyroid Graves' disease (Graves' ophthalmopathy), and in clinically euthyroid patients receiving thyroid hormone replacement for primary hypothyroidism.
(4) A slow rise in TSH following TRH administration (60 min level > 20 min), in conjunction with low plasma basal levels of both thyroid hormone and TSH, is suggestive of hypothalamic disease.

(5) An exaggerated TSH response ($> \sim 40$ mU/l) is seen in primary hypothyroidism. This is of no diagnostic value.

The main diagnostic uses of the TRH test are: (1) confirmation of hyperthyroidism in cases where the static thyroid biochemistry tests are equivocal, (2) differentiation and diagnosis of hypothalamic and hypopituitary hypothyroidism.

### Selection of thyroid function tests

Clinically overt thyroid disease requires the appropriate laboratory tests only to confirm the diagnosis, and perhaps to provide a biochemical baseline prior to treatment. Many of the clinical signs and symptoms of thyrometabolic dysfunction are non-specific (e.g. tiredness or anxiety, weight loss or gain, dislike of hot or cold weather, constipation or diarrhoea), and therefore more subtle clinical presentations require laboratory evidence before a diagnosis can be confidently assigned.

For investigating patients suspected of having thyroid disease a variety of strategies for thyroid function testing have been developed by laboratories. Most schemes employ some form of initial screening test (e.g. $T_4$, FTI, ETR, etc.) followed, either sequentially or in combination, by further tests appropriate to the clinical findings and the result of the initial test. An example of a sequential approach, used by our laboratory, is shown below.

### Tests selected

*Plasma $T_4$*  Defines thyroid status in the majority of cases.

*Plasma FTI*  A test having very good correlation with $fT_4$ and clinical status. Because FTI requires two tests it is reserved for investigating situations where there is reasonable suspicion of TBG abnormalities, i.e. in all cases of (1) a high $T_4$ and/or $T_3$, (2) a low $T_4$ with an inappropriately low TSH. (In the authors' laboratory 10–15% of $T_4$ results require an FTI.)

*Plasma $T_3$*  Sensitive test for hyperthyroidism, but poor discriminator of hypothyroidism.

*Plasma TSH*  The most sensitive test for primary hypothyroidism.

*TRH test*  Valuable for diagnosing hyperthyroidism in cases with equivocal clinical and baseline biochemical findings. Also enables hypothalamic and hypopituitary hypothyroidism to be diagnosed.

Chapter 20

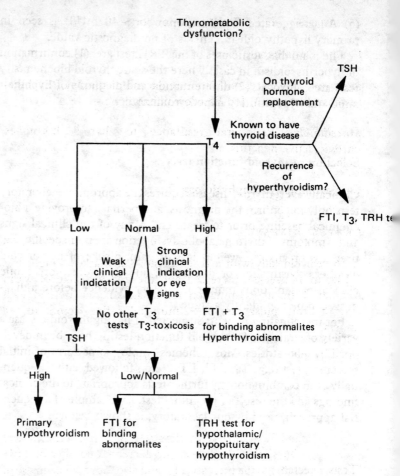

**Fig. 20.7** A biochemical strategy for investigating thyroid disease. Individual decisions are dependent on clinical history and findings.

The TSH standards used for estimating plasma TSH in the following case examples were calibrated against the Medical Research Council WHO standard.

## INCREASED PLASMA THYROID HORMONES

The major causes of a raised level of plasma thyroid hormones are:

*Thyroid disease* (1) Graves' disease, (2) toxic multinodular goitre, (3) toxic adenoma, (4) thyroiditis — acute phase, (5) $T_3$-toxicosis, (6) $T_4$-toxicosis, (7) apathetic thyrotoxicosis.

*Genetic* (1) Congenital excess of TBG, (2) familial dysalbuminaemic hyperthyroxinaemia.

*Physiological* (1) Pregnancy, (2) newborn.

*Drugs* Oestrogens.

*Iatrogenic* (1) Excess hormone replacement therapy, (2) thyrotoxicosis factitia, (3) Jod-Basedow phenomenon.

*Miscellaneous* Peripheral resistance to thyroid hormones, chronic active hepatitis.

## Case examples/pathophysiology

### GRAVES' DISEASE

A 26 year old lady with a recent history of marked weight loss (11 kg in 14 weeks), increased appetite and a dislike of warm weather. Clinical examination found an anxious woman with thyroid bruit, pulse rate of 92/min and a fine tremor of the fingers.

| | | | | |
|---|---|---|---|---|
| Plasma | $T_4$ | 255 | nmol/l | (60–160) |
| | $T_3U$ | 1.28 | | (0.85–1.05) |
| | FTI | 326 | units | (50–150) |
| | $T_3$ | 5.9 | nmol/l | (1.2–2.8) |

Both $T_4$ and $T_3$ are markedly elevated, and confirm the clinical diagnosis. The high $T_3U$ reflects the decreased number of TBG binding sites due to the increased $T_4$, and also the small decrease in TBG that occurs in hyperthyroidism. The raised FTI indicates that the $fT_4$ is also abnormally elevated. Further biochemical tests are unnecessary in such an unequivocal case. Although thyroid stimulating immunoglobulins (TSIg) are found in the plasma of most patients with Graves' disease, tests for their detection have not found a place in the routine investigation of the disease (Fig. 20.8).

### EXOPHTHALMOS

A 54 year old woman with unilateral exophthalmos. No clinical signs or symptoms of hyperthyroidism.

| | | | | |
|---|---|---|---|---|
| Plasma | $T_4$ | 124 | nmol/l | (60–160) |
| | $T_3U$ | 0.83 | | (0.85–1.05) |
| | FTI | 103 | units | (50–150) |
| | $T_3$ | 2.7 | nmol/l | (1.2–2.8) |

Exophthalmos in Graves' disease is due to an autoimmune process, and occurs in about 50% of cases of Graves' disease. As

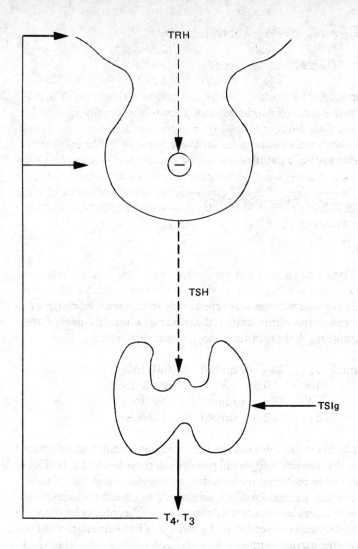

**Fig. 20.8** The hypothalamic-pituitary-thyroid axis in Graves' disease. TSIg, thyroid stimulating immunoglobulins.

Graves' disease may appear biochemically as a $T_3$-toxicosis it is important that any patient with exopthalmos has a full range ($T_4$, FTI, $T_3$) of thyroid function tests performed. The results in the above case suggest that the patient is euthyroid.

### $T_3$-TOXICOSIS

A 43 year old woman, recently diagnosed as having a hyperfunctioning thyroid nodule. Thyroid function tests were performed to provide baseline information prior to treatment.

| Plasma | $T_4$ | 143 | nmol/l | (60–160) |
|---|---|---|---|---|
| | $T_3U$ | 0.86 | | (0.85–1.05) |
| | FTI | 123 | units | (50–150) |
| | $T_3$ | 4.1 | nmol/l | (1.2–2.8) |

At the time of diagnosis only $T_3$ was abnormally raised. The excess $T_3$ is thought to arise from increased secretion by a thyroid nodule(s), rather than from an altered rate of peripheral conversion from $T_4$. $T_3$ toxicosis arises in a small percentage of cases of Graves' disease, toxic adenoma and toxic multinodular goitre. It has been suggested that if they were left untreated, some of these cases would progress to the classic biochemical pattern of hyperthyroidism ($\uparrow T_4, \uparrow T_3$).

$T_3$-TOXICOSIS

A 25 year old woman was referred by her general practitioner to the endocrine clinic with a diagnosis of clinically overt hyperthyroidism. A thyroid function test was performed.

| Plasma | $T_4$ | 143 | nmol/l | (60–160) |
|---|---|---|---|---|
| | $T_3U$ | 0.90 | | (0.85–1.05) |
| | FTI | 129 | units | (50–150) |
| | $T_3$ | 2.3 | nmol/l | (1.2–2.8) |

The above tests did not confirm the diagnosis. It was discovered that the patient had been prescribed propranolol, a β-blocker effective in reducing many of the autonomic symptoms of hyperthyroidism (increased CNS sensitivity to catecholamines). Propranolol can also lower circulating $T_3$ levels, probably by reducing the peripheral conversion of $T_4$ to $T_3$. A TRH stimulation test was therefore carried out.

| Time (min) | TSH mU/l |
|---|---|
| 0 | <2.5 |
| | 200 μg TRH IV |
| 20 | <2.5 |
| 60 | <2.5 |

The exogenous TRH has failed to overcome suppression of the thyrotroph cells caused by the recent excess of circulating hormone. The flat TSH response confirms the original diagnosis. It is presumed that this case represented a $T_3$-toxicosis. Although in this case the circulating $T_3$ was reduced to normal levels for several days by the propranolol, the pituitary often remains suppressed for some time after high hormone levels have been normalized.

## APATHETIC THYROTOXICOSIS

A listless, depressed and obese 55 year old woman was examined in the Accident and Emergency department. She was diagnosed as having congestive cardiac failure and probably myxoedema. A blood sample for thyroid function tests was taken.

| Plasma | $T_4$ | 223 | nmol/l | (60–160) |
|---|---|---|---|---|
| | $T_3U$ | 1.13 | | (0.85–1.05) |
| | FTI | 252 | units | (50–150) |
| | $T_3$ | 3.6 | nmol/l | (1.2–2.8) |
| | TSH | <2.5 | mU/l | (<5.5) |

Apathetic thyrotoxicosis is a rare presentation of hyperthyroidism in which most of the usual clinical features are missing. The $T_3U$ is raised, indicating that more thyroid-binding protein sites are occupied than normal; the high FTI value reflects a raised $fT_4$. Some laboratories advocate that since TSH is the sensitive indicator of primary hypothyroidism it should be the only test used for investigating that situation. However, until TSH assays can distinguish normal from hyperthyroid (↓ TSH) such cases as the above would be missed by such a strategy. The strategy would also misdiagnose hypothyroidism secondary to hypothalamic or anterior pituitary failure (↓ TSH).

## TOXIC ADENOMA

A 50 year old woman was found to have a diffuse enlarged thyroid, and a pulse rate of 98/min. There were no other clinical signs or symptoms of hyperthyroidism.

| Plasma | $T_4$ | 156 | nmol/l | (60–160) |
|---|---|---|---|---|
| | $T_3U$ | 1.12 | | (0.85–1.05) |
| | FTI | 175 | units | (50–150) |
| | $T_3$ | 2.6 | nmol/l | (1.2–2.8) |

The raised $T_3U$ and FTI are suggestive of excess circulating hormone. In view of the borderline nature of the biochemical results and the clinical findings, a TRH stimulation test was carried out.

| Time (min) | TSH mU/l |
|---|---|
| 0 | <2.5 |
| | 200 µg TRH IV |
| 20 | <2.5 |
| 60 | <2.5 |

A flat TSH response to TRH. The exogenous TRH cannot overcome the feedback inhibition exerted by the excess circulating thyroid hormone. This strongly suggests, but does not prove, that the patient is hyperthyroid. TSH may also fail to respond in euthyroid Graves' disease, in patients taking excess thyroid hormone replacement, and in secondary hypothyroidism. The TRH stimulation test is of distinct value in cases where the static biochemical tests are equivocal.

GRAVES' DISEASE

A 24 year old prima gravida (28/52) was found to have a palpable goitre, thyroid bruit and a pulse rate of 100/min.

| Plasma | $T_4$ | 328 | nmol/l | (60–160) |
|---|---|---|---|---|
|  | $T_3U$ | 0.98 |  | (0.85–1.05) |
|  | FTI | 321 | units | (50–150) |
|  | $T_3$ | 6.8 | nmol/l | (1.2–2.8) |

Although the increased oestrogens of pregnancy raise $T_4$ by increasing TBG synthesis, it is unusual for $T_4$ to exceed $\sim 250$ nmol/l. A raised TBG increases thyroid hormone binding sites, which would result in a low $T_3U$ ($\sim 0.60$–$0.85$). In the above case the value of 0.98 suggests that many of the additional hormone binding sites are occupied, i.e. the TBG is more saturated than

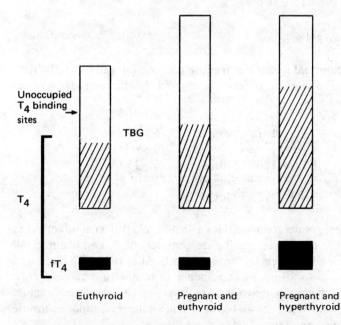

Fig. 20.9 Graves' disease. See cases on pp. 447–449.

normal (Fig. 20.9). FTI indicates that the $fT_4$ is indeed abnormally high. Note that the $T_3$ result in isolation is of no value in the diagnosis, as with $T_4$ it must be interpreted in conjunction with the $T_3U$ value.

PREGNANCY

A 31 week pregnant lady consulted her general practitioner because she found the recent warm weather very uncomfortable. She had a family history of Graves' disease.

| Plasma | $T_4$ | 186 | nmol/l | (60–160) |
|---|---|---|---|---|
| | $T_3U$ | 0.70 | | (0.85–1.05) |
| | FTI | 130 | units | (50–150) |
| | $T_3$ | 3.5 | nmol/l | (1.2–2.8) |

The plasma level of TBG can almost double during pregnancy due to the oestrogen effect on hepatic synthesis; this is reflected by the low $T_3U$ value. The FTI normalizes the high $T_4$, suggesting that the $fT_4$ is normal, and that the subject is euthyroid (see Figs 20.4 and 20.9). The $T_3$ result is of no value on its own, but when considered in association with the $T_3U$ value it suggests that the $fT_3$ would also be normal. Plasma TBG, and therefore total $T_4$, levels return to normal within 4–6 weeks postpartum.

NEWBORN

A 48 hour old infant was feeding poorly, and was clinically 'flat'. Among a variety of tests a thyroid function study was requested.

| Plasma | $T_4$ | 191 | nmol/l | (60–160) |
|---|---|---|---|---|
| | $T_3U$ | 0.94 | | (0.85–1.05) |
| | FTI | 180 | units | (50–150) |
| | $T_3$ | 3.6 | nmol/l | (1.2–2.8) |
| | TSH | 17 | mU/l | (<5.5) |

These results are normal for a 48 hour old full term infant. After birth the plasma $T_4$ and $T_3$ are generally high, and begin to fall towards adult levels by about 1 month. At birth the TSH shows a surge (up to 20 mU/l) in response to postpartum cooling, then returns to normal in two or three days. Typical $T_4$ reference values for newborns are shown below, but laboratories should determine their own age-related reference values.

| Age | $T_4$ nmol/l |
|---|---|
| <7 days | <250 |
| 1–12 weeks | <200 |
| 3 months–~3 yr | <180 |

## CONGENITAL EXCESS OF TBG

A 57 year old man was noted to be extremely anxious and hyperkinetic prior to, and during, a hospital admission for the investigation of hypercalcaemia. There were no other clinical signs or symptoms of thyroid disease.

| Plasma | $T_4$ | 405 | nmol/l | (60–160) |
|---|---|---|---|---|
| | $T_3U$ | 0.65 | | (0.85–1.05) |
| | FTI | 263 | units | (50–150) |
| | $T_3$ | 7.5 | nmol/l | (1.2–2.8) |

$T_4$, FTI and $T_3$ clearly suggest hyperthyroidism, but their gross levels are not really compatible with the clinical picture. The $T_3U$ value suggests that there are many unoccupied hormone sites on the thyroid binding proteins, a situation that can arise from:
(1) Normal thyroid binding proteins, with fewer than normal binding sites occupied by hormone, i.e. ↓ $T_4$.
(2) Normal hormone levels with excess thyroid binding proteins, e.g. pregnancy, oral contraceptives, oestrogen medication, congenital excess of thyroxine-binding globulin, i.e. ↑ TBG.

This patient was obviously not hypothyroid (option 1), and therefore must have a large excess of TBG. In cases of elevated TBG:
(1) The TBG is saturated to its normal level (~30%) with hormone and therefore $T_4$ and $T_3$ are high.
(2) The $fT_4$ and $fT_3$ are maintained at normal levels by the pituitary so that the subject is euthyroid (see Fig. 20.4).

Many of the methods used for determining binding capacity are insensitive to high and low levels of thyroid-binding proteins, and this is reflected in the above case by the failure of the FTI to indicate the normal $fT_4$ by inadequately correcting the high $T_4$ value. Despite this drawback the $T_3U$ and FTI clearly identify the cause of the raised $T_4$ in this patient.

## OESTROGEN EFFECT

A 22 year old woman consulted her general practitioner about recent weight loss. On clinical examination she was an extremely anxious person without clinical evidence of thyroid disease. Blood was taken for thyroid function tests, and the laboratory informed that the subject was taking oral contraceptive medication.

| Plasma | $T_4$ | 178 | nmol/l | (60–160) |
|---|---|---|---|---|
| | $T_3U$ | 0.73 | | (0.85–1.05) |
| | FTI | 130 | units | (50–150) |
| | $T_3$ | 3.8 | nmol/l | (1.2–2.8) |

The oestrogen present in the oral contraceptive preparation has stimulated an increased liver synthesis of TBG. As with the previous case, the thyroid hormone output is increased until the $fT_4$ and percentage saturation of TBG is returned to normal (Fig. 20.4). The $T_3U$ indicates the increase in unoccupied hormone-binding sites, and the FTI reflects the normal $fT_4$. Note that the $T_3$ result is of no diagnostic value unless it is assessed in association with the $T_3U$ result. If clinical evidence was stronger, a TRH stimulation test would be required. Oral contraceptives generally contain less oestrogen than they did a few years ago, and so the problem of misleading $T_4$ values due to the oestrogen effect has also declined.

CHRONIC ACTIVE HEPATITIS

A grossly obese 43 year old woman was admitted to hospital for evaluation of her chronic active hepatitis. Thyroid function tests were requested because of the patient's weight and slow manner.

| Plasma | $T_4$ | 201 | nmol/l | (60–160) |
|---|---|---|---|---|
| | $T_3U$ | 0.86 | | (0.85–1.05) |
| | FTI | 173 | units | (50–150) |
| | $T_3$ | 1.4 | nmol/l | (1.2–2.8) |
| | TSH | 10.0 | mU/l | (<5.5) |

A moderately increased $T_4$ and a slightly elevated TSH appear at first to be inconsistent in a patient who is not receiving $T_4$ replacement therapy. Recent studies have demonstrated that plasma TBG levels can be increased in chronic active hepatitis, and that this is reflected in the elevated or high normal $T_4$ often seen in such cases. Some studies have found that the free $T_4$ is often low and is not truly reflected by an FTI, presumably because of the insensitivity of the $T_3U$ method. The $T_3$ in the above case is low normal and probably reflects a sick euthyroid situation. A significant proportion of subjects with chronic active hepatitis have high titres of thyroid antibodies, so that the increased TSH seen in some of these cases may indicate the presence of a degree of hypothyroidism resulting from an associated autoimmune thyroiditis. As the first-line thyroid function tests are likely to be misleading in these cases, it is advisable to estimate plasma TSH as well.

## DECREASED PLASMA THYROID HORMONES

The major causes of a low level of plasma thyroid hormones are:

*Thyroid disease* (1) Idiopathic (primary) hypothyroidism, (2) autoimmune thyroiditis, (3) hypothalamic or pituitary disease, (4) dyshormonogenesis, (5) infantile hypothyroidism.

*Dietary* Iodine deficiency.

*Iatrogenic* (1) Treatment of hyperthyroidism (a) $^{131}$I, (b) surgical, (c) drugs, (2) inadequate hormone replacement therapy.

*Genetic* Congenital deficiency of TBG.

*Drugs* (1) Triiodothyronine (low $T_4$), (2) lithium, (3) drug interaction.

*Miscellaneous* (1) Nephrotic syndrome, (2) liver disease, (3) acute or chronic illness (low $T_3$), (4) starvation, fasting (low $T_3$).

**Case examples/pathophysiology**

EUTHYROID

A 38 year old man complaining of tiredness and inability to concentrate. Amongst other tests, a thyroid function test was carried out.

| Plasma | $T_4$ | 113 | nmol/l | (60–160) |
|---|---|---|---|---|
| | $T_3U$ | 1.03 | | (0.85–1.05) |
| | FTI | 116 | units | (50–150) |
| | $T_3$ | 2.1 | nmol/l | (1.2–2.8) |
| | TSH | <2.5 | mU/l | (<5.5) |

Many available TSH assays are currently unable to detect low concentrations of TSH, so that up to 30% of normal subjects may have undetectable levels of plasma TSH. Laboratories that use such methods do not quote a lower limit to their reference range and cannot use their assay to demonstrate the low TSH values that might be expected in hypothalamic hypothyroidism (no TRH stimulation), hypopituitary hypothyroidism (loss of TSH secreting cells) and hyperthyroidism (suppression of TSH secretion).

## SICK EUTHYROID (LOW $T_3$) SYNDROME

A 67 year old inpatient noted to have anxiety and a pulse rate of 96/min. The patient had undergone a cholecystectomy 3 days prior to having the thyroid function test done.

| Plasma | $T_4$ | 110 | nmol/l | (60–160) |
|---|---|---|---|---|
| | $T_3U$ | 0.89 | | (0.85–1.05) |
| | FTI | 98 | units | (50–150) |
| | $T_3$ | 0.8 | nmol/l | (1.2–2.8) |

Patients with non-thyroid illness often alter their peripheral metabolism, so that plasma $[T_3]$ ↓ and $[rT_3]$ ↑ (see p. 429, 438).

This pattern ($T_4$ often normal, ↓ $T_3$) can be seen in patients with almost any form of acute or chronic illness, starvation or trauma, and is often referred to as the 'sick euthyroid' or low $T_3$ syndrome. Since it is possible for such a fall in $T_3$ to mask a previously elevated level, it is advisable to delay taking samples for thyroid function tests if the patient is expected to improve in general health in the near future (postsurgery, MI, infections, etc.), and biochemical diagnosis of a possible thyrometabolic problem is not vital to immediate management. Although the circulating level of $T_3$ is reduced in the syndrome the plasma TSH level is not increased, suggesting that the patient remains euthyroid. Although the plasma level of reverse-$T_3$ ($rT_3$) is raised in the sick euthyroid state (↑ production, ↓ metabolic clearance) the estimation of plasma $rT_3$ levels is of no diagnostic value.

## SUBCLINICAL HYPOTHYROIDISM

A 64 year old lady was referred to a general medical clinic for the investigation of her tiredness. Complete blood picture, iron and thyroid studies were performed.

| Plasma | $T_4$ | 73 | nmol/l | (60–160) |
|---|---|---|---|---|
| | $T_3U$ | 0.90 | | (0.85–1.05) |
| | FTI | 66 | units | (50–150) |
| | TSH | 12.0 | mU/l | (<5.5) |
| | $T_3$ | 1.9 | nmol/l | (1.2–2.8) |

The situation of a low normal $T_4$ and FTI, with a small elevation of TSH, is sometimes termed a subclinical or compensated hypothyroidism. Under increasing TSH stimulation, due to a falling $T_4$ level, the failing thyroid tissue manages enough hormone output to maintain the euthyroid state. Patients are usually asymptomatic, and the biochemical condition is discovered inadvertently. Patients are usually not treated but monitored regularly until replacement therapy is required. It is possible that vague symptoms of hypothyroidism in such a patient could be due to the $T_4$ being below the patient's personal reference range, although they are classified as euthyroid in terms of the population reference range.

SUSPECTED HYPOTHYROIDISM

A 50 year old lady sought medical advice about her dry skin and recent dislike of cold weather.

| Plasma | $T_4$ | 51 | nmol/l | (60–160) |
|---|---|---|---|---|
| | $T_3U$ | 0.92 | | (0.85–1.05) |
| | FTI | 47 | units | (50–150) |
| | TSH | 3.2 | mU/l | (<5.5) |

The physician considered the TSH to be inappropriately low for the $T_4$ and FTI values, and investigated the possibility of hypothyroidism secondary to hypothalamic or pituitary disease by arranging a TRH stimulation test.

| Time (min) | TSH mU/l |
|---|---|
| 0 | 4.3 |
| | 200 µg TRH IV |
| 20 | 28.0 |
| 60 | 16.0 |

The response to TRH is normal, indicating an adequate pituitary reserve of TSH. The 20 minute level is greater than that at 60 minute, suggesting that a hypothalamic problem is unlikely. In addition the TSH does not show an exaggerated response (> 30–40 mU/l), as is seen in primary hypothyroidism. The results suggest that the subject is euthyroid. It should be recalled that laboratory reference ranges define the values seen in a selected percentage (usually 95%) of a relevant population, and therefore it is to be expected that some (2.5%) of the population, will lie outside the range at both ends. The above case may represent such a situation for the $T_4$ and FTI values.

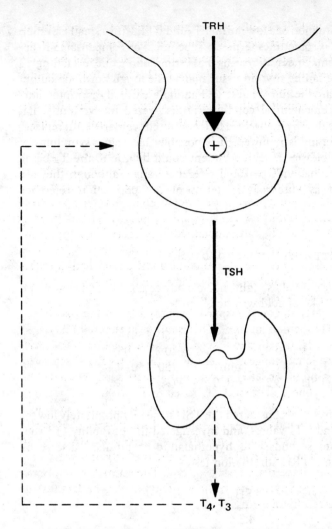

**Fig. 20.10** Hypothalamic-pituitary-thyroid axis in primary hypothyroidism. The low plasma level of thyroid hormone removes feedback inhibition on the thyrotrophs, allowing an increased positive effect by TRH. TSH secretion is also increased.

PRIMARY HYPOTHYROIDISM (Fig. 20.10)

A 64 year old lady consulted a medical practitioner about her general lassitude. Clinical history revealed some weight gain and episodes of constipation. A blood sample was taken for biochemical and haematological investigations; the results of the thyroid function tests are shown below.

| Plasma | $T_4$ | 52 | nmol/l | (60–160) |
|---|---|---|---|---|
| | $T_3U$ | 0.92 | | (0.85–1.05) |
| | FTI | 48 | units | (50–150) |
| | TSH | 28 | mU/l | (<5.5) |
| | $T_3$ | 1.3 | nmol/l | (1.2–2.8) |

The $T_4$ and FTI results are marginally below the lower limit of the population reference range, but since the patient's own normal ranges are not known a biochemical diagnosis cannot be made from these values alone. The plasma TSH is high, indicating that the circulating level of thyroid hormones in this patient, as perceived by *her* pituitary thyrotrophs, is subnormal. This clear evidence of a failing thyroid gland allows a diagnosis of primary hypothyroidism to be made.

The number of unoccupied $T_4$-binding sites on TBG is increased in primary hypothyroidism for two reasons: (1) reduced output of thyroid hormone, (2) slight increase in TBG synthesis. $T_3U$ methods are generally insensitive and often do not reflect these changes until they are substantial. Because of this insensitivity and the low $T_4$ values obtained in primary hypothyroidism the derived FTI values tend to closely match those of $T_4$ and generally do not provide additional diagnostic information unless the biochemical and clinical evidence is contradictory ($\downarrow T_4$, euthyroid, see p. 459).

In the early stages of primary hypothyroidism the plasma $T_3$ level tends to remain normal. This is due to preferential secretion of $T_3$ by the failing thyroid tissue in response to a high TSH level. Recent evidence suggests that the negative feedback effect by thyroid hormone on TSH secretion is at least partly mediated by plasma $T_4$. The $T_4$ is first deiodinated to $T_3$ within the pituitary prior to exerting the inhibitory effect. This mechanism would explain how, in mild primary hypothyroidism, plasma TSH can be elevated in the face of a normal plasma $T_3$ level, despite $T_3$ being the moiety that is believed to express the majority of thyroid hormone action at the peripheral cell level. The estimation of plasma $T_3$, not surprisingly, has proved to be of no value in the diagnosis of primary hypothyroidism.

PRIMARY HYPOTHYROIDISM

A 74 year old lady with a 5 month history of weight gain and lethargy. Clinical examination revealed dry skin, bradycardia and a slow ankle jerk relaxation time.

| Plasma | $T_4$ | 16 | nmol/l | (60–160) |
|---|---|---|---|---|
| | $T_3U$ | 0.94 | | (0.85–1.05) |
| | FTI | 15 | units | (50–150) |
| | TSH | 52 | mU/l | ($<5.5$) |

Plasma $T_4$ is very low with an appropriately increased TSH, both results confirming the diagnosis of primary hypothyroidism. Note that neither the $T_3U$ test nor the FTI provide additional diagnostic information to that provided by the $T_4$ and TSH. Although TSH is the most sensitive biochemical test for primary hypothyroidism a $T_4$ or similar test should always be performed because many TSH assays cannot distinguish between low and normal levels, with the danger of misdiagnosing secondary hypothyroidism as euthyroid.

### PRIMARY HYPOTHYROIDISM

A 66 year old woman with atypical chest pain was provisionally diagnosed as having had a myocardial infarction (no ECG changes), and was admitted to a medical ward for observation and tests. Initial cardiac enzyme results suggested that a small infarct had occurred. Enzyme tests were repeated daily. It was noted that the enzyme levels did not return towards normal although there was no clinical evidence of further cardiac damage. A thyroid function test was requested on day 4.

| Date | CPK U/l (30–140) | HBD U/l (125–250) | $T_4$ nmol/l (60–160) | TSH mU/l (<5.5) |
|---|---|---|---|---|
| 31/7 | 582 | 683 | — | — |
| 2/8  | 544 | 695 | — | — |
| 3/8  | 522 | 654 | — | — |
| 4/8  | —   | —   | <10 | >40 |

The plasma enzyme activities of a number of enzymes, particularly CPK (isoenzyme MM) and LDH (and HBD) are persistently elevated in hypothyroidism. It is thought that the rate of leakage of enzymes from tissue cells, particularly skeletal muscle, is increased in hypothyroidism.

### HORMONE REPLACEMENT THERAPY

A 34 year old woman, initially taking 100 μg $T_4$, and then 200 μg $T_4$, as hormone replacement therapy following $^{131}$I thyroid ablation for Graves' disease.

| Date | $T_4$ nmol/l (60–160) | TSH mU/l (<5.5) |
|---|---|---|
| 18/6  | 122 | >40 |
| 15/7  | 115 | 38 |
| 10/8  | 163 | 2.7 |
| 20/11 | 164 | <2.5 |

Plasma $T_4$ provides a useful monitor for the adequacy of the replacement dosage, a level in the upper half of, and even above, the reference range usually proving clinically satisfactory. A more sensitive biochemical test is to estimate TSH, since its level indicates whether the patient's own pituitary thyrotrophs are 'satisfied' with the circulating level of hormone. However, the clinical evaluation of the patient remains the major assessment of the adequacy of replacement dose.

## HYPOTHALAMIC DISEASE

A 20 year old man had a triple function test performed to investigate his pituitary reserve of hormones. The TSH response to TRH is shown below.

| Time (min) | TSH mU/l (<5.5) |
|---|---|
| 0 | 4.5 |
| | 200 μg TRH IV |
| 30 | 18 |
| 60 | 29 |
| 90 | 36 |

A normal response to a standard TRH test (not part of a triple function test) is for the 20 minute TSH level to be greater than that at 60 minutes. A sluggish TSH response, as shown above, is suggestive of normal pituitary thyrotrophs that have not undergone hypothalamic stimulation for some time, and would be consistent with the presence of a hypothalamic lesion (Fig. 20.6).

## SECONDARY HYPOTHYROIDISM

A 73 year old man was investigated for the cause of his fatigue and general malaise, which he complained of having for 4–5 months. Low values for plasma thyroid and cortisol hormones led to a triple function test. The TSH response is shown below.

| Time (min) | $T_4$ nmol/l (60–160) | TSH mU/l (<5.5) |
|---|---|---|
| 0 | 32 | 3.6 |
| | 200 μg TRH IV | |
| 30 | — | 2.7 |
| 60 | — | 3.6 |
| 90 | — | 3.2 |

The basal TSH level appears to be inappropriately low for the prevailing plasma $T_4$ concentration to be due to primary hypothyroidism. The TSH fails to respond to exogenous TRH, suggesting a lack of pituitary capacity (Figs 20.6 and 20.11). Other pituitary tests also indicated hypopituitarism.

## Chapter 20

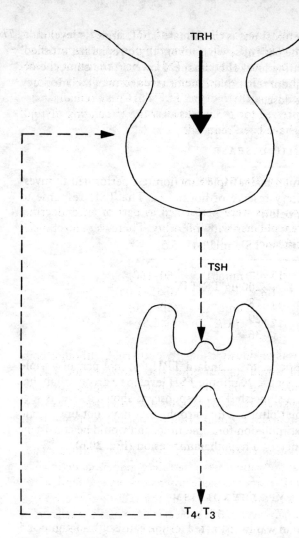

**Fig. 20.11** The hypothalamic-pituitary-thyroid axis in panhypopituitarism.

### THYROID ABLATION

A 43 year old woman given an ablative dose (15 mCi) of $^{131}$I for her Graves' disease. Her thyroid function tests were periodically monitored to assess when hormone replacement therapy should be commenced.

| Date | $T_4$ nmol (60–160) | TSH mU/l (<5.5) |
|---|---|---|
| 12/7 | >400 | <2.5 |
| 30/8 | 265 | <2.5 |
| 4/10 | 175 | <2.5 |
| 31/10 | 89 | <2.5 |
| 7/12 | 29 | 3.3 |

The pituitary thyrotroph cells are sometimes slow to respond to low levels of circulating $T_4$ after a lengthy period of suppression due to thyrotoxicosis, and therefore TSH is not the test of choice for monitoring hormone replacement needs immediately following treatment for hyperthyroidism. The TSH may remain inappropriately suppressed for 2–3 weeks after the circulating thyroid hormone levels have been reduced.

## CONGENITAL TBG DEFICIENCY

Thyroid function tests were performed as part of an endocrine screen on a 32 year old man with infertility. There were no clinical findings of significance.

| Plasma | $T_4$ | 13 | nmol/l | (60–160) |
|---|---|---|---|---|
| | $T_3U$ | 1.85 | | (0.85–1.05) |
| | FTI | 24 | units | (50–150) |
| | TSH | <2.5 | mU/l | (<5.5) |

The subject was obviously clinically euthyroid, although both the $T_4$ and FTI would suggest otherwise. However the $T_3U$ is grossly high, indicating that there are few unoccupied hormone binding sites (Fig. 20.5). This must be due to either: (1) excess hormone secretion and normal TBG, (2) normal hormone level and reduced TBG.

The second option correlates with the subject's clinical state, and suggests that the patient has a harmless congenital deficiency of TBG. Many $T_3U$ methods are insensitive at low (and high) levels of TBG, and this is reflected by the failure of the FTI to indicate that the $fT_4$ level in this subject is normal. In practice this is not important, because the grossly abnormal $T_3U$ clearly indicates the binding protein abnormality.

## LOW THYROID BINDING PROTEINS

A 42 year old man with a short history of swollen ankles and feeling generally unwell. His urinary total protein was found to be 6.2 g/day (<0.15 g/day). Thyroid function tests were also requested.

| Plasma | $T_4$ | 51 | nmol/l | (60–160) |
|---|---|---|---|---|
| | $T_3U$ | 1.31 | | (0.85–1.05) |
| | FTI | 67 | units | (50–150) |
| | TSH | 2.7 | mU/l | (<5.5) |

The high $T_3U$ indicates that the hormone-binding proteins are abnormally low, and allows the $T_4$ to be corrected; the FTI suggests the $fT_4$ is normal. In the above case the TBG and other thyroid hormone-binding proteins are being lost through the damaged glomeruli, thus reducing the plasma level of total $T_4$ (Fig. 20.5).

TREATMENT WITH THYROID HORMONE

Thyroid function tests were requested on a grossly overweight 11 year old boy.

| Plasma | $T_4$ | <10 | nmol/l | (60–160) |
|---|---|---|---|---|
| | TSH | 2.9 | mU/l | (<5.5) |
| | $T_3$ | 2.3 | nmol/l | (1.2–2.8) |

The normal TSH, in association with an undetectable $T_4$ was inappropriate, and suggested a secondary or tertiary hypothyroidism. The laboratory consulted the clinician and learned that the boy was receiving $T_3$ medication for his weight problem. The exogenous $T_3$ suppresses the thyroid ($T_4$ ↓), whilst the $T_3$ supplied is adequate to represent euthyroidism as far as the patient's pituitary is concerned (normal TSH). With $T_4$ medication plasma $T_3$ levels are maintained despite no $T_3$ output by the thyroid because of peripheral conversion from the $T_4$.

DRUG INTERACTION

A 49 year old woman was being treated for primary hypothyroidism (300 μg $T_4$ mane) and for primary hypercholesterolaemia (cholestyramine resin). Despite the high dose of hormone replacement the patient appeared mildly hypothyroid (slow ankle jerk relaxation time). Thyroid function tests also suggested mild hypothyroidism.

| Date | 2/9 | | | |
|---|---|---|---|---|
| Plasma | $T_4$ | 86 | nmol/l | (60–160) |
| | $T_3U$ | 0.84 | | (0.85–1.05) |
| | FTI | 72 | units | (50–150) |
| | TSH | 12.0 | mU/l | (<5.5) |

Two weeks later the patient attended her clinic for further assessment of the problem and was found to be very anxious and hyperkinetic, although she had not changed her $T_4$ dose.

Date    14/9

| Plasma | $T_4$ | 233 | nmol/l | (60–160) |
|---|---|---|---|---|
| | $T_3U$ | 0.96 | | (0.85–1.05) |
| | FTI | 221 | units | (50–150) |
| | TSH | <2.5 | mU/l | (<5.5) |
| | $T_3$ | 1.0 | nmol/l | (1.2–2.8) |

Close questioning of the patient revealed that up to her penultimate consultation she had been taking her $T_4$ and cholestyramine together after breakfast. After that visit she had changed her routine and now took the $T_4$ at breakfast and the cholestyramine after lunch. With the patient's initial routine the basic anion exchange resin (cholestyramine) was absorbing most of the $T_4$ dose in the stomach, but taking the two drugs separately allowed the patient the full effect of 300 μg $T_4$.

# 21 Cortisol

Cortisol is synthesized and secreted by the adrenal cortex, both processes being under the hormonal control of the anterior pituitary and hypothalamus.

**The hypothalamic-pituitary-adrenal (HPA) axis** (also see p. 388)

Neurosecretory cells within the hypothalamus synthesize corticotropin releasing factor (CRF), which passes down the neuronal axons to the median eminence before being released into the hypothalamic-hypophyseal portal system. The secretion of CRF:
(1) Is episodic.
(2) Has a circadian rhythm, which is at least partly entrained by the sleep-wake pattern, with the highest level normally occurring in the morning.
(3) Is increased by stimulation from higher centres in the brain.
(4) Is increased by stress—emotional, physical, biochemical.
(5) Is suppressed by cortisol.
Stress can override the inhibitory effect of cortisol and the circadian influence.

On reaching the anterior pituitary CRF stimulates certain cells, via adenyl cyclase, to synthesize and release various peptides (see p. 390). The peptide that is of major diagnostic interest is adrenocorticotropic hormone (ACTH, corticotropin). ACTH is a polypeptide that comprises 39 amino acid residues. Animal studies have shown that ACTH, prior to its release, forms part of a much larger polypeptide called 'pro-opiocortin'. Pro-opiocortin also contains amino acid sequences that correspond to β-endorphin. As well as ACTH, CRF appears to also simultaneously stimulate the release of these peptides from the precursor. It is likely that a similar polypeptide precursor system will be demonstrated in man.

Under the influence of CRF the release of ACTH from the anterior pituitary is (1) episodic, (2) shows a circadian rhythm, (3) increased by stress.

The secretion of pituitary ACTH is suppressed by cortisol (negative feedback).

ACTH binds to specific receptors on cells in the adrenal cortex and stimulates, via adenyl cyclase, steroidogenesis. The major product is cortisol, which is secreted episodically and provides negative feedback at the hypothalamus and the anterior pituitary (Fig. 21.1).

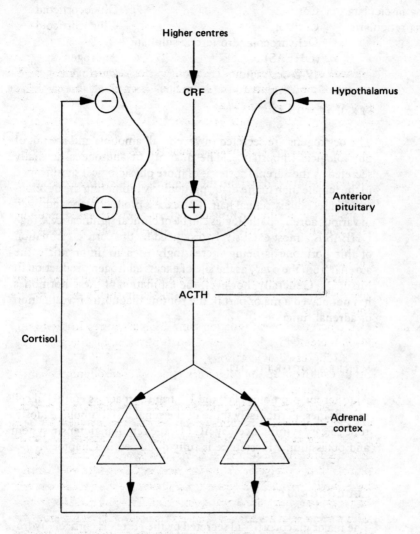

**Fig. 21.1** Hypothalamic-pituitary-adrenal axis in the normal adult.

## The adrenal cortex

The adrenal cortex forms some 80% of the total gland, and comprises three histologically distinct areas, each of which synthesizes and secretes steroid hormones.

| Cortex zone | Major steroid product | Steroid action |
|---|---|---|
| Zona glomerulosa | Aldosterone | Mineralocorticoid |
| Zona fasciculata | Cortisol | Glucocorticoid |
| Zona reticularis | Cortisol | Glucocorticoid |
| | Dehydroepiandrosterone sulphate (DHEAS) | Androgen |
| | 17-β oestradiol | Oestrogen |

## SEX HORMONES

The oestrogens are secreted in very small amounts and are not of physiological importance. The quantity of androgens normally secreted by the adrenal cortex is of more physiological significance in the female than in the male, particularly promoting the development of secondary sexual hair. The excess production of androgen in various adrenal pathologies can be of clinical significance. Since DHEAS is almost exclusively produced by the cortex, estimation of this hormone is being increasingly used to differentiate the adrenal from the ovary as the site of excess androgen production in hirsutism. Generally, however, the estimation of these hormones has not played a major part in the routine diagnostic investigation of adrenal function.

## MINERALOCORTICOID

Aldosterone promotes $Na^+$ and $K^+$ transport across epithelial cell membranes, particularly those of the renal distal tubule. Aldosterone release is influenced by ACTH, the renin-angiotensin system and potassium. Aldosterone is fully discussed in Chapter 7.

## GLUCOCORTICOIDS

The major glucocorticoid secreted by the cortex is cortisol (hydrocortisone). The biosynthetic pathways for cortisol and the other major adrenal steroids are shown in Fig. 21.2. Cortisol is secreted episodically (8–10 bursts/24 h), and provides more than 95% of the glucocorticoid released from the adrenal. Cortisol is secreted with a circadian rhythm (Fig. 21.3) that reflects that of CRF and ACTH, and although this pattern is at least partly sleep-entrained it needs a consistent change in sleep/wake pattern of a week or more to realign the rhythm of cortisol release to the new sleeping period.

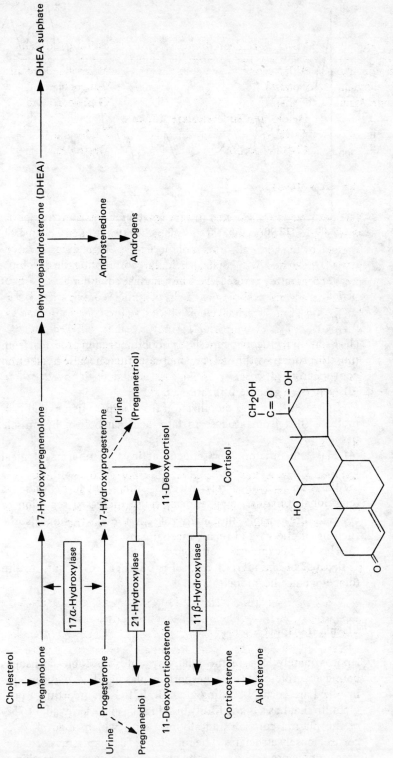

**Fig. 21.2** Major pathways of steroid biosynthesis in the adrenal cortex.

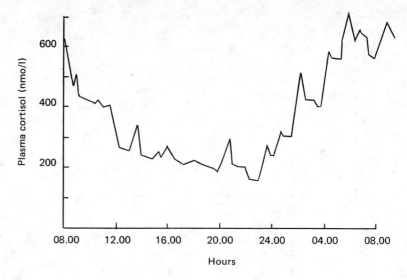

**Fig. 21.3** Plasma cortisol levels in a normal adult during a 24 hour period.

*Cortisol*

(1) Promotes gluconeogenesis by mobilizing amino acids from muscle (protein synthesis ↓) and increases their hepatic uptake and conversion to glucose.
(2) Depresses glucose utilization.
(3) Increases fatty acid mobilization from adipose tissue, and promotes a shift from glucose to fatty acid catabolism by muscle tissue.
(4) Has anti-inflammatory effects, at least in pharmacological doses.
(5) Plays an undefined role in stress. Almost all forms of stress (emotional, physical, biochemical) cause increased secretion of cortisol. Presumably the main role in these situations is the fuelling of energy and repair mechanisms.
(6) Has a weak mineralocorticoid activity.

Because of actions 1–3 cortisol is often called an anti-insulin (diabetogenic) hormone.

**Plasma transport**

Approximately 90% of the relatively water-insoluble cortisol is bound to protein for its transport in plasma. The two proteins involved are cortisol-binding globulin (CBG, transcortin), an $\alpha_2$-globulin that has a high affinity binding site and carries about 75% of the plasma cortisol, and albumin with which about 15% of cortisol loosely associates. The remaining 10% of plasma cortisol is 'free', and is presumed to be the physiologically active fraction at the cell level. CBG is synthesized in the liver and, as with thyroxine-binding globulin, its synthesis is enhanced by increased levels of oestrogen (therapy, oral contraceptives). Alterations to CBG levels are reflected by similar changes in plasma total cortisol, but the level of free cortisol is kept adjusted within the eucorticoid range by a normally functioning HPA axis. CBG is largely saturated with cortisol, and so any increase in adrenal output is quickly reflected by a rise in the level of free cortisol. The plasma half-life of cortisol is approximately 80 minutes, whilst that of aldosterone, which is poorly bound to CBG ($\sim 10\%$), is around 30 minutes.

## LABORATORY INVESTIGATIONS

### PLASMA ELECTROLYTES

Two features of cortisol action may be reflected in plasma $[Na^+]$ and $[K^+]$:
(1) Cortisol has a weak mineralocorticoid action, and therefore excess cortisol (Cushing's syndrome) will generally result in

(a) renal $Na^+$ retention → ↑ ECF osmolality → ↑ water retention
(b) renal $K^+$ loss → hypokalaemia

Because of the concomitant water retention, plasma $[Na^+]$ is either slightly raised, or in the upper half of the reference range. The overall effect is: ↑ total body $Na^+$, ↑ total body $H_2O$, ↓ total body $K^+$.
(2) ADH secretion, and therefore water retention, is inhibited when ECF osmolality falls. The presence of cortisol is required for this inhibition of ADH secretion to occur. Thus in primary adrenal failure (↓ cortisol, ↓ aldosterone) the lack of aldosterone causes:

(a) ↑ renal $Na^+$ loss → ↓ ECF osmolality
(b) ↓ renal $K^+$ loss → ↑ plasma $[K^+]$

The normal response to ↓ ECF osmolality is:

↓ ADH → ↑ water excretion → ↑ ECF osmolality

but if cortisol is lacking ADH release is not inhibited, and therefore water retention continues. Therefore, in primary adrenal failure the plasma electrolyte pattern is generally that of: (a) hyponatraemia, (b) hyperkalaemia.

If hypocortisolism occurs without affecting aldosterone production (i.e. ↓ ACTH due to pituitary dysfunction) hyponatraemia may occur due to dilution (↑ ADH activity), but plasma potassium concentration is usually normal because of the intact aldosterone pathway.

## PLASMA GLUCOSE

Cortisol opposes the action of insulin and conserves glucose, so that excessive glucocorticoid activity (Cushing's syndrome) may result in hyperglycaemia. Lack of cortisol may cause hypoglycaemia, but this is generally only seen in panhypopituitarism, when both 'diabetogenic' hormones (GH, cortisol) are lacking.

## PLASMA CRF

This hypothalamic factor has recently been isolated and shown to be a 41 residue peptide. An assay for CRF is currently unavailable for diagnostic purposes.

## PLASMA ACTH

Sensitive bioassays are available, but are suited more to research rather than to diagnostic use. Radioimmunoassays have been available for a number of years, but many of these have proved technically unsatisfactory, and yield values that are higher than those obtained with bioassays. These problems mainly centre on (1) low plasma concentration ($<100$ pg/ml), (2) rapid degradation of ACTH in blood (proteases), (3) adsorption of ACTH to glass, (4) specificity of antisera used.

Of the 39 amino acid structure of ACTH, the $NH_2$-terminal 1–24 residues express ACTH activity. Antisera to peptides 1–24 (N-terminus), 25–39 (C-terminus), and 1–39 (whole ACTH) are available. A number of ACTH degradation peptides can also cross-react, so that the apparent concentration of ACTH generally exceeds the actual level of biologically active ACTH present in the sample (immunoreactivity $>$ bioactivity).

The effect of the circadian rhythm of ACTH secretion on plasma levels can be reduced by taking blood samples at the same time of day. However, the secretion of ACTH is also episodic which, in conjunction with the effects of stress, can make the interpretation of plasma ACTH levels in single samples very difficult.

Despite these technical and interpretative difficulties, an estimation of plasma ACTH can be of value in a limited number of clinical situations:

(1) Aetiology of Cushing's syndrome (chronic inappropriate hypercortisolism) (a) undetectable ACTH suggests an adrenal tumour (negative feedback by excess cortisol), (b) normal or high normal (ACTH ($\sim$ 100 pg/ml) is compatible with Cushing's disease (pituitary-dependent hypercortisolism), (c) very high ACTH ($>$ 200 pg/ml) suggests ectopic production by a non-endocrine tumour.
(2) Occasionally for assessing Nelson's syndrome ($\uparrow$ ACTH from pituitary tumour expressing itself postadrenalectomy) and other treatment situations.

If requests for ACTH measurement are occasionally received by a laboratory it is wise, because of the problems mentioned above, to have the samples assayed by a specialist centre.

## Plasma ACTH samples

(1) Contact laboratory prior to commencing procedure, to arrange sample preparation.
(2) The patient does not require special preparation.
(3) Samples should be taken between 08.00 and 09.00 hours because of the circadian variation in ACTH secretion.
(4) A sample should be obtained at the first attempt at venepuncture, as stress will markedly elevate ACTH. An indwelling needle in place for 30 minutes prior to sampling is an alternative procedure.
(5) A plastic syringe and plastic containers for blood collection and plasma storage should be used.
(6) The sample requires immediate centrifugation, followed by freezing of the plasma at $-20°C$ if analysis is to be deferred.

### PLASMA CORTISOL

Although the adrenal cortex produces a variety of steroids, the measurement of plasma cortisol forms the basis of tests for assessing adrenocortical function.

Fluorometric techniques are widely used for the estimation of plasma cortisol, as 11-hydroxycorticosteroids (principally cortisol and corticosterone). In careful hands these methods yield satisfactory results, but they require relatively large amounts of plasma, need an extraction procedure and suffer from specific (e.g. spironolactone) and non-specific interferences. Competitive protein binding and radioimmunoassay methods are more robust techniques for estimating total (CBG-bound + free) cortisol, but care is required in ensuring that antisera have low cross-reactivity with other steroids, particularly those used for medication; prednisolone has a 90% cross-reactivity with the cortisol antiserum used in a currently available commercial kit.

Some 75% of plasma cortisol is bound to CBG, and therefore any increase in CBG (oestrogen excess, pregnancy), or decrease in CBG (severe liver disease, nephrotic syndrome, protein-losing enteropathy), will be reflected in the plasma total cortisol, which will give a misleading indication of the free cortisol level. Methods are not available to correct plasma cortisol estimations for these binding abnormalities.

At birth the adrenal cortex predominantly secretes cortisone, so that plasma cortisol is low at first, but increases to adult levels by the first week postpartum. The circadian rhythm of plasma cortisol secretion takes about 3 months to become established in the infant.

The 24 hour pattern, and the episodic nature of cortisol release, causes rapid and wide fluctuations in cortisol plasma levels, so to reduce this effect and allow meaningful comparisons, blood samples should always be taken about the same time of day, preferably 08.00-09.00 hours. Despite this precaution a blood sample may be taken at the peak or trough, rendering random blood samples difficult to interpret (Fig. 21.3).

In Cushing's syndrome the plasma levels of cortisol do not fall during the day, so that the circadian pattern is often lost; a demonstration of this by taking an additional sample at 23.00-24.00 hours is often used as a screening test for Cushing's syndrome. However, a comparison of the midnight cortisol level, with that obtained at 08.00 hours in patients with Cushing's syndrome, discriminates the group very poorly from normals. Studies have shown that comparing the midnight sample with a reference range for normal samples taken at midnight provides a screening test with few false positives or negatives. Laboratories must determine their own reference limit; in the authors' laboratory, normal midnight cortisol levels fall below 140 nmol/l.

The clinical questions that usually underlie a request for a plasma cortisol estimation are:
(1) Is the adrenal output of cortisol low, normal, or high?
(2) Does the adrenal cortex have an adequate reserve of cortisol?
(3) Does the output of cortisol respond normally to its control mechanisms?

A diseased and failing adrenal cortex can for some time secrete sufficient hormone to maintain the plasma cortisol concentration within the population reference range. On the other hand, an acutely stressed patient (i.e. burns, infections, fever, surgery, other acute illness) can secrete cortisol in amounts that are typical of Cushing's syndrome, whilst subjects anxiously awaiting venepuncture may raise their plasma cortisol levels enough to be misleading. Some 20-30% of patients with Cushing's syndrome have been shown to have normal 08.00 hours plasma cortisol levels. For these reasons the taking of single blood samples for the estimation of plasma cortisol is of very limited, and often misleading, use. The proper investigation of any of the above questions requires that capacity and control of the adrenal cortex be evaluated by means of an appropriate dynamic function test.

## Stimulation tests

Protocols vary, and therefore it is always advisable to consult the local laboratory or specialist unit as to requirements before commencing a test.

Hypocortisolism may follow destruction of the adrenal cortex itself, or arise as a result of a lack of appropriate stimulation due to hypothalamic or pituitary disease. The investigation and diagnosis of the secondary and tertiary causes of adrenal hypofunction are discussed in Chapter 19, therefore the following section will concentrate on the tests used to directly evaluate the function of the adrenal gland itself.

### SHORT ACTH STIMULATION (SYNACTHEN) TEST

The biologically active portion of the 39 amino acid ACTH molecule comprises the N-terminal 24 amino acid residues. This active portion is readily available in synthetic form (tetracosactrin; Cortrosyn-Organon; Synacthen-Ciba), and following IM injection, will cause a marked rise in cortisol secretion by the normal adrenal cortex. The response may be slow if the adrenal has remained unstimulated for some time due to pituitary dysfunction, or has been suppressed by steroid medication; in both cases a prolonged stimulation may be required.

Protocol

*Patient preparation*  No special preparation. If the patient has been placed on steroid cover prior to the test ascertain whether the steroid used will interfere with the laboratory method for plasma cortisol estimation. If steroid cover is required, dexamethasone should generally be the steroid of choice until tests and investigations are completed.

*Method*

Time (min)

| | |
|---|---|
| 0 | Basal blood sample for cortisol estimation |
|   | 250 µg tetracosactrin IM |
| 30 | Sample for plasma cortisol |
| 60 | Sample for plasma cortisol |

Interpretation

A normal response is shown if: (1) the cortisol level in a post-stimulation sample exceeds the basal level by at least 200 nmol/l, (2) a poststimulation cortisol level exceeds 550 nmol/l.

Normally the cortisol level in the 30 minute sample exceeds that in the 60 minute one. A complete lack of response is suggestive of adrenal insufficiency, or a sluggish response (60 min value > 20 min value) secondary to pituitary disease or adrenal suppression. Depending on the clinical findings, it may be necessary in the case of a partial response to further investigate the adrenal reserve by means of a prolonged stimulation test.

## PROLONGED ACTH STIMULATION TEST

Protocol

*Patient preparation*   As above.

*Method*   Basal blood sample for plasma cortisol: 1 mg depot (slow release) tetracosactrin IM at 09.00 h daily for 3 days.

*Blood samples*   (1) 5 hours postinjection sample (14.00 h) for plasma cortisol estimation each day. (2) Day 3—short ACTH stimulation test performed as described above.

Interpretation

Same criteria as for the short stimulation test. A slow response over the period of the test is seen in: (1) adrenal failure secondary to hypothalamic or pituitary disease, the cortex not having been stimulated for some time, (2) patients on prolonged steroid therapy, causing a suppression of the HPA axis.

## METYRAPONE TEST

Metyrapone inhibits 11 β-hydroxylase, the adrenocortical enzyme that converts 11-deoxycortisol to cortisol (Fig. 21.2). In the normal subject this reduction of cortisol feedback causes increased secretion of ACTH, which in turn stimulates adrenocortical steroidogenesis as far as the enzyme block. The intact and normally responsive HPA axis is therefore reflected by a low plasma cortisol and a high plasma 11-deoxycortisol (or urinary 17-hydroxycorticosteroids) following metyrapone administration.

By demonstrating the pituitary reserve of ACTH the metyrapone test may be used to elucidate the aetiology of Cushing's syndrome:

*(1) Cushing's disease* The ACTH reserve is either normal or high, so that there is a marked increase in steroidogenesis (↑ 11-deoxycortisol) following the metyrapone-induced reduction of cortisol secretion.

*(2) Adrenal tumour, ectopic ACTH* In both these conditions the high level of feedback inhibition by cortisol causes a low pituitary reserve of ACTH, so that following a metyrapone-induced reduction of cortisol there is a failure to stimulate steroidogenesis.

The simplicity with which the prolonged dexamethasone test can be used to investigate the aetiology of Cushing's syndrome has led to a decline in the use of the metyrapone test.

Protocol

Before undertaking the metyrapone test the ability of the adrenal cortex to respond to ACTH should be demonstrated by performing a short Synacthen test.

*Precautions* Metyrapone blocks cortisol production, so patients in whom there is a possibility of adrenocortical insufficiency (primary or secondary) must not undergo this test, as in such cases metyrapone may precipitate an adrenal crisis.

*Dose* Metyrapone can cause nausea and giddiness and should be given with milk. The effect of the standard dose can be altered by disease or drugs, e.g. an impaired ability to catabolize cortisol due to liver disease increases the plasma half-life of cortisol with a consequent reduction in ACTH response. Anti-convulsant drugs may reduce dose effectiveness by increasing the rate of hepatic metabolism of metyrapone.

*Method*
Day (1) 08.00 hours—samples for plasma 11-deoxycortisol and cortisol estimation followed by 750 mg oral metyrapone 4-hourly for six doses.
Day (2) 08.00 hours—plasma 11-deoxycortisol and cortisol.
Alternatively a 24-hour urine is collected for 2 days prior to (basal) and for 2 days following the dose (test) and 17-hydroxycorticosteroids (17-OHCS) or 11-hydroxysteroids are estimated.

Interpretation

To demonstrate that the dose has been effective the plasma cortisol level should decline to less than approximately 200 nmol/l. A normal response to metyrapone is indicated by plasma 11-deoxycortisol rising from almost undetectable levels to more than 200 nmol/l. If urinary 17-OHCS are measured, a normal response is shown by the level of 17-OHCS in the test sample increasing at least two-fold over that in the basal level.

## Suppression tests

Cushing's syndrome describes the clinical condition that develops from prolonged hypercortisolism; non-iatrogenic causes are due to inappropriate production of either ACTH (pituitary or ectopic tumour), or cortisol itself (adrenal tumour). The morning plasma cortisol levels in patients with Cushing's syndrome often fall within the reference range, and therefore it is very important to be able to clearly distinguish this group from normal subjects. Equally, there is a need to differentiate normal subjects with high plasma cortisol (psychological stress, obesity) from those with Cushing's syndrome. Dexamethasone, a glucocorticoid with an approximately 25-fold potency relative to cortisol, may be administered in a small dose and used to assess the response of the negative feedback mechanism by measuring the effect on the plasma cortisol level at a suitable time interval. Dexamethasone does not interfere with fluorometric assays of cortisol.

### OVERNIGHT DEXAMETHASONE SUPPRESSION TEST

Protocol

*Patient preparation*   No special preparation.

*Method*   Day 1—08.00 hour basal plasma sample (optional). Dose—1-2 mg dexamethasone orally at 23.00 hours. Blood sample: Day 2—08.00 hours for plasma cortisol estimation.

Interpretation

In the normal individual, plasma cortisol is suppressed to less than 50% of basal (if sample taken) or less than 140 nmol/l. Failure to suppress is suggestive of Cushing's syndrome or acute stress, e.g. infection, surgery, acute illness, alcohol abuse, severe depression. Oestrogen therapy enhances the synthesis of CBG and, therefore, increases plasma protein-bound cortisol. The resulting reduction in cortisol clearance rate can lead to false positive results for the dexamethasone suppression test (see p. 488 for other causes of false results).

### PROLONGED DEXAMETHASONE SUPPRESSION TEST

In Cushing's disease (i.e. pituitary dependent) the normal negative feedback mechanism is relatively insensitive to circulating cortisol levels, so that ACTH secretion (and therefore plasma cortisol) is not suppressed by a low dose of dexamethasone. However, a large dose of dexamethasone will overcome the insensitivity of the feedback mechanism and cause a suppression of CRF and ACTH output, which will be reflected in a cessation of cortisol secretion.

Dexamethasone will not suppress plasma cortisol levels in Cushing's syndrome due to ectopic ACTH because the tumour ACTH secretion is autonomous. Similarly, dexamethasone will have no effect on the autonomous cortisol production by an adrenal tumour. In the latter case, pituitary ACTH will be suppressed (Figs 21.6 and 21.7). Therefore, the response of plasma cortisol to a high dose of dexamethasone will differentiate Cushing's disease from other causes of Cushing's syndrome in most cases (see plasma ACTH).

Protocol

*Patient preparation* Nil.

*Method*

| Day | 1 | 2 | 3 | 4 | 5 |
|---|---|---|---|---|---|
| Plasma sample for cortisol (h) | | | 07.00 | | 07.00 |
| Dexamethasone (start at 0.700) | 0.5 mg/6 h | 0.5 mg/6 h | 2 mg/6 h | 2 mg/6 h | |

Daily 24 hour urine collections for 17-OHCS or 17-Oxo. estimations may be taken if plasma cortisol methods are not available.

Interpretation

A failure to suppress plasma cortisol (to <140 nmol/l or by >70% of basal level) at the low dexamethasone dose (2 mg/day) confirms Cushing's syndrome. An inability to cause suppression (by <50% of basal level) at the higher dose (8 mg/day) indicates a pituitary-independent hypercortisolism (ectopic ACTH, adrenal tumour), whilst reduction of cortisol (to<50% of basal level) at this high dose is suggestive of Cushing's disease (pituitary dependent).

## Urine

Adrenal glucocorticoids, mineralocorticoids and androgens undergo various catabolic conversions before they are secreted as water soluble conjugates in the urine. Various chemical methods of estimating these excretory compounds are available, and have been successfully used for many years for the investigation and diagnosis of adrenal dysfunction. The methods are tedious, and require care in their performance and an awareness of the many sources of specific and non-specific interference. An advantage of the methods are that they are performed on 24 hour urine collections, and so results usually reflect a production rate for the steroid measured; however the usual drawbacks to such urine collections are present, and it is often helpful to estimate urinary creatinine to assess the reliability of collection and to express the results in terms of creatinine excretion.

### URINARY 17-HYDROXYCORTICOSTEROIDS (17-OHCS)

This assay utilizes the Porter-Silber reaction, and measures about one-third to one-half of the metabolites of cortisol (tetrahydrocortisol) and cortisone (tetrahydrocortisone). The estimation of 17-OHCS on a 24 hour urine sample provides an adequate screening test in place of a urinary free cortisol; values may be elevated in obese subjects. The test may also be used instead of plasma cortisol estimations for most dynamic function tests.

### URINARY 17-KETOGENIC STEROIDS (17-KG)

Relative to 17-OHCS, this method measures additional metabolites of cortisol (cortol) and cortisone (cortolone). It also measures pregnanetriol, a steroid produced in large amounts in patients with one form of congenital adrenal hyperplasia (21-hydroxylase deficiency). This test is used in similar situations as the 17-OHCS estimation.

### URINARY 17-OXOSTEROIDS (17-OXO., 17-KETOSTEROIDS)

This technique estimates mainly adrenal androgens, and is used for investigating hirsutism and virilization. The method is becoming superseded by the general availability of plasma testosterone and dehydroepiandrosterone sulphate (DHEAS) assays.

### URINARY FREE CORTISOL

Some 10% of plasma cortisol is not protein-bound (free), and is therefore filtered by the glomeruli. Any increase in adrenal output of cortisol quickly saturates the remaining protein-binding sites, and causes an increase in free cortisol. An estimation of urinary free cortisol (RIA or competitive protein binding) provides an accurate reflection of the biologically active plasma free cortisol. Estimations are performed on an aliquot of a 24 hour collection. A lower reference limit is usually difficult to define, and therefore urinary cortisol measurements are useless for the investigation of adrenal insufficiency. The method provides an ideal screening test for Cushing's syndrome.

## Selection of tests

### HYPOCORTISOLISM

An initial clinical suspicion of adrenal hypofunction may be strengthened by the finding of hyperkalaemia and/or a slight to moderate hyponatraemia, and result in a direct investigation of adrenal function. Nowadays patients with Addison's disease rarely present in a crisis with total adrenal failure. Mostly the adrenals in such patients are capable of secreting sufficient cortisol to maintain a normal plasma concentration, although they are unable to adequately respond to stress situations. Therefore random plasma cortisol determinations are valueless. For all cases of suspected adrenal hypofunction the adrenal reserve of cortisol must be demonstrated by means of an ACTH (Synacthen) stimulation test. Depending on clinical findings and drug history (i.e. long term steroid therapy) a partial or lack of response may require further investigation with a prolonged stimulation test.

### HYPERCORTISOLISM

The investigation and diagnosis of hypercortisolism is undertaken in two stages by establishing: (1) that the patient has inappropriate hypercortisolism (Cushing's syndrome), (2) the aetiology of the Cushing's syndrome (Table 21.1).

A wide variety of tests and protocols have been used to answer these questions, and obviously the approach selected depends on the laboratory methods that are locally available. A commonly used strategy is shown below (Fig. 21.4). Because of wide method variations laboratories must determine their own reference values for the static and dynamic tests.

### HYPOCORTISOLISM

Low levels of plasma cortisol may be caused by:

Chapter 21

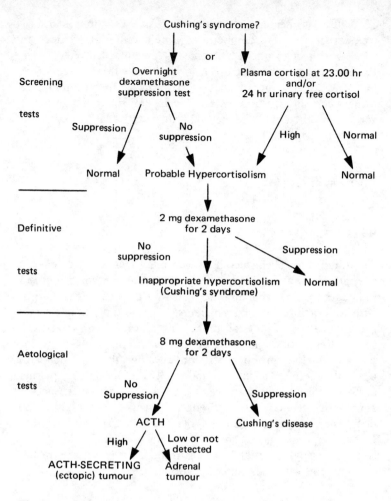

**Fig. 21.4** A commonly used strategy for the biochemical investigation of hypercortisolism.

Deficiency of CRF

*Hypothalamic disease* (see Chapter 19)

Deficiency of ACTH

*Anterior pituitary disease or surgery* (see Chapter 19)

Deficiency of adrenal cortex

*Addison's disease*   (1) Autoimmune, (2) tuberculosis.

*Iatrogenic*   (1) Surgery (adrenalectomy), (2) drugs (e.g. aminoglutethimide), (3) inadequate steroid replacement, (4) cessation of prolonged steroid replacement.

Table 21.1 Plasma tests for the investigation of hypercortisolism

| | Stress | Pituitary dependent (Cushing's disease) | Ectopic ACTH | Adrenal tumour |
|---|---|---|---|---|
| 08.00 hours plasma cortisol | N or ↑ | N or ↑ | N or ↑ | N or ↑ |
| 23.00 hours plasma cortisol | N or high | High | High | High |
| Overnight dexamethasone (DXM) 08.00 hours | May not be suppressed | Not suppressed | Not suppressed | Not suppressed |
| Prolonged DXM | | | | |
| @ 2 mg/day | Suppressed | Not suppressed | Not suppressed | Not suppressed |
| @ 8 mg/day | Suppressed | Suppressed | Not suppressed | Not suppressed |
| Plasma ACTH concentration | N or ↑ | N or ↑ | ↑↑ | ↓ |

*Chapter 21*     Defect of enzyme required for cortisol synthesis

*Congenital adrenal hyperplasia*     Deficiency of (1) 20,22 desmolase, (2) 3 β-hydroxysteroid dehydrogenase, (3) 17 α-hydroxylase, (4) 21-hydroxylase, (5) 11 β-hydroxylase.

Reduced cortisol-binding globulin

*Nephrotic syndrome*

*Severe liver disease*


**Case examples/pathophysiology**

In the following examples cortisol was measured by a solid-phase radioimmunoassay technique.


SEVERE STRESS

A thin, anxious 64 year old man was admitted to hospital with hypotension. Although there was no strong clinical evidence of adrenal disease, a short tetracosactrin stimulation test was performed to exclude adrenal insufficiency.

| Time (min) | Plasma Cort. nmol/l (140–690) |
|---|---|
| 0 | >1600 |
|  | 250 µg Synacthen |
| 30 | >1600 |
| 60 | >1600 |

The grossly elevated levels of cortisol obviously rule out adrenal insufficiency, and are probably due to the general stress of the patient's illness.


NORMAL ADRENAL FUNCTION

A 34 year old lady with insulin-dependent diabetes had a thyroid lump removed 2 months ago. A short stimulation test was performed to assess whether there was any adrenal involvement as part of a generalized autoimmune process.

| Time (min) | Plasma Cort. nmol/l (140–690) |
|---|---|
| 0 | 640 |
| | 250 µg Synacthen |
| 30 | 990 |
| 60 | 1080 |

The poststimulation samples exceeded an absolute value of 550 nmol/l, whilst at least one also rose above the basal level by more than 200 nmol/l. The adrenal response to the stimulation was therefore normal.

## ADRENAL INSUFFICIENCY

A 68 year old man was considered to be clinically and biochemically euthyroid on his current thyroid hormone replacement therapy for primary hypothyroidism. He was now admitted for investigation of extreme tiredness. A 24 hour urinary free cortisol estimation and a short stimulation test were performed.

| Time (min) | Plasma Cort. nmol/l (140–690) | Urine Cort. nmol/l (<280/24 h) |
|---|---|---|
| 0 | 248 | 95 |
| | 250 µg Synacthen | |
| 30 | 331 | |
| 60 | 373 | |

The basal level of cortisol falls within the reference range and gives no clue as to adrenal reserve. Cortisol levels fail to exceed the basal level by 200 nmol/l, or surpass a value of 550 nmol/l in either of the poststimulation samples. This response is typical of adrenal insufficiency. If clinical history was suggestive of pituitary dysfunction or steroid therapy, then a prolonged stimulation test would be required in case the adrenal was slow to respond due to lengthy inactivity or suppression. Note that the urinary cortisol estimation was valueless, as was the basal plasma cortisol, emphasizing the need to perform a stimulation test.

## ADDISON'S DISEASE

A 39 year old woman presented complaining of 4 days of vomiting after a history of tiredness, poor appetite, salt craving and dizziness over the previous 4 months. Examination revealed marked pigmentation in all exposed areas and palmar grooves. A clinical diagnosis of primary adrenal failure was made, and a short adrenal stimulation test performed.

| Time  |         | 08.10 | 08.40 | 09.10 |        |              |
|-------|---------|-------|-------|-------|--------|--------------|
| Plasma| Na      | 128   |       |       | mmol/l | (132–144)    |
|       | K       | 5.2   |       |       | mmol/l | (3.1–4.8)    |
|       | Cl      | 91    |       |       | mmol/l | (93–108)     |
|       | $HCO_3$ | 21    |       |       | mmol/l | (21–32)      |
|       | Urea    | 15.5  |       |       | mmol/l | (3.0–8.0)    |
|       | Creat.  | 0.11  |       |       | mmol/l | (0.06–0.12)  |
|       | Glu.    | 3.8   |       |       | mmol/l | (3.0–5.5)    |
|       | Cort.   | 28    | 97    | 124   | nmol/l | (140–690)    |

The failure of cortisol to respond to stimulation supports the diagnosis of primary Addison's disease. Adrenal failure causes a loss of both cortisol and aldosterone secretion. The consequences of aldosterone loss are:

↓ Aldo. → ↓ renal secretion of $K^+$ → hyperkalaemia
→ ↑ renal excretion of $Na^+$ → ECF Osmo. ↓ →
↓ ADH
→ ↑ renal water loss → ↓ ECF volume (therefore plasma $[Na^+]$ ↑)

However, when ECF volume contracts by more than 10%, this overrides the inhibition of ADH release by low osmolality (see p. 22), so that water is now retained to preserve the ECF volume (circulation maintained).

Also in this patient cortisol is lacking, further reducing the inhibition of ADH release (p. 467), therefore

glucocorticoid ↓ + ECF vol ↓↓ → ↑ ADH
→ ↑ renal water retention + continued renal $Na^+$ loss → hyponatraemia

In the above case the hyponatraemia is slight, probably because of the concomitant loss of water via the vomiting (salt-poor fluid). The pigmentation is due to the MSH activity of β-lipotropin, which is secreted under the same situations as ACTH (p. 391).

Table 21.2 Common clinical signs and symptoms of Addison's disease

|                            | Approx. % of patients |
|----------------------------|-----------------------|
| Muscle weakness/fatigue    | 100                   |
| Excess pigmentation        | 90                    |
| Hypotension                | 90                    |
| Hyponatraemia              | 90                    |
| Hyperkalaemia              | 70                    |
| Gastrointestinal symptoms  | 60                    |

It is uncommon nowadays for patients with Addison's disease to first present in adrenal crisis, instead the usual presentation is one with a history of increasing severity of symptoms. Apart from the very characteristic pigmentation, most of the clinical signs and symptoms are, by themselves, poor markers for the disease and may lead the medical investigation in fruitless directions. A common presentation of Addison's disease is one of increasing tiredness and muscle weakness, gastrointestinal symptoms and increased pigmentation.

ADDISON'S DISEASE

A 61 year old lady had a 3 month history of increasing leg weakness, and a weight loss of 5.5 kg over the last 3 weeks. She had not been vomiting. On examination she was a wasted, deeply pigmented lady with a blood pressure of 80/40 when lying down. Plasma electrolytes were taken, and on the basis of these results, and the provisional clinical diagnosis of primary adrenal failure, the patient was placed on 5 mg prednisolone b.d. and admitted for further investigation. The following morning an adrenal stimulation test was performed.

Admission sample 12/10

| Plasma | Na | 115 | mmol/l | (132–144) |
|---|---|---|---|---|
| | K | 7.2 | mmol/l | (3.1–4.8) |
| | Cl | 85 | mmol/l | (93–108) |
| | HCO$_3$ | 20 | mmol/l | (21–32) |
| | Urea | 12.0 | mmol/l | (3.0–8.0) |
| | Creat. | 0.11 | mmol/l | (0.06–0.12) |
| | Cort. | 97 | nmol/l | (140–690) |

Short stimulation test 13/10

| Time | Plasma Cort. (nmol/l) (140–690) |
|---|---|
| 10.00 | 869 |
| | 250 μg Synacthen |
| 10.30 | 966 |
| 11.00 | 1007 |

The results for the plasma electrolytes (Na$^+$ ↓, K$^+$ ↑) and cortisol taken on admission support the diagnosis of Addison's disease. The results of the stimulation test to demonstrate the adrenal failure are quite unexpected, particularly the markedly raised basal level of cortisol. At this point the laboratory consulted the case records and noted that the patient had been placed on prednisolone for temporary steroid cover. This steroid has an approximate 90% cross-reactivity with the antisera used in the RIA method employed by the laboratory. The steroid medication was changed to dexamethasone, and the stimulation test repeated the following day.

| Time | Plasma Cort. (nmol/l) (140–690) |
|---|---|
| 09.30 | 110 |
|  | 250 µg Synacthen |
| 10.00 | 124 |
| 10.30 | 115 |

This lack of response confirmed adrenal failure.

### ADRENAL SUPPRESSION

A 64 year old man who is a steroid-dependent asthmatic. A prolonged stimulation test was performed to determine whether the steroid dose had allowed his adrenal cortex to remain functional.

|  | Time | Plasma Cort. (nmol/l) (140–690) |
|---|---|---|
| Day 1 | 07.20 | <55 |
|  | 09.00 | 1 mg depot Synacthen IM |
| Day 2 | 09.00 | 1 mg depot Synacthen IM |
| Day 3 | 09.00 | 1 mg depot Synacthen IM |
| Day 4 | 08.30 | <55 |
|  |  | 0.25 mg Synacthen |
|  | 09.00 | 90 |
|  | 09.30 | 135 |

Lack of stimulation of the adrenal cortex, as a result of prolonged suppression of the HPA axis by steroid, results in the start of atrophy of the adrenal within a short period (10–14 days). Suppression may arise from (1) high dose steroid therapy, (2) excess cortisol from an adrenal tumour causing atrophy of the healthy adrenocortical tissue.

In such situations the adrenal has been unstimulated for some time and may be slow to respond, if it is capable. Stimulation is therefore maintained by using depot tetracosactrin over 3 days prior to performing a short stimulation test. The adrenal cortex failed to respond in the above case.

## Congenital adrenal hyperplasia (Adrenogenital syndrome)

Congenital adrenal hyperplasia (CAH) describes a group of inherited conditions in which specific enzymes of the cortisol and aldosterone biosynthetic pathway are deficient. The consequently reduced levels of cortisol raise CRF and ACTH secretion, which then results in increased steroid production and accumulation of intermediates as far as the enzyme block, and bilateral adrenal hyperplasia. The clinical and biochemical findings in these patients vary according to the site and completeness of the enzyme deficiency (Fig. 21.2).

### 21-HYDROXYLASE DEFICIENCY

The most common of the CAH group of disorders, with an occurrence of one in about 5000 births. If the enzyme block is partial, the elevated ACTH may stimulate cortisol output that is adequate until stressful situations are encountered. The high ACTH also causes increased adrenal androgen production (androstenedione); this may result in masculinization in the newborn or virilization around the time of puberty (androstenedione is a weak androgen that is peripherally converted to testosterone). Depending on the severity of the enzyme block, the condition may be obvious at birth (ambiguous female genitalia), or not declare itself until early adult life. The presence of the condition is usually less obvious in males. The excess androgens are reflected by raised urinary 17-oxo-steroids.

The elevated ACTH, in attempting to raise cortisol production, also increases the levels of intermediate steroids, so that plasma 17-hydroxyprogesterone (and its metabolite in urine, pregnanetriol) are usually markedly raised.

If the 21-hydroxylase deficiency is severe then aldosterone production is also inadequate, so that the renal retention of sodium is reduced (*salt-losing form*). This salt-loss is aggravated further by some of the raised steroid intermediates (progesterone, 17-hydroxyprogesterone) which also promote salt loss. This form of CAH appears within a few days of birth as an adrenal crisis, with loss of salt and water leading to dehydration. Hyperkalaemia is also present.

Laboratory investigations should include plasma electrolytes, cortisol and/or ACTH if available, androgens and 17-hydroxyprogesterone, or the urinary equivalents if collections are feasible.

*Example* A two and a half week old boy was admitted to hospital to investigate his failure to gain weight (3.9 kg at birth, 3.5 kg on admission). Clinical examination and history revealed nothing of diagnostic significance. The baby was kept in for microbiological and nutritional study. On the third day of admission the baby became clinically dehydrated. A blood sample was taken for electrolyte and osmolality estimations.

| Plasma | Na | 106 | mmol/l | (132–144) |
|---|---|---|---|---|
| | K | 6.6 | mmol/l | (3.1–4.8) |
| | Cl | 70 | mmol/l | (93–108) |
| | $HCO_3$ | 19 | mmol/l | (21–32) |
| | Urea | 6.5 | mmol/l | (2.0–5.5) |
| | Creat. | 0.05 | mmol/l | (0.03–0.06) |
| | Osmo. | 217 | mmol/kg | (270–285) |

The clear evidence of severe salt loss, dehydration, and potassium retention suggested aldosterone deficiency. A provisional diagnosis of congenital adrenal hyperplasia was made and samples were taken for the measurement of plasma 17-hydroxyprogesterone and urinary pregnanetriol. Treatment comprising IV saline and mineralocorticoid (fluorocortisone) and glucocorticoid (cortisone acetate) replacement were instituted. Plasma electrolytes, osmolality and hydration state normalized over the following nine days.

| Plasma | 17-OH progesterone | 2240 | nmol/l | (<15) |
|---|---|---|---|---|
| Urine | Pregnanediol | 0.62 | μmol/24 h | (<0.5 at <2 weeks) |
| | Pregnanetriol | 6.8 | μmol/24 h | (<0.6 at <2 weeks) |

Either the abnormal plasma level of 17-hydroxyprogesterone or high urinary excretion of pregnanetriol confirms the diagnosis of 21-hydroxylase deficiency (see Fig. 21.2) in the above case. The plasma electrolytes reflect severe salt-loss, indicating that the enzyme deficiency has also compromised aldosterone biosynthesis.

## 11 β-HYDROXYLASE DEFICIENCY

Depending on the severity of the enzyme deficiency cortisol production may or may not be adequate. The block on the aldosterone pathway leads to increased 11-deoxycorticosterone, a steroid that promotes $Na^+$ retention and the hypertension that is seen in these patients. Plasma hydroxycorticosteroids and urinary 17-OHCS are elevated.

OTHER DEFICIENCIES

Deficiencies of 17 α-hydroxylase, 3 β-hydroxysteroid dehydrogenase and 20,22-desmolase are extremely rare and will not be discussed here.

## HYPERCORTISOLISM

**Classification**

*Physiological*   (1) Stress, (2) depression, (3) obesity.

*Excess ACTH*   (1) Pituitary disease (Cushing's disease), (2) malignant tumour (ectopic ACTH), e.g. oat cell carcinoma of the lung, tumours of ovary, thymus.

*Excess cortisol*   (1) Tumours of the adrenal cortex, (2) alcoholism, (3) iatrogenic—excess steroid replacement therapy.

*Excess cortisol-binding globulin*   Oestrogen—replacement therapy, oral contraceptives containing oestrogen, pregnancy.

**Case examples/pathophysiology**

CIRCADIAN RHYTHM

A 10 year old girl was investigated for obesity, which by itself is a poor marker for Cushing's syndrome. Plasma samples for cortisol levels at various times during a 24 hour period were obtained via an indwelling needle.

| Time | 16.00 | 20.00 | 24.00 | 04.00 | 08.00 | |
|---|---|---|---|---|---|---|
| Plasma Cort. | 348 | 58 | 28 | 160 | 578 | nmol/l |

This series demonstrates the normal circadian rhythm of cortisol secretion. A number of studies have shown that the analysis of this rhythm (i.e. percentage relationship between midnight and a.m. value) is a poor diagnostic test for Cushing's syndrome. The absolute value of the midnight sample is a better test, there being little overlap between values found in those with Cushing's syndrome and those who are eucorticoid. However, midnight levels are also raised in alcoholic patients and in those who are acutely ill (severe stress), such patients giving rise to false positive results.

Laboratories must determine their own reference limit for midnight plasma cortisol levels. The above case showed a normal value (<140 nmol/l), and excluded the need for further investigation of adrenal function.

EXCLUSION OF CUSHING'S SYNDROME

An obese 25 year old lady consulted about her recent onset of amenorrhoea. Physical examination revealed a slightly red patchy facial skin, but nothing else of significance. An overnight dexamethasone suppression test was performed to exclude Cushing's syndrome.

| Date | Time | Plasma Cort. (nmol/l) (140–690) |
|---|---|---|
| 14/10 | 23.00 | 320 |
| | | 2 mg dexamethasone |
| 15/10 | 08.00 | <55 |

Dexamethasone causes a negative feedback inhibition of CRF release at the hypothalamus, which in turn cuts off pituitary ACTH and cortisol secretion. This normal suppression (<140 nmol/l) therefore excludes Cushing's syndrome. This test may give false positive results in the following situations:
(1) Severe stress—acutely ill patients, severe depression, alcoholism.
(2) Oestrogen therapy—excess oestrogen stimulates CBG synthesis by the liver, which is reflected by a raised total cortisol level. Despite the suppression of cortisol release, the plasma cortisol level falls more slowly because the increase in protein-bound cortisol slows the clearance rate.
(3) Phenytoin sodium (*Dilantin*) therapy—dilantin can enhance the hepatic metabolism of dexamethasone, so reducing the effectiveness of the dose given.

With these points in mind the overnight dexamethasone suppression test provides a very useful screening test for Cushing's syndrome in ambulatory subjects. (This test is also beginning to be used, in modified form, for the assessment of severe depression, a state of stress in which plasma cortisol levels fail to suppress for a normal duration of time after the dexamethasone dose.)

The response in the above case indicates that the negative feedback mechanism is normal, and also excludes the possibility of an ectopic (autonomous) source of ACTH. This normal suppression test therefore excludes Cushing's syndrome.

## ACUTE ILLNESS

A 24 year old insulin-dependent diabetic was admitted to hospital in a coma. The level of cortisol in his admission sample was measured.

| Plasma | Na | 125 | mmol/l | (132–144) |
|---|---|---|---|---|
| | K | 8.1 | mmol/l | (3.1–4.8) |
| | Cl | 83 | mmol/l | (93–108) |
| | $HCO_3$ | 7 | mmol/l | (21–32) |
| | Urea | 21.8 | mmol/l | (3.0–8.0) |
| | Creat. | 0.40 | mmol/l | (0.06–0.12) |
| | Glu. | 92.0 | mmol/l | (3.0–5.5) |
| | Osmo. | 400 | mmol/l | (281–297) |
| | Cort. | >1600 | nmol/l | (140–690) |

Almost any form of stress can override the negative feedback of cortisol at the hypothalamus and cause increased stimulation of the adrenal cortex; typically an acute illness as in the above case, also the stress of situations such as burns, infections, fever, post-surgery, severe depression, or emotional upset can cause a prolonged, but appropriate hypercortisolism. The functions of cortisol in these situations are unclear, but probably relate to the provision of energy sources, substrates for repair mechanisms and the control of inflammatory processes.

## CUSHING'S DISEASE

A 31 year old man consulted his general practitioner about muscle weakness in his left leg. The doctor noted the man had a 'mooning of the face' and raised blood pressure. Over the next few weeks the patient developed visual field disturbances, headaches and nausea, and recalled that recently he had begun to bruise easily, with poor healing of damaged skin. The patient was admitted to hospital for investigation.

To exclude Cushing's syndrome a 24 hour urine collection was made, after which an overnight dexamethasone suppression test was performed.

| Date | Time | Plasma Cort. (nmol/l) | | Urine Cort. nmol/24 h | |
|---|---|---|---|---|---|
| 16/05 | | — | (140–690) | 1400 | (<280) |
| | 23.00 | 2 mg dexamethasone | | | |
| 17/05 | 08.00 | 1021 | | | |

The plasma cortisol level failed to suppress (<140). A prolonged dexamethasone test was commenced. Urinary cortisols were also determined.

| Date | Time | Plasma Cort. (nmol/l) | Urine Cort. (nmol/24 h) | |
|---|---|---|---|---|
| 19/05 | 08.30 | 800 | >1300 | |
|  |  | 0.5 mg/6 h dexamethasone |  | |
| 20/05 | 08.00 | 690 | 1180 | |
|  |  | 0.5 mg/6 h |  | |
| 21/05 | 08.15 | 966 | >1300 | (<280) |

The failure to suppress the plasma cortisol indicates Cushing's syndrome. To biochemically determine the aetiology the test should have then been continued with the higher dose of 2.0 mg/6 h for 2 days (hypercortisolism due to Cushing's disease will suppress, whilst high cortisol, caused by ectopic ACTH or adrenal tumour (both autonomous) will not). At this point

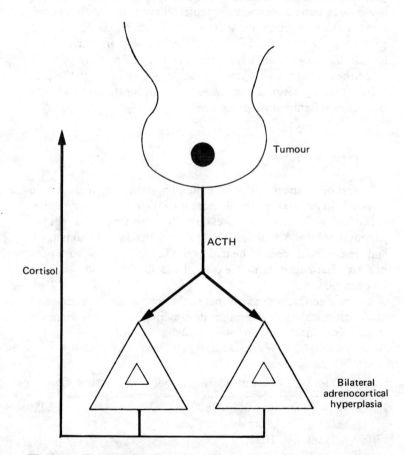

Fig. 21.5 The main features of the HPA axis in an adult with Cushing's disease due to a pituitary tumour.

however, skull X-ray studies clearly revealed the presence of a pituitary tumour. At operation a basophilic pituitary adenoma was removed. Although pituitary tumours appear to be the major cause underlying pituitary-dependent Cushing's syndrome, there is some evidence that the hypothalamus may play a role in the pathogenesis of the disease. The resulting inappropriate secretion of ACTH causes continual over-stimulation of the adrenal cortex, which leads to hypercortisolism and bilateral adrenal hyperplasia (Fig. 21.5). If estimated, plasma ACTH is often found to fall within normal limits, but this apparent normality is clearly inappropriate to the high levels of circulating cortisol.

Approximately 70% of cases of Cushing's syndrome are pituitary-dependent (75% of these are in women), ectopic ACTH underlies some 15% of cases, and the remaining 15% stem from adrenal tumours. Many of the clinical signs and symptoms of Cushing's syndrome are non-specific by themselves, their occurrence depending on the aetiology and the stage at which the disease is detected. Good screening tests are therefore important if the disease is to be diagnosed at an early stage.

**Table 21.3** Percentage occurrence of some clinical features commonly found in Cushing's syndrome

|  | Incidence (%) | Excess cortisol effect |
|---|---|---|
| Truncal obesity | 90 | Altered lipid distribution |
| Hypertension | 80 | Mineralocorticoid action ($Na^+$ & $H_2O$ retention) |
| Plethoric face | 75 | ↓ protein synthesis → weak skin |
| Hirsutism | 70 | ↑ andrenal androgens |
| Muscle weakness | 60 | ↓ protein synthesis, hyperkalaemia |
| Osteoporosis | 60 | ↓ protein synthesis |
| Easy bruising | 50 | ↓ protein synthesis |
| Poor wound healing | 40 | ↓ protein synthesis |

ADRENAL TUMOUR

A 32 year old woman attended the endocrine clinic for the investigation of marked hirsutism, weight gain and a 3 month history of amenorrhoea. The hirsutism was initially studied by taking a plasma sample for testosterone (T+DHT) and dehydroepiandrosterone sulphate estimation.

Plasma T+DHT   8.7 nmol/l   (Female = 0.9–3.6)
       DHEAS   14.7 µmol/l  (Female = 1.2–11.0)

DHEAS is the most abundant plasma steroid and is almost exclusively secreted by the adrenal. The slightly raised level in the above case suggested that the adrenal glands were the source of the excess androgen causing the hirsutism. A 24 hour urine collection for free cortisol estimation and a plasma sample for electrolytes were taken.

| Plasma | Na | 138 | mmol/l | (132–144) | Urine Cort. (nmol/l) |
|---|---|---|---|---|---|
| | K | 3.3 | mmol/l | (3.1–4.8) | 1740 (<280) |
| | Cl | 99 | mmol/l | (93–108) | |
| | HCO₃ | 26 | mmol/l | (21–32) | |
| | Urea | 4.2 | mmol/l | (3.0–8.0) | |
| | Creat. | 0.07 | mmol/l | (0.06–0.12) | |

The elevated urinary free cortisol was suggestive of Cushing's syndrome, and led to a prolonged suppression test.

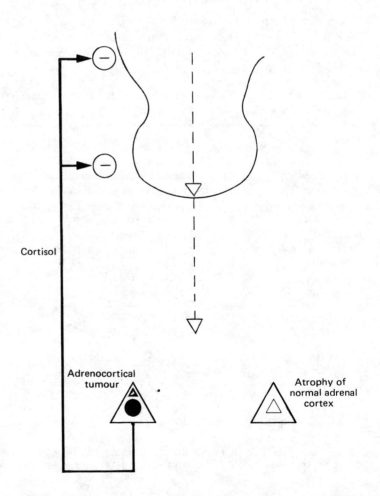

**Fig. 21.6** The main features of the HPA axis in a patient with an adrenal tumour.

| Date | Time | Plasma Cort. (nmol/l) |
|------|------|----------------------|
| 13/01 | | 0.5 mg/6 h dexamethasone |
| 14/01 | 08.30 | 795 |
| | | 0.5 mg/6 h dexamethasone |
| 15/01 | 08.30 | 660 |
| | | 2.0 mg/6 h dexamethasone |
| 16/01 | 08.15 | 730 |
| | | 2.0 mg/6 h dexamethasone |
| 17/01 | 08.30 | 890 |

Both low (2 mg/day) and high (8 mg/day) doses of dexamethasone failed to suppress the level of plasma cortisol, suggesting the presence of either an ectopic ACTH secreting tumour or an adrenal tumour. X-ray studies indicated the presence of a tumour in the right adrenal. Following right adrenalectomy, the left adrenal failed to respond to stimulation tests, suggesting complete atrophy. The patient therefore remained on glucocorticoid and mineralocorticoid medication (Fig. 21.6).

ECTOPIC ACTH SECRETING TUMOUR

A 72 year old lady was admitted to hospital with a 2 week history of epigastric pain and abdominal distension. Physical examination revealed marked hepatomegaly, oedema, widespread skin pigmentation, scaly red plaques on the back and legs and a blood pressure of 180/90. Plasma tests done on admission are shown below.

| Plasma | Na | 142 | mmol/l | (132–144) |
|--------|-----|------|--------|-----------|
| | K | 2.1 | mmol/l | (3.1–4.8) |
| | Cl | 93 | mmol/l | (93–108) |
| | $HCO_3$ | >40 | mmol/l | (21–32) |
| | Urea | 6.0 | mmol/l | (3.0–8.0) |
| | Creat. | 0.10 | mmol/l | (0.06–0.12) |
| | Glu. | 16.4 | mmol/l | (3.0–5.5) |

During the following week investigations revealed the presence of an oat cell carcinoma of the bronchus with metastases throughout the liver. A plasma sample was taken for cortisol and ACTH estimation.

| Plasma | Cort. | >1600 | nmol/l | (140–690) |
|--------|-------|-------|--------|-----------|
| | ACTH | 450 | pg/ml | (<100) |

Many tumours produce peptides that have hormone activity. Some of these peptides show immunoreactivity with ACTH antisera, and it is probable that they are similar in structure, if not identical with, the native product secreted by the anterior pituitary. Two current theories of the mechanism of ectopic hormone production are:

(1) Derepression. Genes coding for the hormone are present but normally repressed following differentiation of cells, but are derepressed by unknown factors.

(2) Arrested differentiation. Some cells forming the tissue fail to fully differentiate, so that genes inappropriate to those cells retain the ability to express themselves in certain conditions.

ACTH activity is commonly associated with oat cell carcinoma of the bronchus, and with tumours of various other tissues. The excess ACTH is autonomously produced, and stimulates the adrenal cortex, causing hyperplasia. The resulting hypercortisolism suppresses pituitary ACTH secretion (Fig. 21.7). The hypercortisolism underlies the biochemical abnormalities seen in this patient.

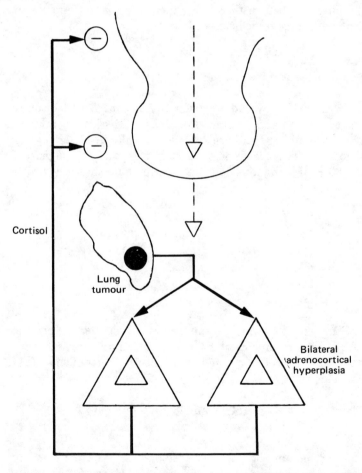

**Fig. 21.7** Main features of the HPA axis in a patient with ectopic ACTH secretion from a lung tumour.

## Pathophysiology

↑ Cort. → ↑ mineralocorticoid activity → (1) ↑ $Na^+$ retention → hypernatraemia → ↑ water retention → *hypertension*
(2) ↑ $K^+$ loss → hypokalaemia → *alkalosis*
(3) → ↑ glucocorticoid activity → anti-insulin effect → *hyperglycaemia*

Other Cushingoid features that occur in the pituitary-dependent disease may not appear in the pituitary-independent types, probably because of the often short survival times in the latter situation (ectopic ACTH secreting tumours).

β-lipotropin (which contains the melanocyte stimulating hormone sequence, MSH) is released in the same situations as ACTH, and although not measured in this case, is presumed to be responsible for the excess pigmentation (p.391).

# 22 Catecholamines and hypertension

The catecholamines are a group of compounds with the following general structure:

|  | $R_1$ | $R_2$ |
|---|---|---|
| Dopamine | $-H$ | $-H$ |
| Adrenalin | $-OH$ | $-CH_3$ |
| Noradrenalin | $-OH$ | $-H$ |

$$R_1-CH-CH_2-NH-R_2$$

(attached to a catechol ring with HO– and –OH substituents)

The major catecholamines synthesized and released from the nerve-endings of the sympathetic nervous system and the adrenal medulla are:

|  | Production site | Major action |
|---|---|---|
| Noradrenalin | Sympathetic nerve-endings and adrenal medulla | ↑ blood pressure<br>↓ heart rate<br>↑ sweating<br>↑ blood glucose |
| Adrenalin | Adrenal medulla | ↓ blood pressure<br>↑ heart rate<br>↑ sweating<br>↑ blood glucose<br>causes apprehension |
| Dopamine | Precursor of adrenalin produced in large quantities by the CNS | ↑ blood pressure<br>↓ prolactin production by the anterior pituitary |

## Metabolism of adrenalin and noradrenalin

The synthesis of the catecholamines is shown in Fig. 22.1. Adrenalin and noradrenalin are then converted to metanephrines (metadrenalin, normetadrenalin), and hydroxymethoxymandelic acid (HMMA, VMA), and excreted in the urine (see Fig. 22.2). A small amount of free adrenalin and noradrenalin is also excreted.

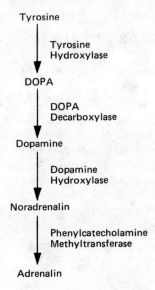

**Fig. 22.1** Synthesis of adrenalin. DOPA, dihydroxyphenylalanine.

## Urinary products

(1) HMMA—<30 µmol/day
(2) metanephrines—<5 µmol/day (mainly conjugated as sulphate and glucuronide)
(3) free adrenalin + free noradrenalin—<0.6 µmol/day (as much again excreted as conjugates)

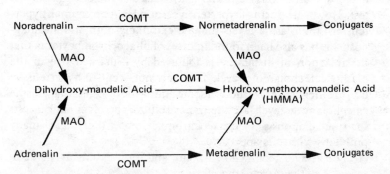

**Fig. 22.2** Metabolism of adrenalin and noradrenalin. COMT, catechol-O-methyltransferase; MAO, monoamine oxidase.

Chapter 22  **Metabolism of dopamine**

The main excretory product of dopamine is homovanillic acid. Homovanillic acid (HVA) is excreted in large amounts by patients with a neuroblastoma. Increased secretion may also occur in phaeochromocytoma, particularly if the tumour is poorly differentiated and malignant.

## LABORATORY INVESTIGATION

Tests commonly used for investigating disorders of catecholamine metabolism are:
(1) Plasma catecholamines (adrenalin and noradrenalin).
(2) Urinary catecholamines (adrenalin and noradrenalin).
(3) Urinary metanephrines (breakdown products of adrenalin and noradrenalin—not catecholamine).
(4) Urinary hydroxy-methoxymandelic acid (HMMA, VMA) (breakdown product of adrenalin and noradrenalin—not catecholamine).
(5) Urinary homovanillic acid (HVA) (metabolite of DOPA).

The major application of these tests is in the diagnosis and evaluation of phaeochromocytoma. The above tests will be discussed with this point in mind.

### PLASMA CATECHOLAMINES

For the investigation of suspected phaeochromocytoma plasma catecholamines have been measured under the following conditions: (1) resting, (2) during a hypertensive episode, (3) after a histamine injection, (4) after clonidine administration.

Resting plasma catecholamines are increased in the majority of cases of phaeochromocytoma, but normal values do occur and the diagnosis can be missed if this assay is used as the only diagnostic test. However, during hypertensive episodes (paroxysmal hypertension) the plasma levels may be extremely high and are diagnostic. It has also been found in cases of phaeochromocytoma that the injection of histamine is followed by a marked rise in the plasma catecholamine level. This does not occur in hypertension due to other causes. Although some authorities suggest that plasma catecholamine levels are the most useful of the biochemical tests for the diagnosis of phaeochromocytoma, there are many problems with this approach:
(1) The assay is technically difficult and is often only available in reference or research laboratories.
(2) Blood levels are unstable, and may decrease during collection and storage.

(3) Levels are influenced by stress (e.g. venepuncture) and exercise.
(4) Levels may be normal in a number of cases of phaeochromocytoma.
(5) Many patients with essential hypertension may have raised plasma catecholamine levels.

A test recently introduced to aid in the diagnosis of phaeochromocytoma is the clonidine suppression test. Clonidine is an antihypertensive agent which decreases resting plasma catecholamine levels by inhibiting central sympathetic activity; it has no effect on the release of catecholamines by phaeochromocytomas. Thus this test, which involves measurement of plasma catecholamine levels before and after administration of 300 µg of clonidine, is useful in the differentiation of patients with phaeochromocytoma (no suppression of plasma catecholamine), from those patients with essential hypertension and high plasma catecholamine levels (will suppress).

A useful application of plasma catecholamine levels is the determination of the site of the tumour. Blood samples are taken from various levels of the *vena cava* to determine the site of drainage of the neoplasm.

## URINARY CATECHOLAMINES

Catecholamines are excreted in both the free state and as conjugates with both sulphate and glucuronic acid. Patients with sustained hypertension due to a phaeochromocytoma excrete large quantities of catecholamines in their urine, and most patients with paroxysmal hypertension will have increased urinary levels. The estimation is technically difficult, but has proved to be a reliable test for phaeochromocytoma, provided other causes of increased excretion are excluded.

Causes of increased urinary catecholamines

(1) Severe stress, e.g. myocardial infarct, severe psychological stress, (2) severe exercise, (3) foods, e.g. bananas, (4) drugs, e.g. adrenalin, (5) phaeochromocytoma.

## URINARY METANEPHRINES

The estimation of urinary metanephrines (metadrenalin, normetadrenalin) is a relatively simple procedure. The urinary levels of metanephrines are markedly elevated in most, if not all, cases of phaeochromocytoma. Some false negatives have been recorded, but this problem can usually be overcome by estimating urinary levels on several consecutive days.

False positive results may be found following the ingestion of certain substances: (1) foods—vanilla containing substances (ice cream, bananas), (2) drugs—monoamine oxidase inhibitors, phenacetin, phenothiazides, chlorpromazine.

When all factors are considered (e.g. false positives, false negatives, technical ease of the assay, etc.) the estimation of urinary metanephrines is the investigation of choice for the diagnosis of phaeochromocytoma.

### URINARY HYDROXY-METHOXYMANDELIC ACID (HMMA)

HMMA or vanillylmandelic acid (VMA) is the major catecholamine metabolite excreted in normal urine. Increased levels occur in cases of phaeochromocytoma. The test is not technically difficult, but it is time consuming and subject to interference by a variety of substances, e.g. (1) foods—coffee, tea, citrus fruits, vanilla containing compounds, (2) drugs (a) increased levels occur with ingestion of: chlorpromazine, methyldopa, naladixic acid, phenolsulphophthalein, (b) decreased levels occur with phenothiazines, reserpine, laevodopa.

Compared with urinary metanephrines this test has a similar diagnostic efficiency, but it is technically more difficult and liable to interference by a greater number of substances.

### URINARY HOMOVANILLIC ACID (HVA)

HVA is the main excretory product of dopamine; excretion is increased in neuroblastoma and in some phaeochromocytomas, especially the poorly differentiated variety. The estimation is of little use in the evaluation of phaeochromocytoma, but it may be useful in monitoring patients following surgery for neuroblastoma. HMMA estimations usually provide similar information.

## Selection of tests

In the investigation of suspected phaeochromocytoma the most reliable tests are urinary catecholamines, HMMA and metanephrines. Most routine laboratories are able to measure HMMA and metanephrines, and either of these is suitable, but it is not necessary to measure both as they give similar information. When using any of these three tests it is necessary to observe the restrictions with regard to certain types of food and drugs (see above) and to carry out the estimations on several consecutive days (symptoms of the disease are usually of a paroxysmal nature).

If neuroblastoma is suspected, an estimation of HMMA (or metanephrines) and HVA is recommended.

# PHAEOCHROMOCYTOMA

Prevalence

Less than 0.5% of patients with hypertension.

Pathology

Secretion of catecholamines by tumours of the sympathetic nervous system (adrenal and extra-adrenal).

*Site* 99% in abdomen, and of these 90% are located in the adrenal. 10% of the tumours are malignant.

The tumours secrete large quantities of noradrenalin, whilst a few also secrete large amounts of adrenalin (Table 22.1).

**Table 22.1** Major clinical effects of the catecholamine producing tumours.

|  | Noradrenalin | Adrenalin |
| --- | --- | --- |
| Blood pressure | ↑ | Slight ↑ or ↓ |
| Heart rate | Slow | Rapid |
| Sweat glands | ↑ sweat | ↑ sweat |
| Nervous system | — | Apprehension |
| Glucose metabolism | ↑ blood glucose | ↑ blood glucose |

Chemical pathology

↑ plasma adrenalin and noradrenalin (during acute attacks)
↑ urinary excretion of adrenalin and noradrenalin (free and conjugated)
↑ urinary excretion of HMMA
↑ urinary excretion of metanephrines

Clinical

The classical presentation is that of a patient with paroxysmal bouts of hypertension associated with vasomotor symptoms (anxiety, sweating, palpitations), but usually the commonest clinical presentation is one of a sustained hypertension associated with intermittent increases in the blood pressure:

(1) Paroyxsmal hypertension with vasomotor symptoms.
(2) Persistent hypertension.
(3) Hypotension (?due to adrenalin) has been described, but orthostatic hypotension is quite common.
(4) Disorders of glucose metabolism, (a) 60% have a raised fasting blood glucose level, (b) 40% have glycosuria, (c) 30% have a diabetic glucose tolerance test.

Diagnosis

Diagnosis depends on the clinical features (hypertension) associated with one of the following: (1) increased urinary catecholamines, (2) increased urinary HMMA, (3) increased urinary metanephrines.

Plasma levels of catecholamines are not necessary for diagnosis, and they may be misleading if the sample has not been collected during a bout of hypertension; however, they are of value in localizing the tumour by taking blood samples collected at various levels in the *vena cava* to detect the site of drainage of the tumour.

*Example* Male, aged 11 years, admitted with headache, vomiting and convulsions. On examination he was found to have a blood pressure of 155/80, and a pulse rate of 122 per minute. In hospital his blood pressure ranged from 125/80 to 185/140 throughout the day. Retinal examination revealed small haemorrhages and slight hypertensive changes.

Radiology revealed a mass in the hilum of the left kidney.

| Plasma | Na | 138 | mmol/l | (132–144) |
|---|---|---|---|---|
| | K | 4.7 | mmol/l | (3.2–4.8) |
| | Glu. | 10.7 | mmol/l | (3.0–5.5) |

Urine metanephrines 5.0, 9.8, 8.3 µmol/l on 3 consecutive days (ref. range <5)

At operation a phaeochromocytoma was found in the left adrenal.

# NEUROBLASTOMA

These tumours, like the phaeochromocytomas, are tumours of tissue derived from the embryonic neural crest and are located along the sympathetic chain and in the adrenal gland. Again, like the phaeochromocytomas, they are often associated with increased secretion of catecholamines, however hypertension is rare. Neuroblastomas are very malignant tumours, mainly found in children.

Chemical pathology

Increased secretion of: (1) noradrenalin (but not adrenalin), (2) HMMA and metanephrines, (3) dopamine (main product, rarely causes sustained hypertension), (4) homovanillic acid (HVA)—this is the major excretory product of dopamine and is characteristic of neuroblastoma.

GANGLIONEUROMA

These are benign tumours that are found in both adults and children. They originate in the sympathetic chain and may be associated with increased secretion of catecholamines and hypertension.

# HYPERTENSION

Hypertension is difficult to define, but for the purposes of treatment it is based on the diastolic blood pressure level. In the adult population the normal diastolic pressure is generally agreed to be less than 90 mmHg. About 40% of the population over the age of 40 years have a diastolic pressure greater than 90 mmHg, 25% greater than 95 mmHg, and 15% greater than 100 mmHg. Some authorities suggest that all patients with a diastolic pressure greater than 100 mmHg should be treated, and others suggest that treatment should be instituted when the pressure rises above 90 mmHg.

Causes

There are two broad groups of causes. Patients are said to have primary or 'essential' hypertension if, after extensive investigation, no cause can be found. This group comprises 70–90% of the hypertensive population. Renal disease is the commonest of the secondary causes, with phaeochromocytoma and primary hyperaldosteronism constituting less than 1% of this group.

Aetiology

*Primary* Essential.

*Secondary* (1)Renal disease (a) reno-vascular, parenchymal renal disease, (2) endocrine (a) adrenal cortex—Cushing's syndrome, congenital adrenal hyperplasia, primary hyperaldosteronism, (b) adrenal medulla—phaeochromocytoma, (3) toxaemia of pregnancy, (4) coarctation of aorta, (5) miscellaneous (a) contraceptive/oestrogen therapy, (b) licorice ingestion, (c) carbenoxolone therapy, (d) renin secreting tumour.

**Laboratory investigation**

The biochemical investigation of hypertension can be divided into two groups of tests: (1) those to evaluate the hypertension (all hypertensives), (2) specific tests for causes of secondary hypertension (when indicated).

Evaluation

Plasma electrolytes and creatinine, glucose, urate, calcium, cholesterol and triglyceride.

Specific tests

(1) Catecholamines, (2) adrenocortical function.

**Evaluation**

PLASMA ELECTROLYTES AND CREATININE

These should be measured to:
(1) Assess renal function—hypertension may cause renal insufficiency (nephrosclerosis) and renal disease may result in hypertension.

(2) Screen for mineralocorticoid excess—the classical presentation of primary hyperaldosteronism is hypertension associated with hypokalaemic alkalosis. However there are many other causes of this presentation: (a) essential hypertension treated with diuretics (common), (b) Cushing's syndrome, (c) congenital adrenal hyperplasia, (d) renin secreting neoplasm, (e) renovascular hypertension, (f) licorice and carbenoxolone ingestion, (g) essential malignant hypertension associated with secondary aldosteronism.

PLASMA GLUCOSE

Diabetes mellitus is commonly associated with hypertension (accelerated atherosclerosis producing renovascular disease). Cushing's syndrome, phaeochromocytoma and Conn's syndrome often have an associated hyperglycaemia.

Acromegaly.

PLASMA CALCIUM

Approximately one third of patients with primary hyperparathyroidism have hypertension. In most cases this is due to renal disease (nephrocalcinosis).

PLASMA URATE

Hyperuricaemia (gout) can produce renal insufficiency (urate nephropathy), whilst high plasma urate levels are often associated with essential hypertension.

PLASMA CHOLESTEROL AND TRIGLYCERIDES

Plasma lipid levels should be estimated to identify factors other than hypertension which predispose to ischaemic heart disease.

## Specific tests

CATECHOLAMINES

See Phaeochromocytoma, p. 502.

ADRENOCORTICAL FUNCTION

(1) Investigation of suspected Cushing's syndrome, (2) investigation of suspected congenital adrenal hyperplasia, (3) investigation of suspected Conn's syndrome.

## Pathophysiology

ESSENTIAL HYPERTENSION

This disease is diagnosed by the following criteria: (1) hypertensive (diastolic pressure $>90$ mmHg), (2) there is no clinical, biochemical or radiological evidence to support a secondary cause of the increased blood pressure.

This group comprises 70-90% of all hypertensive patients, and within this group there are two types: (1) essential benign hypertension (major type), (2) essential malignant hypertension which comprises only a small number of cases, and is characterized by a rapid increase in blood pressure which may result in (a) papilloedema, (b) hypertensive encephalopathy, (c) rapid deterioration of renal function (nephrosclerosis).

10-30% of patients with benign essential hypertension have low (suppressed) plasma renin levels. In these cases, called 'low renin hypertension' (LHR), the plasma aldosterone level is usually within reference range. The significance of these findings is unclear (see also p. 140).

In essential malignant hypertension there may be renin overproduction (?decreased renal blood perfusion due to renal pathology), and secondary aldosteronism accompanied by hypokalaemic alkalosis.

### RENO-VASCULAR HYPERTENSION

Stenosis of a main renal artery or one of its branches results in activation of the renin-angiotensin system. The angiotensin increases the blood pressure by direct vasoconstriction and stimulation of aldosterone secretion (sodium retention). Hypokalaemic alkalosis may occur and the condition be confused with primary aldosteronism, but in the latter case the plasma renin activity is depressed. The diagnosis of renovascular hypertension can usually be made on clinical and radiological grounds.

### RENAL PARENCHYMAL DISEASE

Acute and chronic renal failure is often accompanied by hypertension. This increased blood pressure may be due to a variety of mechanisms: (1) activation of renin-angiotensin system, (2) production of unidentified vasopressor agents, (3) failure to inactivate circulating vasopressor agents, (4) sodium retention.

### CUSHING'S SYNDROME

See p.

### CONGENITAL ADRENAL HYPERPLASIA

Hypertension due to excessive mineralocorticoid production may be due to the following enzyme deficiencies: (1) $C_{11}$ hydroxylase, (2) $C_{17}$ hydroxylase.

The hypertension will have an associated hypokalaemic alkalosis.

## PRIMARY HYPERALDOSTERONISM

See p. 141.

## PHAEOCHROMOCYTOMA

See p. 501.

## ORAL CONTRACEPTIVE/OESTROGEN THERAPY

The oestrogen component of oral contraceptives may occasionally produce hypertension by stimulating the hepatic synthesis of the renin substrate angiotensinogen.

## LICORICE/CARBENOXOLONE

Licorice contains glycyrrhizinic acid which has an aldosterone-like effect. Ingestion of large amounts may produce hypertension and hypokalaemic alkalosis. Carbenoxolone, a drug used in the management of peptic ulcer, has a similar effect.

## RENIN SECRETING TUMOUR

Haemangiopericytomas (juxtaglomerular cell tumour) of the kidney have been associated with renin secretion, hypertension and hypokalaemic alkalosis.

## PREGNANCY

Toxaemia is a complication of late pregnancy, usually in primigravidas. It is characterized by hypertension, oedema, proteinuria and occasionally convulsions. The cause of the hypertension is unknown.

# 23 Tumours: biochemical syndromes

Tumours, both benign and malignant, are capable of producing a variety of substances which may be detected in the blood. These substances often have biological activity and may provide the earliest evidence of the presence of a neoplasm, or produce severe and distressing biochemical abnormalities in the later stages of the disease. These substances may be: (1) amines, (2) hormones, (3) enzymes, (4) proteins.

## TUMOURS AND AMINES

The major amines produced are 5-hydroxytryptamine (serotonin) and the catecholamines (adrenalin, noradrenalin, dopamine). Serotonin is associated with the carcinoid syndrome which is described below, while the catecholamines are secreted by phaeochromocytomas, ganglioneuromas and neuroblastomas which are discussed in Chapter 22.

### CARCINOID SYNDROME

Carcinoid tumours (neoplasm of argentaffin cells) are associated with diarrhoea, flushing and excessive production of serotonin.

*Pathology* The tumours may arise in the large bowel, the appendix, the small bowel, the stomach or the bronchus. Most carcinoid syndromes are due to tumours of the ileum. These tumours are metastatic which, if sited in the small bowel, involve the liver. Tumours of the appendix rarely metastasize, and those of the large gut metastasize but do not produce the carcinoid syndrome.

*Chemical pathology* Serotonin (5-hydroxytryptamine, 5HT) is synthesized in the argentaffin cells of the gut from tryptophan, and metabolized to 5-hydroxyindole acetic acid (5HIAA) before excretion in the urine (Fig. 23.1). In the carcinoid syndrome there is increased production of serotonin and excretion of 5HIAA. This overproduction of serotonin may give rise to tryptophan deficiency, and result in hypoproteinaemia. Carcinoid tumours de-

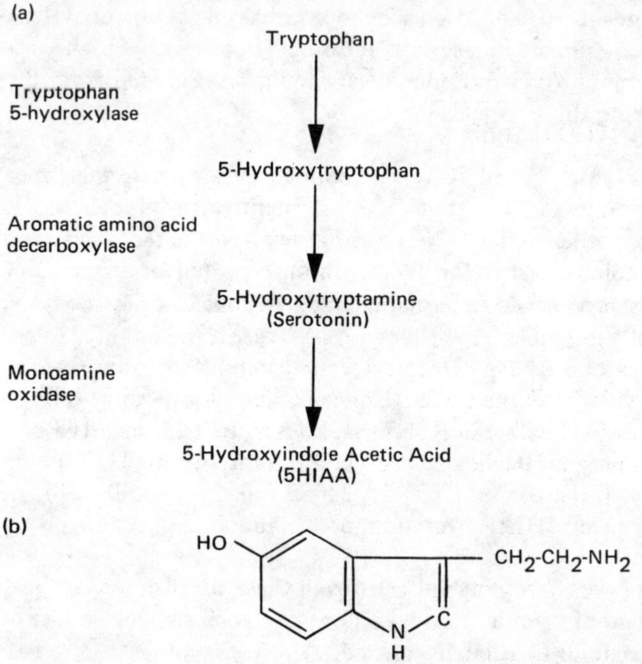

**Fig. 23.1** (a) Tryptophan metabolism. (b) Structure of 5-hydroxytryptamine (serotonin).

rived from the bronchus and stomach frequently have a deficiency of the enzyme aromatic amino acid decarboxylase, and therefore release large amounts of 5-hydroxytryptophan rather than serotonin.

*Clinical* Ileal carcinoids do not usually produce clinical signs and symptoms of the carcinoid syndrome until they metastasize to the liver; this may be due to inactivation of the tumour products as they pass through the liver via the portal vein. The common clinical manifestations of the carcinoid syndrome are: (1) flushing—erythematous, involving head and neck, (2) gut symptoms—severe diarrhoea and hypermobility, occasional malabsorption, (3) cardiac symptoms—right-sided heart failure, (4) pulmonary symptoms—bronchospasm, (5) skin—occasionally pellagra-like rash (tryptophan deficiency).

Serotonin has been implicated in the diarrhoea and heart symptoms (endocardial fibrosis), but the cause of the other clinical features is unclear.

## Diagnosis

Diagnosis is made on the clinical features, and is supported by the finding of an increased urinary output of 5HIAA. If the carcinoid syndrome is suspected, then a 24 hour urinary excretion of 5HIAA should be estimated on several occasions. Foods with a high serotonin content (bananas, tomatoes) should be excluded from the diet during the test period.

*Example* Male, aged 72 years, complaining of severe diarrhoea and flushing. The diarrhoea was persistent, with up to 12 stools being passed each day. The flushing involved the face, was deep red in colour, lasted for 1 or 2 minutes, often occurred after breakfast, and could be brought on by alcohol or excitement.

| Plasma | | | | |
|---|---|---|---|---|
| | Na | 136 | mmol/l | (132–144) |
| | K | 3.2 | mmol/l | (3.2–4.8) |
| | Cl | 112 | mmol/l | (98–108) |
| | $HCO_3$ | 20 | mmol/l | (23–33) |
| | AP | 240 | U/l | (30–120) |
| 24 hour urine 5HIAA excretion | | 680 | µmol | (<60) |

There are three notable features in these results:
(1) Hyperchloraemic (normal anion gap) acidosis, due to loss of bicarbonate in the diarrhoea fluid.
(2) High plasma alkaline phosphatase, possibly due to tumour metastases in the liver.
(3) High urinary excretion of 5HIAA.

## TUMOURS AND HORMONES

Hormone-producing tumours may be derived from (1) endocrine tissue, (2) non-endocrine tissue (*ectopic* hormone production).

# Tumours of endocrine tissue

| Tissue | Tumour | Hormone(s) secreted |
|---|---|---|
| Anterior pituitary | Adenoma | Prolactin |
| | | Growth hormone |
| | | ACTH |
| Thyroid | Medullary carcinoma | Calcitonin |
| Parathyroid | Adenoma, carcinoma | PTH |
| Adrenal cortex | Adenoma, carcinoma | Cortisol |
| | | Androgens |
| Adrenal medulla | Phaeochromocytoma | Catecholamines |
| Pancreas (islet cells) | Insulinoma ($\beta$-cells) | Insulin |
| | Gastrinoma ($\delta$-cells) | Gastrin |
| | Glucagonoma ($\alpha$-cells) | Glucagon |
| | VIPoma ($\alpha$-cells) | Vasoactive intestinal polypeptide (VIP) |
| Multiple | Multiple endocrine neoplasia (MEN) | |
| | $MEN_1$—adenoma of (1) pituitary, (2) parathyroid, (3) pancreas | |
| | $MEN_2$—(1) parathyroid adenoma, (2) medullary carcinoma (thyroid), (3) phaeochromocytoma. | |

## MEDULLARY CARCINOMA OF THYROID

This tumour, derived from the parafollicular cells of the thyroid, secretes calcitonin. Despite the high plasma levels of calcitonin, which would be expected to lower plasma calcium, the patient is usually normocalcaemic, probably due to a compensatory increase in parathyroid hormone secretion. These tumours may also secrete ACTH (ectopic ACTH syndrome) and serotonin, and may also be associated with phaeochromocytomas (see MEN below).

## GASTRINOMA

These tumours produce large quantities of gastrin, and are usually found in the pancreas. The increased plasma gastrin causes hypersecretion of acid (HCl) by the stomach, which may cause the following: (1) ulceration of stomach and upper small bowel, (2) diarrhoea, (3) malabsorption (steatorrhoea) due to neutralization of lipase activity by HCl.

The above symptom complex is known as the 'Zollinger-Ellison syndrome'. It is diagnosed by finding gastric hyperacidity associated with increased plasma gastrin levels. It should be noted that plasma gastrin levels are also increased in: (1) atrophic gastritis, (2) pernicious anaemia, (3) antacid therapy.

*Example* Male, aged 46 years, presented with a 10 year history of peptic ulceration and gastrointestinal bleeding. A gastrojejunostomy, vagotomy and anthrectomy were performed on separate occasions without controlling his symptoms. Over the previous 4 years he had nine admissions to hospital, and on several occasions has required blood transfusion because of recurrent intestinal bleeds. The plasma gastrin level was found to be $>1000$ pg/ml (r.r$<$150).

A secretin stimulation test (4 U/kg IV) showed the following results:

| Time (min) | Plasma gastrin (pg/ml) |
|---|---|
| $-30$ | 820 |
| 0 | 725 |
| $+10$ | $>1000$ |
| $+20$ | $>1000$ |
| $+30$ | $>1000$ |
| $+40$ | $>1000$ |

An abdominal ultrasound revealed a large pancreatic mass which was subsequently removed at operation. Histology revealed a benign tumour consistent with a gastrinoma. In patients with gastrinomas the administration of secretin (IV) produces a marked rise in the plasma gastrin level, whereas in normal subjects there is little or no response. The mechanism of this phenomenon is unknown.

GLUCAGONOMA

Glucagon producing tumours arise from the α-cells of the pancreatic islets. They are associated with glucose intolerance, and a curious bullous skin lesion known as necrolytic migratory erythaemia.

VIPoma

These rare tumours arise from pancreatic islet cells, and produce the gastrointestinal hormone called vasoactive intestinal polypeptide (VIP). This peptide increases intestinal motility, therefore these tumours are associated with profuse watery diarrhoea. The associated syndrome is known as the Werner-Morrison or WDHA (Watery Diarrhoea, Hypokalaemia, Achlorhydria) syndrome.

## MULTIPLE ENDOCRINE NEOPLASIA (MEN)

In this rare condition there are hormone producing tumours of two or more endocrine tissues present at the same time. They are familial disorders inherited as autosomal dominant traits. Two distinct groups of syndromes have been described:
(1) $MEN_1$ involves two or more tumours of the following tissues: parathyroid gland, pancreatic islets (insulinoma, gastrinoma), anterior pituitary.
(2) $MEN_2$—Medullary carcinoma of thyroid, phaeochromocytoma, parathyroid adenoma.

### Hormone production by tumours of non-endocrine tissue

Synthesis (and secretion) of a hormone by tissues that do not normally produce that hormone has been defined as 'ectopic production'. These ectopic hormones are invariably polypeptides, and are usually produced by malignant or potentially malignant tissue. In most cases the ectopic polypeptides produced have yet to be demonstrated as being identical to the natural hormone. Various theories have been put forward to explain this ectopic hormone synthesis.

The derepression theory suggests that all cells have the potential to produce any peptide normally produced in the body, but during tissue development and differentiation most of these mechanisms are repressed or rendered functionless. Thus a cell only produces those peptides necessary for its own function. The theory suggests that as a cell undergoes neoplastic change some of these mechanisms are derepressed, thus allowing peptides not normally produced by that cell to be synthesized. An alternative theory proposes that some cells fail to fully differentiate from their primitive level, leaving some genes, not normally functional in the differentiated tissue, able to express themselves under certain conditions (arrested differentiation).

For the definitive diagnosis of ectopic hormone production the following criteria would have to be satisfied:
(1) Demonstration of the presence of the hormone in the non-endocrine tissue.
(2) Disappearance of the hormone from the plasma and loss of clinical manifestations after removal of the tumour.
(3) Production of the hormone by the tumour tissue *in vitro*.

Under normal circumstances it is difficult to satisfy these criteria, and a diagnosis is usually suspected on clinical and biochemical grounds alone, i.e. ectopic hormone production is commonly suspected but rarely proved.

Ectopic hormone syndromes

Tumours that produce substances that have the biological effects of the following hormones have so far been described:
(1) Parathyroid hormone (PTH), (2) antidiuretic hormone (ADH), (3) corticotropin (ACTH), (4) calcitonin, (5) growth hormone, (6) prolactin, (7) gonadotropins, (8) placental lactogen (HPL), (9) tumour hypoglycaemia, (10) erythropoietin.

Tumour hypercalcaemia

10–20% of patients with malignant tumours have hypercalcaemia, particularly those with tumours of the lung, breast, kidney or lymphoid tissue. In many of these patients ($\sim 80\%$) bony metastases are demonstrably present. However, in some ($\sim 10\%$) there is no evidence of bone secondaries, and the biochemical features resemble those of primary hyperparathyroidism (i.e. hypercalcaemia associated with hypophosphataemia). For this syndrome the term 'pseudohyperparathyroidism' has been applied, as it was thought to be due to ectopic production of parathyroid hormone.

Investigation of these cases by measuring plasma PTH has not produced a clear answer. In some patients there was an increased circulating level of PTH, but in others the PTH was undetectable. In those cases where plasma PTH was increased the polypeptide appeared to have immunological characteristics different from the native PTH found in primary hyperparathyroidism. This suggests that the hypercalcaemia in these cases of pseudohyperparathyroidism may be due to the elaboration by the tumour of a PTH-like substance rather than native PTH. Until the question is resolved it is better to refer to this syndrome as 'humoral hypercalcaemia of malignancy'.

The possible causes of hypercalcaemia of malignancy are manifold: (1) ectopic PTH production, (2) elaboration of a PTH-like substance, (3) bone secondaries, (4) elaboration of a vitamin D-like substance, (5) production of prostaglandins (PGE will cause dissolution of bone), (6) concurrent primary hyperparathyroidism. For examples see p. 181.

## Ectopic ADH secretion

The inappropriate secretion of ADH results in: (1) hyponatraemia, (2) plasma hypo-osmolality, (3) urine/plasma osmolality $>1$, (4) urine [sodium] $>20$ mmol/l.

This syndrome of inappropriate secretion of ADH (SIADH) is often associated with carcinoma of the bronchus, but may also occur with neoplasms of the gut, pancreas and lymphoid tissue. SIADH may also arise in other non-neoplastic conditions, e.g. diseases of the lung, brain (see p. 51).

## Ectopic ACTH syndrome

Tumours of the lung, pancreas and thymus are often associated with corticotropin production, leading to adrenal hyperplasia and hypercortisolism. This syndrome is usually manifested by hypokalaemic alkalosis, and mucosal and skin pigmentation. The symptoms and signs of Cushing's syndrome are not usually present.

The hypokalaemia reflects excessive mineralocorticoid activity, and is more likely to be due to excessive production of corticosterone and deoxycorticosterone rather than to the action of the massive amounts of circulating cortisol. The hyperpigmentation is due to the MSH (melanocyte-stimulating hormone) activity of β-lipotropin, which is secreted along with ACTH. The clinical features of Cushing's syndrome are rarely encountered in this disorder because they take some months to develop, and often patients with these ectopic secreting tumours do not survive long enough for this to occur. Other neoplasms which may be associated with ectopic ACTH syndrome include medullary carcinomas of the thyroid, tumours of the pancreatic islets and phaeochromocytomas (p. 493).

## Calcitonin

Elevated plasma levels of calcitonin do not cause recognizable clinical manifestations, and so measurement of the neoplastic tissue levels of this hormone would be required to demonstrate ectopic production of this hormone. Increased plasma levels of immunoreactive calcitonin have been reported to occur in oat cell carcinomas of the lung and in carcinomas of the breast.

### Growth hormone

Increased plasma levels of immunoreactive growth hormone have been observed in patients with tumours of the lung, and in some with gastric carcinomas. The lung tumours have been associated with hypertrophic osteoarthropathy, a condition which resembles the bone disorders caused by the growth hormone excess of acromegaly. The bone disorders have regressed after resection of the tumour, and therefore it has been suggested that the tumour had been producing growth hormone.

### Prolactin

Increased plasma levels of immunoreactive prolactin have been detected in some cases of breast carcinoma, renal cell carcinoma and oat cell carcinoma of the lung. In all such cases the plasma level has decreased after specific therapy. Although prolactin was elevated, not all those cases had associated galactorrhoea.

### Gonadotropins

Human chorionic gonadotropin (hCG) is normally detectable in plasma only during pregnancy. High plasma levels have been reported in males with testicular tumours, and in females with ovarian tumours, hydatiform moles and choriocarcinomas. However, in these cases the hormone production could not be considered to be ectopic in nature. Ectopic production, often resulting in gynaecomastia, has been reported in patients with tumours of the stomach, pancreas or liver.

### Placental lactogen (HPL)

HPL, normally produced by the placenta, has been detected in some tumours of the ovary, testis and lung.

### Tumour hypoglycaemia

The hypoglycaemia due to non-islet cell tumours is of the fasting type, and is indistinguishable from that caused by pancreatic islet cell tumours. The tumours that have been implicated are: (1) mesenchymal tumours (mesotheliomas, fibrosarcomas), (2) hepatomas, (3) adrenocortical carcinomas, (4) carcinomas of the gastrointestinal tract.

The mechanism of the hypoglycaemia is unclear but several possibilities have been proposed: (1) ectopic insulin production (unlikely), (2) production of an insulin-like substance (likely), (3) increased glucose utilization by the tumour, (4) inhibition of glycogenolysis by an unidentified tumour substance.

Erythropoietin

Erythropoietin is a hormone, normally produced by the kidney, which stimulates bone marrow production of erythrocytes. Ectopic production of this hormone has been described in association with hepatomas, uterine fibromas, bronchogenic carcinomas and haemangioblastomas.

## TUMOURS AND ENZYMES

Malignant disease is often associated with increased plasma levels of a wide variety of enzymes. The origin of these enzymes is two-fold:
(1) Release from malignant cells due to increased cell turnover, increased cell membrane permeability or cell necrosis.
(2) Release from the cells of the invaded tissue due to cell necrosis, or enzyme induction, or regurgitation into the plasma due to obstruction of duct systems.

**Enzymes released from neoplastic cells**

The enzymes released from neoplastic cells fall into three categories.

Glycolytic enzymes

Neoplastic cells have an increased rate of aerobic and anaerobic glycolysis compared with that of normal tissue. Thus enzymes such as lactate dehydrogenase, aldolase and phosphohexokinase are often released into the plasma in large amounts.

Organ specific enzymes

This group includes bone alkaline phosphatase (osteogenic sarcoma), acid phosphatase (carcinoma of the prostate) and possibly amylase (carcinoma of the pancreas).

Enzymes not normally associated with the particular tissue

*Alkaline phosphatase* The isoenzyme of alkaline phosphatase known as the Regan enzyme may be found in patients with carcinoma of the lung, ovary, pancreas or colon.

*Amylase* Increased plasma levels of this enzyme have been described in cases of carcinomas of the lung, ovary and colon.

It is possible that these two enzymes are normally present in many tissues and that the increased levels found in malignancies of various tissues merely reflects increased production, i.e. they are not examples of ectopic production of enzymes.

**Enzymes released due to invasion of tissue**

The enzymes released from organs involved in malignant infiltrations include the following:

*Liver* Alkaline phosphatase, gamma-glutamyltransferase, transaminases, 5'-nucleotidase, lactate dehydrogenase, aldolase.

*Bone* Alkaline phosphatase.

*Pancreas* Amylase.

No enzyme has yet been found to be specific for carcinoma, and therefore increased plasma levels are not useful diagnostic indicators for malignancy. However, they are occasionally useful in the detection of secondaries (e.g. alkaline phosphatase for liver metastases), and in following the results of therapy (e.g. acid phosphatase in carcinoma of the prostate).

## TUMOURS AND PROTEINS

Most neoplastic diseases are associated with a decrease in the total plasma protein concentration, due mainly to a decreased albumin concentration. In some cases there is an increase in the total protein due to the presence of paraproteins. In many cases there are increases in the acute phase reactants. Other proteins that may be increased are the tumour associated antigens.

### IMMUNOGLOBULINS

See p. 251.

## ACUTE PHASE REACTANTS

Proteins which make up the $\alpha_1$- and $\alpha_2$-globulins on the electrophoretic pattern all tend to increase when there is acute tissue damage. These include $\alpha_1$-antitrypsin, haptoglobins and caeruloplasmin, and are increased in inflammations, trauma and autoimmune diseases as well as malignancy. The other acute phase proteins such as albumin, prealbumin and transferrin tend to decrease, while fibrinogen and C-reactive protein tend to increase.

## TUMOUR ASSOCIATED ANTIGENS

### Carcinoembryonic antigen (CEA)

CEA is a glycoprotein that is synthesized by cells of fetal intestinal tissue. In normal adults it may be detected in low concentration ($<2.5$ µg/l) in the plasma. High plasma levels are often found in adult patients with carcinomas of the gastrointestinal tract (70–80%), and occasionally in other malignancies, e.g. carcinoma of the breast, leukaemia. Increased levels may also be found in the plasma of patients with non-malignant conditions such as pancreatitis, ulcerative colitis and cirrhosis of the liver. Approximately 20% of heavy cigarette smokers also have increased plasma levels. Because of the wide range of conditions in which elevated plasma levels of CEA occur its estimation is useless as a test for the diagnosis of intestinal carcinoma. However, CEA measurement is of value in the detection of intestinal tumour recurrence in treated patients. In such patients several plasma CEA levels should be estimated to establish a post-treatment baseline, followed by regular monitoring.

### α-Fetoprotein (αFP)

This protein is synthesized by the developing fetal liver and probably fulfils the role of albumin. Synthesis ceases at about the beginning of extra-uterine life, so that αFP is almost undetectable in adult plasma. Raised plasma levels in adults are often associated with hepatomas (60–90%) and teratomas, but again the value of measurement in these conditions has been in the ongoing assessment of treatment, rather than in the initial diagnosis. Plasma αFP levels may also be raised in various non-malignant conditions such as hepatitis, liver cirrhosis and ulcerative colitis.

# Further reading

## GENERAL

Brown S.S., Mitchell F.L. & Young D.S. (eds) (1979) *Chemical Diagnosis of Disease.* Elsevier/North Holland, Biochemical Press, Amsterdam.

Zilva J.F. & Pannall P.R. (1979) *Clinical Chemistry in Diagnosis and Treatment*, 3rd edn. Lloyd-Luke (Medical Books), London.

Tietz N.W. (ed.) (1976) *Fundamentals of Clinical Chemistry*, 2nd edn. W.B. Saunders, Philadelphia.

Whitby L.G., Percy-Robb I.W. & Smith A.F. (1980) *Lecture Notes on Clinical Chemistry*, 2nd edn. Blackwell Scientific Publications, Oxford.

Guyton A.C. (1981) *Textbook of Medical Physiology*, 6th edn. W.B. Saunders, Philadelphia.

Galen R.S. & Gambino R.S. (1975) *Beyond Normality: the Predictive Value and Efficiency of Medical Diagnoses.* Wiley, New York.

## FLUID BALANCE, ELECTROLYTES, ACID BASE

Maxwell M.H. & Kleeman C.R. (eds) (1980) *Clinical Disorders of Fluid and Electrolyte Metabolism*, 3rd edn. McGraw-Hill, New York.

Beck L.H. (ed.) (1981) Symposium on fluid and electrolyte disorders. *Med. Clin. N. Amer.* **65**, 2.

Brenner B.M. & Stein J.H. (eds) (1978) *Acid-base and Potassium Homeostasis.* Churchill Livingstone, New York.

## RENAL DISEASE

Black D.K. & Jones N.F. (1979) *Renal Disease*, 4th edn. Blackwell Scientific Publications, Oxford.

## ENDOCRINOLOGY

De Groot L.J. (ed.) (1979) *Endocrinology.* Grune and Stratton, New York.

Williams R.H. (ed.) (1981) *Textbook of Endocrinology*, 6th edn. W.B. Saunders, Philadelphia.

YEN S.S.C. & JAFFE R.B. (eds) (1978) *Reproductive Endocrinology.* W.B. Saunders, Philadelphia.

## METABOLISM

STANBURY J.B., WYNGAARDEN J.B. & FREDRICKSON D.S. (eds) (1978) *The Metabolic Basis of Inherited Disease*, 4th edn. McGraw-Hill, New York.
BONDY P.K. & ROSENBERG L.E. (eds) (1980) *Metabolic Control and Disease*, 8th edn. W.B. Saunders, Philadelphia.

## GASTROENTEROLOGY

SHERLOCK SHEILA (1981) *Diseases of the Liver and Biliary System*, 6th edn. Blackwell Scientific Publications, Oxford.
DUTHS H.L. & WORMSLEY K.G. (eds) (1979) *Scientific Basis of Gastroenterology.* Churchill Livingstone, Edinburgh.

## LIPIDS-LIPOPROTEINS

LEWIS B. (1977) *The Hyperlipidaemias.* Blackwell Scientific Publications, Oxford.

## ENZYMES

WILKINSON J.H. (1976) *The Principles and Practice of Diagnostic Enzymology.* Edward Arnold, London.

## DIABETES

NATIONAL DIABETES DATA GROUP, Classification and diagnosis of diabetes mellitus and other categories of glucose intolerance. *Diabetes* (1979) **28**, 1039–1057.
WHO Expert Committee on Diabetes Mellitus. WHO Technical Report Series, (1980), No. 646.

# Conversion factors for SI units

| Analyte | SI units | Conversion (×) | Traditional units |
|---|---|---|---|
| **Plasma** | | | |
| Albumin | g/l | 0.1 | g/dl |
| Bicarbonate | mmol/l | 1 | meq/l |
| Bilirubin | μmol/l | 0.058 | mg/dl |
| Blood gases | | | |
| $P\text{co}_2$ | kPa | 7.5 | mmHg |
| $P\text{o}_2$ | kPa | 7.5 | mmHg |
| Calcium | mmol/l | 4 | mg/dl |
| Chloride | mmol/l | 1 | meq/l |
| Cholesterol | mmol/l | 38.6 | mg/dl |
| Cortisol | nmol/l | 0.036 | μg/dl |
| Creatinine | mmol/l | 11.3 | mg/dl |
| Glucose | mmol/l | 18 | mg/dl |
| Iron | μmol/l | 5.6 | μg/dl |
| Iron binding capacity | μmol/l | 5.6 | μg/dl |
| Lactate | mmol/l | 9 | mg/dl |
| Magnesium | mmol/l | 2.4 | mg/dl |
| Oestradiol | μmol/l | 27.24 | μg/dl |
| Osmolality | mmol/kg | 1 | mosmol/kg |
| Phosphate | mmol/l | 3.1 | mg/dl |
| Potassium | mmol/l | 1 | meq/l |
| Protein (total) | g/l | 0.1 | g/dl |
| Sodium | mmol/l | 1 | meq/l |
| Testosterone | nmol/l | 28.8 | ng/dl |
| Thyroxine | nmol/l | 0.078 | μg/dl |
| Triiodothyronine | nmol/l | 65.1 | ng/dl |
| Triglyceride | mmol/l | 88.5 | mg/dl |
| Urate | mmol/l | 16.8 | mg/dl |
| Urea | mmol/l | 6 | mg/dl |

*Conversion factors for SI units*

| Analyte | SI units | Conversion (×) | Traditional units |
|---|---|---|---|
| **Urine** | | | |
| Calcium | mmol/24 h | 40 | mg/24 h |
| Cortisol | nmol/24 h | 0.36 | µg/24 h |
| Creatinine | mmol/24 h | 113 | mg/24 h |
| Creatinine clearance | ml/s | 60 | ml/min |
| 5 HIAA | µmol/24 h | 0.2 | mg/24 h |
| Metanephrines | µmol/24 h | 0.24 | mg/24 h |
| Phosphate | mmol/24 h | 31 | mg/24 h |
| Urate | mmol/24 h | 168 | mg/24 h |
| Urea | mmol/24 h | 0.06 | g/24 h |

# Index

Abetalipoproteinaemia 333

Acetoacetate 210, 211

Acetone 210, 211

Acidaemia 98

Acid-base
    acidosis: metabolic 121
        metabolic, and respiratory
          acidosis 130
        metabolic, and respiratory
          alkalosis 133
        renal tubular 83, 123
        respiratory 130
    actual bicarbonate 110
    alkalosis: metabolic 126
        metabolic, and respiratory
          acidosis 132
        metabolic, and respiratory
          alkalosis 132
        respiratory 124
    anion gap and 90, 112
    base excess 111
    blood gases 108
    bone buffer 107
    carbon dioxide metabolism 104
    definitions: acid 95
        acidaemia 98
        acidosis 98, 99
        alkalaemia 98
        alkali 96
        alkalosis 98, 99
        base 96
        buffer 96
        compensation 99
        metabolic component 98
        respiratory component 98
    hydrogen ion metabolism 100
    interpretation of blood gases 113
    intracellular buffer 106
    laboratory investigation of 107
    standard bicarbonate 110

Acidosis
    metabolic: causes 121
        compensation 118
        definition 98, 99
        diabetes mellitus and 121
        diarrhoea and 124
        hyperchloraemic 83, 112
        increased anion gap and 90, 112
        normal anion gap and 90, 112
        renal failure and 122
        renal tubular acidosis and 83, 123
        respiratory acidosis and 130
        respiratory alkalosis and 133
    respiratory: acute 119
        causes 130
        chronic 119
        compensation 119
        emphysema and 130
        metabolic acidosis and 130
        metabolic alkalosis and 132
        respiratory arrest and 131

Acid phosphatase (ACP), plasma 288
    increased, causes of 288
    prostatic disease and 288, 289

Acromegaly 422

ACTH (see adrenocorticotropic hormone)

Acute intermittent porphyria (AIP) 372

Acute phase proteins 236

Acute tubular necrosis 156, 161

Addison's disease
    cortisol and 481, 483
    hyperkalaemia and 70, 482, 483
    hyponatraemia and 46, 482, 483

ADH (see antidiuretic hormone)

Adrenal cortex
    hormones of 464
    steroid biosynthesis and 465

# Index

Adrenal hyperplasia, congenital (CAH) 485
Adrenalin 496
  phaeochromocytoma and 501
Adrenal medulla
  hormones of 496
Adrenal tumour
  cortisol and 491
  testosterone and 491
Adrenocorticotropic hormone (ACTH) 391, 399, 462, 468, 511, 515
  ectopic 87, 493, 515
  metyrapone test and 472
  plasma electrolytes and 400
  stimulation test 401, 471, 472
Adrenogenital syndrome (see congenital adrenal hyperplasia)
Agammaglobulinaemia 254
Airway disease, chronic obstructive
  acidosis and 130, 132
  hyperkalaemia and 69
Alanine aminotransferase (ALT), plasma 298
  alcoholic liver disease and 314, 315
  cholestatic jaundice and 299, 306
  cirrhosis and 311, 312, 315
  differential diagnosis of jaundice and 302, 309
  hepatitis and 302, 305, 306, 307, 309
  liver anoxia and 310
  liver disease and 298, 302
  liver infiltrations and 313
Albumin, plasma 242 (see also hyper- and hypoalbuminaemia)
  chronic hepatitis and 307
  cirrhosis and 312
  diagnosis of jaundice and 298, 302
  function of 243
  hyperalbuminaemia 243
  hypoalbuminaemia 244
  liver function tests and 298
Alcoholic liver disease
  cirrhosis and 315
  fatty liver and 314
  hepatitis and 315
  hypoalbuminaemia and 245, 315
  plasma enzymes and 314
Alcoholism
  γ-glutamyltransferase and 287
  magnesium and 202
  plasma lipids and 330
Aldosterone
  ACTH and 135
  action of 138
  control of 135
  deficiency of 139, 147
  enzyme defects and 147
  excess of 138, 142
  heparin and 148
  laboratory investigation of 139
  metabolism 134
  plasma electrolytes and 139
  plasma volume and 137
  plasma renin activity (PRA) and 140
  potassium and 61, 136
  renal disease and 148
  sodium homeostasis and 24, 138
  spironolactone and 148
  structure of 134
  synthesis of 134
Aldosteronism, primary
  alkalosis and 127
  hypokalaemia and 77, 141, 143
  laboratory investigation of 141
  pathophysiology of 143
  plasma sodium and 41, 139, 143
Aldosteronism, secondary
  causes of 142, 144
Alkalaemia 98
Alkali ingestion
  alkalosis and 128
  hypokalaemia and 73
  magnesium and 201
Alkaline phosphatase, plasma
  alcoholic liver disease and 314
  bone disease and 283
  calcium (plasma) and 281
  carcinoma and 284, 285
  children and 282
  cholestasis and 283, 308
  cirrhosis and 312
  diagnosis of jaundice and 309
  γ-glutamyltransferase and 280
  hepatitis and 283, 305
  hyperparathyroidism and 284
  increased, causes of 281
  isoenzymes of 280
  laboratory investigation of 280

# Index

liver disease and 299
liver infiltrations and 313
5'-nucleotidase and 280
Paget's disease and 283
pregnancy and 282
renal failure and 285
Alkalosis
  metabolic: aldosteronism
    and 127
    classification of 126
    compensation of 118
    definition of 98, 99
    hypophosphataemia and 193
    respiratory acidosis and 132
    saline responsive 127
    saline unresponsive 127
    vomiting and 127
  respiratory: acute 119, 125
    causes of 124
    chronic 120
    compensation of 119, 120
    metabolic acidosis and 133
    metabolic alkalosis and 132
    septicaemia and 125
Allopurinol
  uric acid and 383, 386
δ-Aminolaevulinic acid (ALA) 361
Ammonia
  urinary buffer and 104
Amylase 263 (see also
  hyperamylasaemia)
  plasma 263
  urinary 264
Amylase: creatinine ratio 265
Angiotensin converting enzyme (ACE),
  plasma 291
  sarcoidosis and 291
Angiotensin 136, 137
Anion gap 89
  acidosis and 90
  calculation of 89
  decreased, causes of 92
  definition of 89
  error of 90
  hypoalbuminaemia and 93
  increased, causes of 90
  ketoacidosis and 90, 92
  laboratory error and 91, 93
  laboratory investigation of 90
  myeloma and 94
  renal failure and 93
Anterior pituitary gland
  control of 388

hormones, investigation of 398
hormones of 390
Antidiuretic hormone 22
  control of 22
  ectopic 51, 515
  inappropriate secretion of
    (SIADH) 51
  osmolality and 22
$\alpha_1$-Antitrypsin, plasma
  liver disease and 301
Aspartate aminotransferase (AST),
  plasma 276
  elevated, causes of 276
  liver disease and 276, 298
  muscle disease and 276
  myocardial infarction and 276
Atheroma
  lipoproteins and 319
Bartter's syndrome
  aldosterone and 145
Base excess 111
Bence-Jones protein 240, 255, 256
Bicarbonate, plasma
  actual bicarbonate 110
  aldosteronism and 127, 143
  alkali ingestion and 128
  cardiopulmonary arrest and 131
  consumption of 100, 104
  control of renal reabsorption 103
  decreased plasma levels,
    causes of 109, 120
  decreased plasma levels,
    investigation of 107, 109
  diabetic ketoacidosis and 121
  diarrhoea and 124
  increased plasma levels,
    causes of 108, 125
  increased plasma levels,
    investigation of 107, 108
  renal conservation of 103
  renal failure and 122
  renal regeneration of 104
  renal tubular acidosis and 123
  respiratory acidosis and 119, 130
  respiratory alkalosis and 119, 124
  salicylate ingestion and 133
  standard bicarbonate 110
Bile salt $^{14}CO_2$ test 342
Bile salts, plasma
  liver disease and 300
Bilirubin (see also hyperbilirubinaemia)
  metabolism 294

# Index

plasma 297
urinary 297
Blind loop syndrome (see stagnant loop syndrome)
Blood gases 108
    interpretation of 113, 114, 115, 116
Broad beta disease 331
Bromsulphthalein (BSP) excretion
    liver disease and 300
    liver infiltrations and 313
Buffer
    definition 96
    extracellular 100
    pH of a system 96
Caeruloplasmin, plasma
    liver disease and 301
Calciferol (see Vitamin D)
Calcitonin
    calcium homeostasis and 170
Calcium 165 (see also hyper- and hypocalcaemia)
    albumin and 166, 173
    alkaline phosphatase and 174, 281
    bicarbonate and 176
    calcitonin and 170
    cholecalciferol and 168
    kidney and 165
    laboratory investigation of 172
    metabolism 165
    phosphate and 174, 177
    plasma levels 166, 173
    PTH and 167, 175
    regulation of plasma level 166, 167–172
    steroid suppression test and 176
    tumour hypercalcaemia 178, 180, 181, 514
    urinary 165, 177
    urinary phosphate and 177
    vitamin D and 168
Carbohydrate
    digestion and absorption 205
Carbon dioxide 104
    excretion 105
    production 104
    renal response to 106
    transport in blood 104
Carbonic anhydrase
    erythrocytes and 105
    renal tubules and 101

Carcinoembryonic antigen (CEA)
    malignancy and 519
Carcinoid syndrome 508
    5-hydroxyindole acetic acid and 508
Cardiac enzymes 271
Carotenes, plasma 337
Catecholamines 496
    action of 496
    clonidine suppression test and 499
    DOPamine 496
    homovanillic acid (HVA) 500
    laboratory investigation of 498
    metabolism 497
    metanephrines 499, 502
    neuroblastoma and 502
    phaeochromocytoma and 501
    production site 496
    plasma 498
    urinary 499
    vanillylmandelic acid (VMA, HMMA) 500
Cerebrospinal fluid (CSF) 258
    chloride and 260
    glucose and 217, 260
    Guillain-Barré syndrome and 261
    immunoglobulins and 259
    increased protein and 259
    laboratory investigation of 258
    meningitis and 260, 261
    production of 258
    protein and 259
Chloride (see also hyper- and hypochloraemia)
    absorption 80
    excretion 80
    intake 80
    laboratory investigation of 82
    plasma-erythrocyte shift 80, 105
    plasma levels 81
    urinary 82
Cholestatic jaundice
    alkaline phosphatase and 283, 308
    carcinoma of pancreas and 309
    causes of 308
    cholelithiasis and 308
    extrahepatic 308
    γ-glutamyltransferase and 287, 299

intrahepatic 309
lipids, plasma and 328
liver function tests and 302, 309
Cholesterol (see also hypercholesterolaemia)
dietary 319
endogenous 319
hyperlipidaemia and 322, 325, 326
laboratory investigation of 321
metabolism 318
Cholinesterase 290
dibucaine number 290
liver disease and 291
organophosphates and 291
suxamethonium sensitivity and 290
Chylomicrons 317
Cirrhosis
albumin and 311
alcoholic 315
ALT and 312
alkaline phosphatase and 312
bilirubin and 312
calcium and 186
causes of 311
definition 310
Clearance, renal
creatinine 154
urea 154
Clomiphene test 410
Clonidine suppression test 499
Coagulation factors
liver disease and 300
Coeliac disease 346
Congenital adrenal hyperplasia (CAH) 146, 147, 485
Congenital erythropoietic porphyria 378
Corticotropin-releasing-factor (CRF) 389, 468
Cortisol 462 (see also hyper- and hypocortisolaemia)
Addison's disease and 481, 483
circadian rhythm and 462, 466, 487
control of 462
Cushing's disease and 489
effects of 466
glucose and 468
investigation, selection of tests 477
laboratory investigation of 467
plasma 400, 469
plasma electrolytes and 467
sex hormones and 464
stimulation tests 471
suppression tests 474
transport, plasma 466
urinary 401
Cortisol-binding globulin (CBG) 466, 470
C-peptide 216
Creatine kinase (CK), plasma 272
elevated, causes of 272
hypothyroidism and 275
isoenzymes 275
muscle disease and 274
myocarditis and 274
myocardial infarct and 273
Creatinine 150
decreased plasma level, causes of 164
increased plasma level, causes of 156
laboratory investigation of 151, 157
metabolism 150
plasma level 153
renal clearance 154
renal disease and 156
urinary 154
Crigler-Najjar syndrome 302, 304
Cushing's disease 469, 489
Cushing's syndrome 146, 469, 491, 493
Cystic fibrosis 344
sweat electrolytes and 342, 344
Dehydration
consequences of:
hypertonic 42
hypotonic 54
isotonic 58
laboratory investigation of 36
plasma osmolality and 31
urine osmolality and 31
Dexamethasone suppression test 402
Cushing's syndrome and 488, 489, 491
prolonged 474
short 474
Diabetes insipidus
hypernatraemia and 37, 39
Diabetic ketoacidosis 220, 223

Diabetes mellitus
    acidosis and 86, 92, 121, 206, 220, 223, 225
    amylase and 270
    anion gap and 86, 92
    classification 218
    coma and 219
    definition of 218
    glucose tolerance test and 213
    hyperchloraemia and 86, 223
    hyperkalaemia and 69, 222
    hyperosmolar coma and 224
    hyperphosphataemia and 196, 223
    hypertriglyceridaemia and 329
    hypokalaemia and 73
    hyponatraemia and 47, 49
    hypophosphataemia and 193, 223
    ketoacidosis and 220, 223
    laboratory investigation of 211
    lactic acid and 206, 225
    magnesium and 203
    osmotic diuresis and 37, 47
    plasma glucose and 212
    plasma lipids and 329
Diarrhoea
    acidosis and 124
    hyperchloraemia and 84
Dibucaine number 290
Digestion and absorption
    bile salts, of 335
    calcium, of 336
    carbohydrates, of 335
    iron, of 336, 353
    lipids, of 334
    magnesium, of 336
    proteins, of 335
    vitamins, of 335
1,25-Dihydroxycholecalciferol 169
Discretionary testing 16
Disaccharidases 335, 340
    deficiency of 339, 350
Diuretics, thiazide
    hypokalaemia and 76
    hypophosphataemia and 194
    plasma calcium and 184
DOPamine 496
Drugs
    allopurinol 383
    cholestyramine 460
    effect on test interpretation 6
    glibenclamide 231
    oestrogens 5, 432, 449
    phenformin 228
    phenytoin sodium 432
    prednisolone 6
    salbutamol 74
    salicylate 133
    spironolactone 71
    suxamethonium 290
    thiazides 76, 184, 194, 203
Dubin-Johnson syndrome 302, 304
Ectopic hormones 513
    adrenocorticotropin (ACTH) 87, 493, 515
    antidiuretic hormone (ADH) 51, 87, 515
    calcitonin 515
    erythropoietin 517
    gonadotropins 516
    growth hormone (GH) 516
    hypoglycaemia and 516
    parathyroid hormone (PTH) 175, 180, 514
    prolactin 516
EDTA (ethylenediamine tetra-acetate)
    plasma alkaline phosphatase and 9
    plasma calcium and 9
    plasma potassium and 9
Electrophoresis
    plasma lipids and 322
    plasma proteins and 238
Emphysema
    acidosis and 130
Enzymes (see also individual enzymes)
    acid phosphatase (ACP) 288
    alanine aminotransferase (ALT) 298, 305, 308, 309, 312
    alkaline phosphatase (AP) 279, 299, 302, 309
    amylase (Amy.) 263
    angiotensin converting enzyme (ACE) 291
    aspartate aminotransferase (AST) 276, 298
    cardiac 271
    creatine kinase (CK) 272
    γ-glutamyltransferase (GGT) 286, 299, 314, 315
    hepatic 298, 299
    hydroxybutyrate dehydrogenase (HBD) 278
    isoenzymes 262

lactate dehydrogenase (LD) 277
macroamylase 271
Error(s)
　patient preparation and 5
　sample collection and 7, 8
　sample contamination and 7, 8
　sample handling and 4
　sample preservation and 9
　sample storage and 7
Erythrocyte
　carbonic anhydrase 105
　porphyrins 366
Erythropoietic coproporphyria 379
Faecal fat
　steatorrhoea and 337
False positives 13
False negatives 13
Familial hypocalciuric
　hypercalcaemia 185
Ferritin 354
　plasma 357
α-Fetoprotein
　tumours and 519
　plasma 301
Folic acid
　malabsorption and 341
Follicle stimulating hormone (FSH) 396
　plasma 409
Fractional excretion of sodium 33
Fredrickson classification (hyperlipidaemia) 322
Free fatty acids (FFA) 318
Free thyroxine index (FTI) 436
Gastrin, plasma 511
Gastrinoma
　gastrin and 511
Gilbert's disease
　liver function tests and 304
Glibenclamide
　hypoglycaemia and 231
Globulins, plasma 247 (see also immunoglobulins, hyper- and hypogammaglobulinaemia)
　$\alpha_1$-antitrypsin 247
　caeruloplasmin 248
　chronic active hepatitis and 251, 307
　chronic infection and 251
　cirrhosis and 311
　$\alpha_1$-globulins 247
　$\alpha_2$-globulins 248

β-globulins 249
γ-globulins 249
haptoglobin 248
immunoglobulins 249
$\alpha_2$-macroglobulins 248
myeloma and 242, 252
nephrotic syndrome and 255, 256
rheumatoid arthritis and 241, 252
Glomerular filtration rate
　creatinine clearance and 154
　sodium excretion and 24
Glucagon 208
Glucagonoma 512
Glucocorticoids 464
Glucose (see also hyper- and hypoglycaemia)
　adrenalin and 208
　C-peptide and 216
　CSF and 217
　glucagon and 208
　glucocorticoids and 209, 468
　gluconeogenesis 207
　glycogenesis 206
　glycogenolysis 206
　glycolysis 206
　growth hormone and 209
　haemoglobin $A_{1C}$ and 217
　homeostasis 207
　insulin and 207, 216
　laboratory investigation of 211
　lipid synthesis and 207
　metabolism 205
　plasma, fasting 212
　tolerance tests 213
　renal handling 205
　urinary 211
Glucose tolerance test 213
　diabetic response 219
　flat response 220
　impaired response 219
　lag storage response 220
　intravenous 214
　oral 213
γ-Glutamyltransferase 286, 299
　alkaline phosphatase and 280
Gluten-sensitive enteropathy (see coeliac disease)
Gonadotropins (see follicle stimulating and luteinizing hormones)

Gout
    investigation of 384
    primary 385
    secondary 385
    treatment 383
Graves' disease 443, 447
Growth hormone 393
    acromegaly and 422
    arginine load test and 406
    deficiency of 415, 416, 417, 418, 420
    dynamic function tests of 403
    ectopic production 516
    exercise test and 404
    glucose and 209
    glucose suppression of 407
    insulin hypoglycaemia and 406
    plasma 403, 405
    sleep and 404
Guillain-Barré syndrome
    CSF and 261
Haem
    biosynthesis of 361
Haemochromatosis
    plasma iron and 359
Haemoglobin $A_{1C}$ 217
Haemosiderin 354
Heart disease, ischaemic
    lipoproteins and 319
    enzymes and 271
Henderson-Hasselbalch equation 96
Hepatitis, acute
    anicteric 306
    cholestatic 306
    liver function tests and 302, 305, 309
    typical 305
Hepatitis B antigen
    liver disease and 301
Hepatitis, chronic
    chronic active 307
    chronic persistent 307
    hypergammaglobulinaemia and 251
    liver function tests and 302, 307
    thyroid function tests and 450
Hepatocellular jaundice
    causes of 305
    hepatitis: acute and 305
        cholestatic and 306
        chronic and 307
    liver function tests and 302, 309

Hereditary coproporphyria 375
Homovanillic acid (HVA), urinary 500
Humoral hypercalcaemia of malignancy 514
Hydrocortisone suppression test 176
Hydrogen ion (see also acid-base)
    extracellular buffering 100
    metabolism 100
    production 100
    renal excretion 101
    renal secretion 101
    respiratory compensation and 101
Hydroxybutyric dehydrogenase (HBD) 278
17-Hydroxycorticosteroids, urinary 476
5-Hydroxyindole acetic acid (5HIAA)
    metabolism 509
    carcinoid syndrome and 508
21-Hydroxylase deficiency 485
Hydroxymethoxymandelic acid (HMMA, VMA) 500
Hyperalbuminaemia 243
    causes 243
    dehydration and 241, 243
Hyperalimentation
    hypokalaemia and 73
    hypophosphataemia and 193
Hyperamylasaemia
    alcohol and 267
    biliary disease and 268
    causes of 267
    diabetes mellitus and 270
    drugs and 271
    hyperlipidaemia and 266, 268, 329
    mumps and 270
    pancreatic carcinoma and 268
    renal insufficiency and 270
    tumours and 270
Hyperbilirubinaemia
    classification of 303
    differential diagnosis of 309
    Gilbert's disease and 304
    haemolysis and 303
    hepatitis: actue and 305
        cholestatic and 306
        chronic and 307
    liver anoxia and 310

Hypercalcaemia
    causes  178
    dehydration and  179
    diuretics and  184
    familial hypocalciuric hypercalcaemia and  185
    immobilization and  184
    malignancy and  180, 514
    myeloma and  181
    primary hyperparathyroidism and  182
    renal failure and  183
    vitamin D excess and  178
Hyperchloraemia  83
    acidosis and  83, 84, 85, 86
    causes  83
    dehydration and  83
    diabetes mellitus and  86, 223
    diarrhoea and  84, 124
    renal failure and  85
    renal tubular acidosis and  83, 123
    respiratory alkalosis and  86, 124
    uretero-sigmoidostomy and  85
Hyperchloraemic acidosis  83, 84, 85, 86
Hypercholesterolaemia
    broad beta disease and  331
    causes of  322, 325, 326
    dysglobulinaemia and  328
    familial  326
    hypertriglyceridaemia and  331
    hypothyroidism and  327, 332
    laboratory investigation of  321
    nephrotic syndrome and  327
    obstructive jaundice and  328
    primary  326
    secondary  327
Hypercortisolism  487
    adrenal tumour and  491
    causes of  487
    Cushing's disease and  489
    Cushing's syndrome and  491
    ectopic ACTH and  493
    laboratory investigation of  467, 477, 478, 479
    stress and  462, 470, 480, 489
Hypergammaglobulinaemia
    autoimmune disease and  252
    benign paraproteinaemia and  253
    causes of  251

chronic infection and  251
chronic liver disease and  251
essential paraproteinaemia and  254
heavy chain disease and  253
macroglobulinaemia and  253
malignant paraproteinaemia and  252
sarcoidosis and  252
Hyperglycaemia  217
    causes of  218
    diabetes mellitus and  218
    hyponatraemia and  49
Hyperkalaemia  67
    acidosis and  61, 62
    body redistribution of K and  60, 67, 69, 70
    causes of  67
    consequences of  71
    decreased renal excretion and  70, 71
    diabetes mellitus and  69, 222
    factitious  67, 68
    haematological disorders and  68
    hyperkalaemic periodic paralysis and  70
    hyporeninaemic hypoaldosteronism and  147, 149
    improper blood collection and  67
    increased input and  67, 68
    insulin deficiency and  69
    laboratory investigation of  65
    mineralocorticoid deficiency and  70
    potassium EDTA preservative and  9
    renal failure and  70
    renal transport defects and  71
    spironolactone therapy and  71
    therapy, principles of  71
Hyperlipidaemia (see also hypercholesterolaemia and hypertriglyceridaemia)
    causes  322, 325
    combined  326, 331
Hypermagnesaemia  201
    antacid ingestion and  201
    causes of  201
    renal failure and  201
Hypernatraemia  35
    causes of  35
    consequences of  42

*Index*

533

decreased sodium and water and  35, 36, 38
diabetes insipidus and  39
inadequate water intake and  40
increased sodium intake and  36, 37
laboratory investigation of  34
osmotic diuresis and  37
primary aldosteronism and  41
pure water depletion and  37
therapy, principles of  43
vomiting and  36, 38

Hyperosmolar coma  224

Hyperparathyroidism, primary  182, 183
alkaline phosphatase and  174, 284
calcium and  182
phosphate and  182
secondary  188

Hyperpituitarism  421

Hyperprolactinaemia  423

Hyperproteinaemia  240
causes of  240
haemoconcentration and  241
hypergammaglobulinaemia and  241, 242

Hyperphosphataemia
causes of  195
diabetes mellitus and  196
growth hormone and  198
hypoparathyroidism and  197
malignancy and  196
phosphate therapy and  195
renal failure and  197
vitamin D excess and  178

Hypertension  144, 503
causes of  503
congenital adrenal hyperplasia and  506
essential  503, 505
hyperaldosteronism and  143, 144
laboratory investigation of  141, 504
licorice and  507
oral contraceptives and  144, 507
phaeochromocytoma and  501
pregnancy and  507
renal disease and  144, 506
renin secreting tumour and  145, 507
renovascular disease and  144, 506

Hyperthyroidism
apathetic thyrotoxicosis and  446
clinical signs and symptoms  430
exophthalmos and  443
Graves' disease and  443, 447
plasma lipids and  333
toxic adenoma and  446
$T_3$ toxicosis and  444, 445

Hypertonicity  29, 42
consequences of  42
therapy, principles of  43

Hypertriglyceridaemia
acute pancreatitis and  266, 268, 329
alcoholism and  330
broad beta disease and  331
causes of  322, 325, 328, 329
diabetes mellitus and  322, 329
glycogen storage disease and  330
hypercholesterolaemia and  331
hypothyroidism and  332
increased VLDL and  325
laboratory investigation of  321
lipoprotein lipase deficiency and  328
oestrogens and  330
primary  328
renal failure and  330
secondary  329

Hyperuricaemia
causes of  385
gout and  385
Lesch-Nyhan syndrome and  387
treatment of  383

Hyperventilation
alkalosis and  125
hyperchloraemia and  86

Hypoalbuminaemia  244
acute illness and  247
anion gap and  93
causes of  244
haemodilution and  242, 244
liver disease and  245, 298
malabsorption and  245
malignancy and  247
nephrotic syndrome and  246
pregnancy and  247
protein-losing enteropathy and  246

Hypocalcaemia
acute pancreatitis and  189, 266, 267
causes of  185

EDTA preservative and 9
hypoparathyroidism and 187
magnesium deficiency and 188
plasma albumin and 186
pseudohypoparathyroidism and 189
renal failure and 188
vitamin D deficiency and 186
Hypochloraemia 87
cause of 87
hyponatraemia and 87
metabolic alkalosis and 87
respiratory acidosis and 88
vomiting and 88
Hypocortisolism 477
Addison's disease and 481, 483
adrenal insufficiency and 481
adrenal suppression and 484
causes of 477
congenital adrenal hyperplasia and 485
laboratory investigation of 467, 477
Hypogammaglobulinaemia 254
causes of 254
primary 254
protein loss and 255
transient 254
Hypoglycaemia 230
alcohol and 234
causes of 230
children and 235
diabetes mellitus and 233
endocrine disorders and 232
functional 233
gastrectomy and 233
glibenclamide and 231
infancy and 235
insulin and 231
insulinoma and 232
leucine and 234
liver disease and 232
tumours and 230, 516
Hypogonadotropic hypogonadism 419
Hypokalaemia 72
alkalosis and 61, 62, 73
causes of 72
consequences of 78
diarrhoea and 75
disturbed internal distribution and 73, 74
diuretic therapy and 76

haemodialysis and 78
hypokalaemic periodic paralysis and 74
inadequate intake and 72
increased renal excretion and 76, 77
insulin therapy and 73
laboratory investigation of 66
loss from gut, and 75
mineralocorticoid excess and 77
renal tubular acidosis and 76
salbutamol therapy and 74
therapy, principles of 78
vomiting and 75
Hypolipoproteinaemia
abetalipoproteinaemia 333
hyperthyroidism and 333
α-lipoprotein deficiency (Tangier disease) 332
malnutrition and 333
Hypomagnesaemia
alcohol and 202
aldosterone and 203
causes of 202
diuretics and 203
hypercalcaemia and 203
neonatal 204
osmotic diuresis and 203
Hyponatraemia 44
Addison's disease and 46
causes of 45
consequences of 54
decreased body sodium and 45
diabetes mellitus and 47
factitious 48
hyperlipidaemia and 48
increased body sodium and water and 45, 53
increased body water and 45
increased extracellular solute and 49
laboratory investigation of 35
nephrotic syndrome and 53
normal body sodium and 48
oedematous states and 53
postoperative state and 52
renal disease and 47, 54
SIADH and 51
therapy, principles of 55
vomiting and 45
water overload and 45

*Index*

Hypoparathyroidism
    plasma calcium and   187
    plasma phosphate and   197
Hypophosphataemia
    alkalosis and   193
    alkali therapy and   192
    causes of   192
    diabetes mellitus and   193
    diuretic therapy and   194
    hyperalimentation and   193
    hyperparathyroidism and   182
    hypophosphatasaemic rickets and   195
    liver disease and   194
    vitamin D deficiency and   186
    vomiting and   192
Hypophosphatasaemic rickets   195
Hypopituitarism
    causes of   412
    tests for   411
Hypoproteinaemia   241
    causes of   241
    haemodilution and   242
Hyporeninaemic hypoaldosteronism   147, 149
Hypothalamus   388
    hormones of   389
Hypothalamic disease
    anterior pituitary hormones and   415
    causes of   413
Hypothalamic-pituitary-adrenal axis   392, 462
Hypothalamic-pituitary-thyroid axis   425
Hypothyroidism
    CK and   275
    clinical signs and symptoms   430
    plasma lipids and   327, 332
    primary   454, 455, 456
    secondary   457
    subclinical   452
    hypothalamus and   457
    thyroid ablation and   458
Hypotonicity
    consequences of   54
    therapy, principles of   55
Hypouricaemia, causes   387
Idiogenic molecules   43
Imprecision of a test   10
Immobilization
    plasma calcium and   184

Immunoglobulins, plasma   249
    IgA   249
    IgD   250
    IgE   250
    IgG   249
    IgM   250
    liver disease and   298
    paraproteins   250
    structure of   249
Inaccuracy of a test   10
Indicans, urinary   342
Insulin   207, 216
    hypoglycaemia and   231
    hypoglycaemic test   401
Insulinoma
    hypoglycaemia and   232
Iodine metabolism   425
Iron   351
    absorption   353
    binding capacity   356
    body stores   354
    cycle   354
    deficiency   357
    ferritin and   354, 357
    haemosiderin and   354
    homeostasis   351
    laboratory investigation of   355
    plasma   355
    total binding capacity (TIBC)   356
    transport   353
    overload   358
Isoenzymes   262
    acid phosphatase   288
    alkaline phosphatase   280
    amylase   264
    creatine kinase   272, 275
    lactic dehydrogenase   277
Jaundice (see also hyperbilirubinaemia, cholestatic and hepatocellular jaundice)
    differential diagnosis of   309
Ketoacidosis   209, 211
    amylase and   270
    anion gap and   92
    blood gases and   121
    hyperchloraemia and   86, 223
    hyperkalaemia and   69
    hypokalaemia and   73
Ketogenic steroids, urinary   476
Ketones   209, 210, 211, 215
    diabetes mellitus and   209

ketosis, causes of   211
metabolism   209
Ketosteroids, urinary   476
Laboratory investigation of
   acid-base disorders   108, 109, 114, 115, 116
   aldosteronism   141
   alkaline phosphatase   280
   amylase   263
   anion gap   90
   bicarbonate, decreased plasma   107, 109
   bicarbonate, increased plasma   107, 108
   calcium   172
   carcinoid syndrome   509
   catecholamines   498, 500
   cortisol   467, 477
   dehydration   36
   gastrinoma   511
   glucose   211
   hyperkalaemia   65
   hypernatraemia   34
   hypertension   504
   hypokalaemia   66
   hyponatraemia   35
   iron   355
   jaundice   296, 309
   lipids, plasma   321
   liver disease   296
   oliguria   32, 153
   pituitary disease   397
   porphyrias   364
   renal failure, acute   151, 157
   proteins, plasma   237
   steatorrhoea   336
   thyroid disease   430
   urate   384
Lactate dehydrogenase (LD), plasma   277
   elevated, causes of   277
   isoenzymes of   277
   liver disease and   279
   muscle disease and   279
   myocardial infarction and   277
Lactic acid   215, 225
   acidosis and   227
   biguanide therapy and   228
   diabetes mellitus and   228
   drugs and   229
   hypoxia and   228
   liver failure and   229
   metabolism of   225
   pulmonary embolism and   229
   sepsis and   229
Lactose tolerance test   339
Lead poisoning
   porphyria and   380
Lesch-Nyhan syndrome   387
Liddle's syndrome   147
Lipids, digestion and absorption   317, 319, 334
Lipids, plasma   317 (see also cholesterol, triglycerides, hyper- and hypolipidaemia)
   cholesterol   318, 321
   free fatty acids   318
   laboratory investigation of   321
   lipoproteins   317, 320
   triglycerides   317, 321
Lipoprotein lipase deficiency   328
Lipoproteins
   chylomicrons   317
   Fredrickson classification   322, 323
   function   317
   high density (HDL)   317
   laboratory investigation   321
   low density (LDL)   317
   very low density (VLDL)   317
Liver anoxia
   liver function tests and   310
Liver disease
   alcoholic   314
   ALT and   298, 302, 307, 309, 312
   anoxia   310
   alkaline phosphatase and   282, 299, 302, 309, 312
   AST and   276
   carcinoma and   313
   classification of   302
   cirrhosis   310
   $\gamma$-glutamyltransferase and   299, 314
   hepatitis   304
   infiltrations   313
   obstructive   308, 328
   phosphate and   194
Liver function tests   293, 296, 302, 309
   albumin, plasma   298
   alkaline phosphatase, plasma   299, 302, 309, 312

$\alpha_1$-antitrypsin, plasma   301
bile salts, plasma   300
bilirubin, plasma   294, 297, 302
bilirubin, urine   295, 297
BSP excretion   300, 313
caeruloplasmin, plasma   301
coagulation factors   300
α-fetoprotein, plasma   301
globulins, plasma   298
γ-glutamyltransferase, plasma   299, 314
haemolysis and   303
hepatitis B antigen   301
immunoglobulins   298
interpretation of   302, 309
transaminases, plasma   298, 302, 309, 312
urobilinogen   294, 297
Luteinizing hormone   395
   clomiphene test and   410
   GnRH test and   409
   LHRH test and   409
   plasma   409
Macroamylasaemia   271
Macroglobulinaemia   253
Magnesium   198 (see also hyper- and hypomagnesaemia)
   distribution   198
   metabolism   198
   mineralocorticoids and   200, 203
   plasma   200
   PTH and   199
   renal excretion   199
   vitamin D and   200
Magnesium deficiency
   causes of   202
   plasma calcium and   188
Malabsorption
   bile salt $^{14}CO_2$ test and   342
   carotenes and   337
   classification of   343
   disaccharidase deficiency and   350
   faecal fat and   337
   faecal pH and   339
   faecal trypsin and   341
   folic acid and   341
   glucose tolerance and   339
   indicanuria and   342
   laboratory investigation of   336
   lactose tolerance test and   339
   optical density (plasma) and   337
   protein-losing enteropathy and   350
   secretin-pancreozymin test and   341
   transport defects and   350
   triglyceride $^{14}C$ test and   338
   vitamin A absorption and   338
   vitamin $B_{12}$ absorption and   340
   xylose absorption and   338
Malignancy (see also tumours)
   acid phosphatase and   288
   alkaline phosphatase and   284, 285
   calcium and   180
   hyperphosphataemia and   196
   magnesium and   203
   uric acid and   385, 386
Medullary carcinoma of thyroid   511
Meningitis
   CSF and   260, 261
Metanephrines, urinary   499
Metyrapone test   472
Milk alkali syndrome   179
Mineralocorticoids   134
   cortisol and   146
   11-deoxycorticosterone and   146
Multiple endocrine neoplasia (MEN)   513
Muscle dystrophy
   AST and   276
   CK and   274
Myeloma
   anion gap and   94
   hypergammaglobulinaemia and   252
   plasma calcium and   181
   plasma proteins and   242
Myocardial infarct
   AST and   276
   cardiac enzymes and   271
   CK and   273
   CK isoenzymes and   275
   HBD and   278
   LD and   277
   LD isoenzymes and   278
Myxoedema (see hypothyroidism)
Natriuretic hormone   25
Nephrotic syndrome
   hypoalbuminaemia and   246
   hypogammaglobulinaemia and   255
   hyponatraemia and   53

plasma lipids and   327
    urinary protein and   256
Neuroblastoma
    catecholamines and   502
Nitrogen, faecal
    malabsorption and   338
Noradrenalin   496
    phaeochromocytoma and   501
Normal anion gap acidosis   90, 112
Normonatraemia
    dehydration and   56
    overhydration and   56
    vomiting and   57
5′-Nucleotidase, plasma
    alkaline phosphatase and   280
    liver disease and   280, 299
Oedema   145
Oral contraceptives   5, 432, 449
Osmolality   28 (see also hyper- and hypotonicity)
    calculation of   29
    definition of   29
    plasma and urine   28
Osmolal gap   30
Osmolarity
    definition of   29
Osmotic diuresis
    diabetes mellitus and   37, 47
    plasma sodium and   37
Overhydration
    plasma osmolality and   31
    urine osmolality and   31
Oxogenic steroids, urine   476
Oxygen, blood gases and   112
Paget's disease
    alkaline phosphatase and   283
Pancreatitis, acute
    alcohol and   267
    amylase (plasma) and   263, 267
    amylase (urinary) and   264
    amylase: creatinine ratio and   265
    biliary disease and   268
    calcium and   189, 266
    glucose and   266
    hyperlipidaemia and   266, 268, 329
    laboratory investigation of   263
    lipase and   265
Pancreatic carcinoma
    amylase and   268
Panhypopituitarism   412, 416
Paraproteinaemia   250

benign   253
essential   254
malignant   252
Parathyroid hormone (PTH)   167
    action   168
    assay   175
    ectopic   175, 180, 514
    hyperparathyroidism and   183
    metabolism of   167
    renal failure and   183, 188
    secretion of   168
$P_{CO_2}$
    compensation of: metabolic acidosis   101, 118
                    metabolic alkalosis   118
    diabetic ketoacidosis and   121
    diarrhoea and   124
    lung disease and   130
    renal failure and   122
    renal response to   106
    renal tubular acidosis and   123
    respiratory arrest and   131
    salicylate ingestion and   133
    septicaemia and   125
pH
    definition   96
    Henderson-Hasselbalch equation and   96
    relationship to [H$^+$]   110
Phenformin
    lactic acidosis and   215, 228
Phenytoin sodium
    metyrapone test and   473
    thyroid function tests and   432
Phaeochromocytoma   501
    metanephrines and   499
Phosphate   190
    distribution   190
    metabolism   190
    plasma   190
Phosphate, plasma (see also hyper- and hypophosphataemia)
    concentration   190
    hyperphosphataemia   195
    hypophosphataemia   192
    plasma calcium and   191
    plasma urea and   191
Pituitary disease
    investigation of   397
Pituitary tumour
    anterior pituitary hormones

## Index

and  416, 417, 418
Porphobilinogen (PBG)  362, 365
Porphyrias  369
    acute intermittent, (AIP)  372
    classification of  369
    congenital erythropoietic  378
    erythropoietic  370
    hepatic  370, 371
    hereditary coproporphyria  375
    laboratory investigation of  364
    lead poisoning and  380
    porphyria cutanea tarda (PCT)  376
    protoporphyria  379
    variegate porphyria (VP)  373
Porphyrins  360
    faecal  365
    haem biosynthesis and  361
    laboratory investigation of  364
    red cells and  366
    screening tests for  365
    structure  361
    urinary  364, 365
Potassium  59 (see also hyper- and hypokalaemia)
    acid-base and  61, 62
    aldosterone and  61, 63
    body distribution  59, 60
    catecholamines and  61
    concentration in body fluids  59
    Cushing's syndrome and  467
    excretion of  61
    gut excretion of  61
    homeostasis  59
    insulin and  61
    intake  59
    laboratory investigation of  63, 65, 66
    plasma bicarbonate and  64
    plasma, improper collection of  7, 9, 67
    plasma levels  63
    renal excretion  62, 70, 71, 76
    salbutamol and  74
    spironolactone and  71
    urinary chloride and  65
    urinary  64
Predictive value of a test  17
Prevalence of a disease  17
Profile testing  16
Prolactin, plasma  394
    control of  394
    dynamic function tests and  408
    levels  407
    suppression of  409
    TRH test and  408
Pro-opiocortin  391
Prostate, transuretheral resection
    hyponatraemia and  53
Prostatic carcinoma
    acid phosphatase and  288
Protein-losing enteropathy
    hypoproteinaemia and  246
Proteins  236 (see also hyper- and hypoproteinaemia)
    acute phase  236
    albumin  242
    CSF  258
    estimation of  237
    function  236
    globulins  247
    hyperproteinaemia  240
    hypoproteinaemia  241
    laboratory investigation of  237
    serosal fluid  257
    tumours and  518
    urinary  255
Proteinuria  256
    causes  256
    nephrotic syndrome and  256
    orthostatic  256
Protoporphyria  379
Pseudohyperaldosteronism  147
Pseudohyperparathyroidism  514
Pseudohypoaldosteronism  148
Pseudohypoparathyroidism  189
Purine metabolism  381
Reference range, definition  12
Renal disease (see also renal failure)
    creatinine clearance and  154
    hyperuricaemia and  385
    plasma creatinine and  153
    plasma electrolytes and  151
    plasma urea and  152
    urine creatinine and  154
    urine osmolality and  153
    urine urea and  154
Renal failure, acute  156
    plasma electrolytes and  161
    postrenal failure  156, 162
    prerenal failure  156, 158
    urine electrolytes and  158
    urine/plasma osmolality and  153, 158

Renal failure, chronic  155
    acidosis and  122
    alkaline phosphatase and  285
    amylase and  270
    anion gap and  92
    calcium and  160, 183, 188
    hyperchloraemia and  85
    hyperkalaemia and  70
    hyperphosphataemia and  160, 197
    magnesium and  201
    plasma electrolytes and  159
    plasma lipids and  330
    urea and  152, 160
Renal tubular acidosis
    blood gases and  123
    hyperchloraemia and  83
    hypokalaemia and  76
Renin-angiotensin system  24, 136
Renin, plasma activity (PRA)  140
Respiratory arrest
    acidosis and  131
Reverse $T_3$  438
Salicylate overdose
    acid-base and  133
Salt-losing nephritis
    hyponatraemia and  47
Sarcoidosis
    angiotensin converting enzyme and  291
    plasma calcium and  179
Secretin-pancreozymin test  341
Sensitivity of a test  17
SIADH (see antidiuretic hormone)
Sick euthyroid  430, 438, 452
Sodium  23 (see also hyper- and hyponatraemia)
    aldosterone and  24
    content of body fluids  37
    Cushing's syndrome and  467
    fractional excretion of  33
    homeostasis  23
    laboratory investigation of  27, 34, 35
    plasma concentration  25
    renal excretion  23
    third factor and  25
Sodium overload
    hypernatraemia and  36, 37, 41
Specificity of a test  17
Stagnant loop syndrome  345
Standard bicarbonate  110

Steatorrhoea
    bile salt $^{14}CO_2$ test and  342
    bile salt deficiency and  345
    bowel resection and  345
    carcinoid syndrome and  349
    carotenes (plasma) and  337
    causes of  343
    chronic pancreatitis and  345
    coeliac disease and  346
    cystic fibrosis and  344
    diabetes mellitus and  349
    faecal fat and  337
    faecal nitrogen and  338
    faecal trypsin and  341
    folic acid and  341
    gastrectomy and  348
    glucose tolerance and  339
    indicans, urinary and  342
    infiltrations, gut wall and  347
    intestinal defects and  346
    jaundice and  345
    laboratory investigation of  336
    maldigestion and  344
    optical density (plasma) and  337
    parasitic infestations and  349
    secretin-pancreozymin test and  341
    stagnant loop and  345
    sweat electrolytes and  342
    triglyceride $^{14}C$ test and  338
    triglyceride (plasma) and  337
    tropical sprue and  347
    vitamin A absorption and  338
    vitamin $B_{12}$ absorption and  340
    Whipple's disease and  349
    xylose absorption test and  338
    Zollinger-Ellison syndrome and  348
Steroid suppression test  176
Stress
    plasma cortisol and  462, 470, 480, 489
Suxamethonium sensitivity
    cholinesterase and  290
Sweat electrolytes  342
Synacthen stimulation test
    adrenal insufficiency and  481, 483, 484
    normal response  472, 480
    prolonged  472
    short  471
Tangier disease  332

Third factor  25
Thyroid gland  425
Thyroid function tests (see also thyroid hormones)
    drug interaction and  460
    hormone replacement and  456, 460
    newborn and  448
    nephrotic syndrome and  459
    pregnancy and  447, 448
    selection of  441
    sick euthyroid and  452
Thyroid hormones
    biosynthesis of  425
    control of  426
    decreased levels, causes  451
    deiodination of  429
    half life, plasma  429
    increased levels, causes  442
    laboratory investigation of  430
    metabolic effect of  430
    replacement therapy  456
    transport in plasma  428
Thyroid stimulating hormone (TSH)  396, 425
    estimation (plasma)  410, 438
    TRH test and  411
    plasma levels  410
Thyrotoxicosis (see hyperthyroidism)
Thyrotropin (see Thyroid stimulating hormone)
Thyrotropin-releasing hormone (TRH)
    test for thyroid function  411, 439, 445, 446
Thyroxine (see also thyroid hormones)
    displacement by drugs  432
    free in plasma  428, 433
    peripheral metabolism of  429
    total, plasma  430
Thyroxine-binding globulin (TBG)  428, 437
    congenital deficiency  459
    congenital excess  449
    measurement of  437
    oestrogens and  449
    pregnancy and  447, 448
Thyroxine-binding prealbumin  428
Tonicity of ECF  29
Total iron binding capacity (TIBC)  356
Transaminases (see aspartate and alanine aminotransferases)

Transferrin  353
Triglycerides, plasma (see also hypertriglyceridaemia)
    hyperlipidaemia and  322, 325
    lipoprotein and  317
    laboratory investigation of  321
    metabolism  317
Triiodothyronine ($T_3$), plasma
    estimation  437
    free  433
    half-life  429
    levels in health and disease  437
    sick euthyroid and  430, 438, 452
    $T_3$-toxicosis  444, 445
Triiodothyronine uptake ($T_3U$) test  433
Triple function test  411
Trypsin, faecal activity  341
Tumours
    alkaline phosphatase and  518
    amines and  508
    amylase and  518
    biochemical syndromes of  508
    calcium and  180, 182, 514
    ectopic hormones and  513
    enzymes (plasma) and  517
    proteins and  518
Ultracentrifugation
    plasma lipids and  324
Uraemia
    acute renal failure and  156, 161
    chronic renal failure and  155, 160
    prerenal  156, 158
    postrenal  156, 162
Urate  381 (see also hyper- and hypouricaemia)
    allopurinol and  383
    excretion of  381
    laboratory investigation of  384
    metabolism of  381
    plasma levels  384
    uricosuric agents and  383
    urinary  384
Urate nephropathy
    urinary urate: creatinine ratio and  384
Urea  150
    decreased plasma, causes  163
    increased plasma, causes  156
    increased production  157
    laboratory investigation of  157

metabolism of 150
   plasma levels 152
   renal clearance 154
Ureterosigmoidostomy 85
Urinary chloride 33, 65, 82
Urinary free cortisol 476
Urinary potassium 64, 66
Urinary retention
    urea and 156, 162
    plasma electrolytes and 162
Urinary sodium 32, 34, 35, 36
    hyponatraemia and 32, 35
    oliguria and 32
Urobilinogen
   liver disease and 297
   synthesis of 294, 296
Vanillylmandelic acid (VMA, HMMA) 500
Variegate porphyria (VP) 373
Vasoactive intestinal peptide (VIP) 512
VIPoma 512
Vitamin A absorption test 338
Vitamin $B_{12}$
   malabsorption and 340
Vitamin D
   deficiency 186
   intoxication 178

metabolism 168
Vomiting
   alkalosis and 127
   hypernatraemia and 38
   hypochloraemia and 88
   hypokalaemia and 75
   hyponatraemia and 45
   hypophosphataemia and 192
Water
   balance 20
   dehydration 36, 43, 45, 56, 58
   distribution 21
   effect of drugs on renal clearance 27
   extracellular, control of 22
   homeostasis 21
   intake 22
   loss from body 21
   membrane shifts 21
   renal clearance of free 26
   urinary excretion 22, 26
WDHA syndrome (Verner-Morrison) 513
Wilsons disease
   liver and 311
Xylose absorption test 338
Zollinger-Ellison syndrome 512
   malabsorption and 348